# Vorticity and Turbulence Effects in Fluid Structure Interaction

## An Application to Hydraulic Structure Design

# International Series on Advances in Fluid Mechanics

## Objectives

The field of fluid mechanics is rich in exceptional researchers worldwide who have advanced the science and brought a greater technical understanding of the subject to their institutions, colleagues and students.

This book series has been established to bring such advances to the attention of the broad international community. Its aims are achieved by contributions to volumes from leading researchers by invitation only. This is backed by an illustrious Editorial Board who represent much of the active research in fluid mechanics worldwide.

Volumes in the series cover areas of current interest and active research and will include contributions by leaders in the field.

Topics for the series include: Bio-Fluid Mechanics, Biophysics and Chemical Physics, Computational Methods for Fluids, Experimental & Theoretical Fluid Mechanics, Fluids with Solids in Suspension, Fluid-Structure Interaction, Geophysics, Groundwater Flow, Heat and Mass Transfer, Hydrodynamics, Hydronautics, Magnetohydrodynamics, Marine Engineering, Material Sciences, Meteorology, Ocean Engineering, Physical Oceanography, Potential Flow of Fluids, River and Lakes Hydrodynamics, Slow Viscous Fluids, Stratified Fluids, High Performance Computing in Fluid Mechanics, Tidal Dynamics, Viscous Fluids, and Wave Propagation and Scattering.

---



---

# Vorticity and Turbulence Effects in Fluid Structure Interaction

## An Application to Hydraulic Structure Design


EDITORS

**M. Brocchini**
*University of Genoa, Italy*

**F. Trivellato**
*University of Trento, Italy*



WIT*PRESS* Southampton, Boston

# Vorticity and Turbulence Effects in Fluid Structure Interaction

An Application to Hydraulic Structure Design

**Editors:**

**M. Brocchini**
*University of Genoa, Italy*

**F. Trivellato**
*University of Trento, Italy*

---

Published by

**WIT Press**
Ashurst Lodge, Ashurst, Southampton, SO40 7AA, UK
Tel: 44 (0) 238 029 3223; Fax: 44 (0) 238 029 2853
E-Mail: witpress@witpress.com
http://www.witpress.com

For USA, Canada and Mexico

**WIT Press**
25 Bridge Street, Billerica, MA 01821, USA
Tel: 978 667 5841; Fax: 978 667 7582
E-Mail: infousa@witpress.com
http://www.witpress.com

British Library Cataloguing-in-Publication Data

A Catalogue record for this book is available from the British Library

ISBN: 1-84564-052-7
ISSN: 1353-808X

Library of Congress Catalog Card Number: 2005937242



Printed in Great Britain by Cambridge Printing.

# Contents

# FOREWORD

This book is the collection of 11 chapters that have been contributed by each research unit joining a MIUR (Italian Ministry of University and Research) project, devoted to the topic of fluid structure interaction. The subject matter is divided into chapters covering a wide spectrum of recognized areas of research, such as: wall bounded turbulence; quasi 2-D turbulence; canopy turbulence; large eddy simulation; lake hydrodynamics; hydraulic hysteresis; liquid impacts; flow-induced vibrations; sloshing flows; transient pipe flow; and air entrainment in dropshaft.

The purpose of each chapter is to summarize the main results obtained by the individual research unit. As a result, the main feature of the book is to bring the state of the art on fluid structure interaction to the attention of the broad international community.

Each chapter has been reviewed by leading fluid mechanics scientists. Part of the material completes what already is published in international journals. This has been briefly reviewed in some of the book's chapters for clarity's sake and presented along with original results to give an exhaustive picture of each single topic. The basic mathematical formulations, the physical as well as the numerical modeling of interaction problems, are discussed.

This book is mainly aimed at fluid mechanics scientists, but it can be of value also as a reference volume to postgraduate students and practitioners in the field of fluid structure interaction.

The Editors and the Authors are grateful to Professor Carlos Brebbia, Director of the Wessex Institute of Technology, United Kingdom, and to the AFM Series Editor, Professor Matiur Rahman, Dalhousie University, Canada, for the kind invitation to publish the present book in the AFM series of the prestigious WIT Press. The generous support of the many referees who revised the chapters is gratefully acknowledged. Their considerate advices have improved the final quality of the book.

This work has received financial support by the Italian Ministry of University and Research project "Influence of vorticity and turbulence in interactions of water bodies with their boundary elements and effects on hydraulic design".

May the Editors finally add their wish, which after all is shared by any scientist, that the present book might advance this complex branch of Fluid Mechanics because, as Virgilio (Georgiche, lib.II, v.490) vividly stated: *Felix qui potuit rerum cognoscere causas* (He who succeeded in understanding the reasons of phenomena is a happy person).

The Editors
Maurizio Brocchini and Filippo Trivellato
2006

# CHAPTER 1

# Techniques of research and results in the field of coherent structures of wall-bounded turbulence

G. Alfonsi
*Dipartimento di Difesa del Suolo, Universita della Calabria, Rende (Cosenza), Italy.*

## Abstract

Coherent structures of turbulence represent a widely-used viewpoint in describing turbulence in which categories like coherency and intermittency (associated in this context with the process of evolution of the coherent structures) are implied. In the present work the issue of the coherent structures developing in wall-bounded turbulent flows is considered. After a short historical synthesis, some basic concepts and various research methods and techniques for the scientific investigation of turbulent flows are reviewed. Some emphasis is given to the description of the available approaches to the numerical simulation of turbulent flows and to the problem of the construction of a turbulent-flow database. Then the phenomena occurring in the inner- and in the outer region of the turbulent boundary-layer are considered, mainly with reference to the large amount of experimental research existing on the subject. The flow phenomena are described in terms of: *i)* events occurring in the inner region, *ii)* large-scale motions developing in the outer region and, *iii)* dynamics of vortical structures. The method of the Proper Orthogonal Decomposition for the eduction of the coherent structures of turbulence is then presented. This technique permits the analysis of a turbulent-flow database in terms of dynamics of mathematically-defined coherent structures, allowing the calculation of properties of turbulent flows with precise physical meaning.

## 1 Introduction

A still unresolved problem in fluid sciences is turbulence. In the last decades a particularly intense effort has been produced by researchers in this field and several new concepts have been generated. Nevertheless, still there is a lack of a *general theory* of turbulence. New concepts based on results obtained with the use of continuously evolving research techniques of both numerical and experimental nature are often in conflict with formerly developed ideas on the phenomenon of turbulence, many of which have become obsolete. The aim of this work is to present a review of the techniques of investigation and of the current knowledge of turbulence (restricted to the still relatively wide area of wall-bounded incompressible flows). Appropriate categories to be applied to the description of turbulent flows appear to be *coherency* and *intermittency*, where the latter has mainly to be interpreted as the manifestation of the evolution processes of the coherent structures of the flow.

## 1.1 Historical synthesis

The modern era in turbulence research begins with Osborne Reynolds [1]. Reynolds decomposition and averaging consists in: *i)* separating the dependent variables of the Navier-Stokes equations into a mean and a fluctuating part, *ii)* substituting into the equations, *iii)* taking the average of the equations themselves. Owing to the nonlinear character of the system of the governing equations, the result is that a new term in the momentum equations arises, the Reynolds stress term (or turbulent stress term), a non-zero correlation between the fluctuating components of the velocity (case of the incompressible fluid, index notation, summation convention for repeated indices applies):

$$\partial_i \overline{u}_i = 0 \tag{1}$$

$$\partial_t \overline{u}_i + \partial_j(\overline{u}_i \overline{u}_j) + \partial_j(\overline{u'_i u'_j}) = -\frac{1}{\rho}\partial_i \overline{p} + \nu \partial_j \partial_j \overline{u}_i \tag{2}$$

where overbars denote (time) averaging, primes denote the fluctuating velocity components and $\nu$ and $\rho$ are the fluid kinematic viscosity and density, respectively. Much work has been made in order to devise appropriate models for the Reynolds stress term, to be expressed as a function of the averaged quantities in order to arrive to the algebraic closure of the system of the governing equations. Many ideas, producing several classes of turbulence models of technical use and involving specific concepts like that of the *eddy viscosity* were put forward for this scope. The early times in turbulence research (the years 1920s and 1930s) are characterized by a picture in which turbulence appears as a completely stochastic phenomenon in which a randomly fluctuating portion of the velocity field is superimposed on the average part. Within the highly complex conceptual framework of many randomly interacting turbulent scales, the semi-empirical theory of Prandtl [2] was formulated, together with simplified and abstract concepts like the *homogeneous* and *isotropic* turbulence (Taylor [3]). The statistical viewpoint in describing turbulence was dominant up to the years 1940s, a period during which many researchers realized remarkable progress. Among others, Kolmogorov [4] and Heisenberg [5]. A review of the state of the knowledge on turbulence up to those times can be found in Batchelor [6]. Of all the ideas developed in those years, the most relevant are: *i)* turbulent flows at sufficiently high Reynolds numbers generate energy-containing flow structures that are similar at all higher values of the Reynolds number; *ii)* zones of production and dissipation of turbulent energy are well separated in wavenumber space and the condition of *locally isotropic equilibrium* of the small turbulent scales holds (see Batchelor [6] and Sreenivasan and Antonia [7]); *iii)* the coupling between the small-scale and the large-scale motions is weak and the small eddies behave universally in all flows. These ideas are nowadays subject to ongoing discussion, following both experimental measurements and calculations that started to reveal the non-isotropic character of the small turbulent scales (see Shen and Warhaft [8] and references therein). The first perception of the intermittent character of turbulence can be attributed to Townsend [9], Corrsin and Kistler [10] and Klebanoff [11]. New interpretative categories are introduced like the *superlayer* (the turbulent/non-turbulent interface), together with the idea that the large eddies exhibit quasi-deterministic structures. The process of formation of the contemporary vision of turbulence started in the 1960s. Since then, a large amount of research work has been produced with the use of both experimental and computational techniques and based on the principle that the transport properties of a fluid flow are governed by large scale motions while small scale motions are mainly responsible for the dissipation processes. The concepts of coherency and evolution of coherent structures in the boundary-layer of wall-bounded turbulent flows offer the possibility of devising a better clarification of the physical mechanisms through which turbulent energy of mechanical nature is dissipated into heat. The understanding of these mechanisms brings new perspectives on two important objectives in modern fluid technology, namely the *control of turbulence* and the development of new *predictive models* for the numerical calculation of high-Reynolds-number turbulent flows. Important implications of turbulence control are represented, among others, by reduction of skin friction, delay of separation in wake flows, enhancement of mixing in free shear turbulent flows and controlled sediment transport in the case of multiphase flows.

## 1.2 Research methods and approaches

Research techniques in turbulence are of both experimental and numerical nature. Experimental methods have a long tradition in fluid mechanics and turbulence, ranging from one-point probes for the measurement of mean quantities to multi-point probes for the evaluation of instantaneous values of the velocity and the simultaneous acquisition of entire velocity fields. Laboratory techniques include *HWA* (Hot Wire Anemometry, see Comte-Bellot [12] for a review), *LDA* (Laser Doppler Anemometry, see Buchhave and George [13] for a review), *UDV* (Ultrasonic Doppler Velocimetry, see Alfonsi [14] and references therein) and flow visualization, both qualitative and quantitative (*PIV* in particular, Particle Image Velocimetry, see Adrian [15] for a review of the method and related techniques). The second class of methods involves the numerical simulation. Various numerical techniques, ranging from finite differences, spectral methods (see Canuto *et al* [16] for a review), finite elements (see Glowinsky [17] for a review work), high-order finite elements (see Karniadakis and Sherwin [18]) and also appropriate combinations of the basic methods in mixed techniques (see among others, Alfonsi *et al* [19, 20] and Passoni *et al* [21]), are possible. Each time a new computational code is developed, the reliability of the algorithm has to be assessed by performing fundamental algorithmic tests like the behavior with respect to hydrodynamic stability theory in both linear and nonlinear fields of the computational code. In solving the Navier-Stokes equations with the aim of obtaining a precise correlation with turbulence physics, the accuracy of the calculations has to be deeply monitored and the equations have eventually to be further manipulated, by following one of the existing approaches to the numerical simulation and/or modeling of turbulence. There are three main approaches to the numerical simulation and modeling of turbulent flows: *RANS* (Reynolds Averaged Navier-Stokes equations), *LES* (Large Eddy Simulation) and *DNS* (Direct Numerical Simulation of turbulence, see Speziale [22], Lesieur and Métais [23], Moin and Mahesh [24], respectively, for review works of the three approaches). For the *RANS* approach, Reynolds averaging is performed – eqn. (1) – and the problem of the closure of the system of the Navier-Stokes equations has to be faced. Different types of models, the majority of them incorporating the concept of eddy viscosity, have been introduced for this purpose including: *i)* "zero-equation" models, in which the eddy viscosity is directly related to the mean velocity field, *ii)* "one-equation" models, in which one additional differential equation is added to the system of the governing equations typically for the turbulent kinetic energy $\kappa$, *iii)* "two-equation" models, in which two additional differential equations are added governing the turbulent kinetic energy $\kappa$ and the rate of dissipation of turbulent kinetic energy $\varepsilon$ (the $\kappa - \varepsilon$ models), *iv)* "stress-equation" models, involving a number of additional partial differential equations for the evolution of different terms of the Reynolds stress tensor. In following, the *LES* approach, one wants to simulate the larger scales of the flow and use a model for the smallest scales, based on their isotropic and purely dissipative character. A filter is applied to the Navier-Stokes equations for scale separation and a model is sought (the *subgrid-scale* model, *SGS*) for the term of the momentum equation that is not a function of the resolved variables. This is the so-called *subgrid-scale* stress term. For the other terms, including the Leonard tensor and the cross terms, suitable expressions in terms of the resolved variables can be found. After the application of the filter to the Navier-Stokes equations (case of the incompressible fluid, index notation), one obtains:

$$\partial_i \overline{u}_i = 0 \tag{3}$$

$$\partial_t \overline{u}_i + \partial_j (\overline{\overline{u}_i \overline{u}_j} + \overline{\overline{u}_i u'_j} + \overline{u'_i \overline{u}_j} + \overline{u'_i u'_j}) = -\frac{1}{\rho} \partial_i \overline{p} + \nu \partial_j \partial_j \overline{u}_i \tag{4}$$

where overbars now denote filtering and primes denote subgrid-scale components. Several *SGS* models have been devised. Among others, there are the Smagorinsky's model (Smagorinsky [25]), the Scale Similarity model (Bardina *et al* [26]), the Spectral Eddy Viscosity group of models (Kraichnan [27]), the Structure-Function model (Métais and Lesieur [28]), the *RNG* model (based on the Renormalization Group theory, Yakhot *et al* [29]) and the Dynamic Model (Germano [30]). Besides these, there are both non-eddy viscosity *SGS* models and non-isotropic closures that have also started to appear, the latter incorporating the hypothesis of non-isotropy for the smallest turbulent scales. In the *DNS* approach, the attitude of directly simulating all turbulent scales is followed by considering the Navier-Stokes equations with no modifications of any kind (case of the incompressible fluid, index notation):

$$\partial_i u_i = 0 \tag{5}$$

$$\partial_t u_i + \partial_j (u_i u_j) = -\frac{1}{\rho}\partial_i p + \nu \partial_j \partial_j u_i \tag{6}$$

The critical aspect in following this approach is the accuracy of the calculations, that in theory should be sufficiently high to resolve the Kolmogorov microscales in both space and time. Research work has been performed in order to devise less stringent – though reliable – criteria for the accuracy of *DNS* calculations (see Grötzbach [31]). In all the aforementioned approaches, the major difficulty in performing calculations at Reynolds numbers of practical interest lies in the remarkable amount of computational resources required for fluid flow simulations in terms of both memory and computational time. For a long time the consequence has been that only simple flow cases at relatively low values of the Reynolds number have been analyzed. The advent of the high-performance computing techniques has changed this scenario, opening new perspectives in using vector and parallel computers for computational fluid dynamics (see Passoni *et al* [32, 33] and references therein). Whether experimental or numerical, modern techniques of investigation have the potential of greatly increasing the amount of information gathered during the study of a particular flow. From a condition in which a relatively scarce amount of data was measured and processed by using concise statistical methods, the continuous effort in studying turbulence in its full – three-dimensional and unsteady – complexity, has enabled researchers to manage very large amounts of data. A typical turbulent-flow database includes all three components of the fluid velocity (and pressure) at all points of a three-dimensional domain, gathered for an adequate number of time steps of the turbulent statistically steady state. Such databases contain much information about the character of a given turbulent flow but in the formation of the value of each variable, all turbulent scales have contributed and the effect of each scale is nonlinearly combined with all other scales. It is also recognized that not all scales contribute to the same degree in determining the physical properties of a turbulent flow. Methods have been devised to extract the *relevant information* from a turbulent-flow database, which has permitted the separation of the effect of appropriately defined modes of the flow from the background flow, or finally, has enabled the coherent motions of the flow to be extracted, whatever the definition of coherent structure may be. A general definition of coherent structure is reported (from Robinson [34]) as an introductory concept: "... region of the flow field in which flow variables exhibit significant correlations with themselves or other variables over space/time intervals remarkably higher with respect to the smallest scales of the flow ...". Works dealing with coherent turbulent motions in different kinds of flows are due to Robinson [34], Cantwell [35] and Panton [36].

This work is organized as follows. In Section 2 studies and methods dealing with the inner region of turbulent shear flows are reviewed. A subsection is devoted to the description of the streaks of the boundary-layer that constitute the first perception of forms of organized motions in a turbulent flow. In another subsection the so-called *burst phenomenon* is depicted together with a summary of the *event-detection* techniques, mainly in the framework of conditional sampling and averaging methods. The criticism that has developed around the concept of burst, its evolution and better definition are discussed. Section 3 deals with the outer region. One of the issues (still open) is how the large-scale motions of the outer layer are influenced by the turbulent events occurring in the inner layer. Section 4 is devoted to the description of vortical structures. Vortices of all sizes and strengths are present in both inner- and outer region of a turbulent shear flow and undergo processes of evolution that have to be understood with the use of appropriate investigative techniques. Quasi-streamwise, ring, hairpin (horseshoe), arch and other kind of vortices are considered, together with their dynamical connections with previously discovered structures of the boundary-layer. In Section 5 the method of the Proper Orthogonal Decomposition for the extraction of the coherent structures of a turbulent flow is presented. The unambiguous extraction of turbulent-flow structures from the background flow is related to the mathematical procedure upon which the definition of coherent structure is based. Concluding remarks are at the end.

### 1.3 Mean flow properties

Some remarks have to be made on the mean flow properties of wall-bounded flows often described in terms of wall units, *i.e.* normalized in length by the viscous length scale $(\nu/u_*)$ and in velocity by $u_*$. One has:

$$u^+ = \frac{\overline{u}}{u_*}; \qquad u_* = \sqrt{\frac{\tau_w}{\rho}}; \qquad \tau_w = \mu \frac{\partial \overline{u}}{\partial y}|_{wall}$$

$$x^+ = \frac{x u_*}{\nu}; \qquad y^+ = \frac{y u_*}{\nu}; \qquad z^+ = \frac{z u_*}{\nu}; \qquad t^+ = \frac{t u_*^2}{\nu}$$

$$Re_* = \frac{u_* L}{\nu}$$

where $u_*$ is the friction velocity, $\tau_w$ is the mean shear stress at the wall ($\overline{u}$ denotes the averaged $x$-velocity) and $Re_*$ is friction Reynolds number. For what the mean velocity profile is concerned, various regions can be distinguished:

*i)* viscous sublayer $(0 \leq y^+ \leq +7)$, where:

$$u^+ = y^+ \tag{7}$$

*ii)* buffer layer $(7 \leq y^+ \leq +50)$, the region of maximum average production of turbulent kinetic energy. Several different expressions are available for this region;

*iii)* overlap (logarithmic) region $(y^+ > 50)$, characterized by the logarithmic law:

$$u^+ = \frac{1}{k} \ln y^+ + C \tag{8}$$

where $k$ and $C$ are empirical constants;

*iv)* outer region (strictly), in which the expression due to Coles [37], can be adopted.

Traditionally, the *inner region* includes the viscous sublayer, the buffer layer and the overlap layer in part. The *outer region* includes the rest of the layers. In this sense, the terms inner- and outer region have been introduced in the turbulent-flow terminology in a context of description of the properties of the mean flow. In eqn. (5), $k$ and $C$ (typical values are $k = 0.41$ and $C = 5.0$) have been considered for a long time to be universal constants, independent of the Reynolds number. Actually, the process of derivation of eqn. (5) is based, among other hypotheses, on the boundedness assumption in the Karman-Prandtl argument (see Barenblatt [38] and Barenblatt and Prostokishin [39] for more details). More recently, the use of mathematical tools like incomplete similarity and intermediate asymptotics have shown that the well-known pipe-flow data of Nikuradse [40] are satisfactorily interpreted in the overlap layer by a power law in which the relation between $u$ and $y$ is Reynolds number dependent:

$$u^+ = \left(\frac{1}{\sqrt{3}} \ln Re + \frac{5}{2}\right) \left(y^+\right)^{3/(2 \ln Re)} \tag{9}$$

where $Re$ is the Reynolds number based on the mean velocity averaged over the cross section. Pipe-flow experiments in which the insufficiency of law eqn. (5) is demonstrated can be found in Zagarola *et al* [41] and Alfonsi *et al* [42].

## 2 Inner region

### 2.1 The streaky structure of the boundary-layer

One of the first results of studying the structure of the turbulent boundary-layer is due to Kline *et al* [43]. Using hydrogen bubbles as visualization medium they showed that very near to the wall ($y^+ = 2.7$) the flow organizes itself in alternating unsteady arrays of high- and low-speed regions aligned in the streamwise direction, called *streaks* (*low-speed streaks*). The fluid actually migrates laterally from regions of instantaneous high-speed velocity ($+u'$) with respect to the mean streamwise velocity, toward low-speed ($-u'$) regions. The streaky structure of the boundary-layer actually interacts with the outer portion of the flow through a sequence of events like gradual outflow, liftup, sudden oscillation and breakup. For this sequence of events, the term *burst* (*bursting process*) started to be used. Since then, to the bursting phenomenon in the whole has been associated an essential role in the turbulent energy production and in the energy transfer process between inner and outer regions of the boundary-layer. Introducing the definition of streak spacing in the spanwise direction $\Delta z^+$, it was found $\Delta z^+ = 100$ in the mean, ranging from instantaneous values of 50 to 300. In the streamwise direction the streaks extend up to 1000 $\nu/u_*$ units. The formation of wall-layer streaks has also been associated by some authors with the presence of pairs of counter-rotating vortices aligned in the streamwise direction but other viewpoints exist on this issue (see Guezennec *et al* [44]). According to Panton [36] "... the current concept is that relatively short streamwise vortices are convected over the wall, bring up the low-speed fluid and leave it behind in the long trails...". The idea of coherent structure was then associated with streaks, to emphasize the more ordered structure of the boundary-layer in contrast with its previously accepted random character. The streaky structure of the boundary-layer is not persistent throughout. Moving outward from the wall, many vortices with different scales, strengths and orientation appear. This transition suggests the presence of an *inner layer* with a persistent streaky structure and an *outer region* dominated by vortex motions of various sizes. Timewise, there is a continuous transfer of vortical structures from the wall layer to the outer region during *quiescent periods*, a process that guarantees the maintenance of the double structure. In other brief temporal phases, inner and outer layer interact more strongly during a sequence of turbulent events, and the streaky structure is no longer recognizable. The wall layer undergoes a cyclic (though not periodic) process of streak development and disruption. Other authors contributed to these observations. Corino and Brodkey [45] observed *ejection* (at $5 < y^+ < 15$) and *sweep* events near the wall in a fully turbulent pipe flow. Kim *et al* [46] showed that almost all the net production of turbulent energy occurs within $0 < y^+ < 100$ during bursts. Grass [47] studied the structure of the boundary layer in a free-surface channel flow with both smooth and rough walls, observing that ejections and inrushes happen with both types of wall characteristics. These studies (mainly performed with flow visualizations) represent the initial phase of a long cycle of researches in which the process of production and maintenance of the mean turbulent properties are connected to rather ordered, repeatable and unsteady motions occurring in the near-wall region. The bursting phenomenon is thought to scale with inner variables by some authors, with outer variables by some other authors. Among others, Rao *et al* [48] report that the mean bursting period actually scales with the outer variables of a boundary-layer flow, *i.e.* free-stream velocity and boundary-layer thickness.

## 2.2 Event-detection techniques

In a subsequent phase of research development, experimental methods for velocity and pressure measurements associated with *Conditional Sampling and Averaging* techniques, started to be used. Conditional sampling and averaging represent a group of techniques for quantitatively distinguishing particular regions of a flow, including, but not limited to, coherent stuctures (see Antonia [49]). A conditional average can be seen as a special type of cross-correlation:

$$R(x, \Delta x, \tau_j) = \frac{1}{N} \sum_{i=1}^{N} c(x, t_i) f(x, \Delta x, t_i + \tau_j) \tag{10}$$

where $c(x, t_i)$ is the conditioning function at a point $x$ in space and a time $t_i$. Once the condition is met, a measurement at a possibly different location and later in time is added to the averaged ensemble. An early example of application of these methods in the framework of coherent structures is represented by the work of Willmarth and Wooldridge [50]. They performed a study in which space-time correlations are used to investigate pressure fluctuations in the wall layer. Introducing the so-called vector field of correlations, they plotted a field of vectors with components $R_{pu}$ and $R_{pv}$ ($p$ is pressure, $u$ is $x$-velocity, $v$ is $y$-velocity). Note that in general and in terms of velocity one has (index notation):

$$R_{ij} = \langle u_i(x_k, t) u_j(x'_k t) \rangle \tag{11}$$

where $\langle\ \rangle$ denotes averaging. The correlation maps of [50] represent one of the first realizations of conditional averaging of velocity field with respect to the background flow, *i.e.* an attempt to represent the flow field associated with organized structures in the turbulent boundary-layer. A useful tool for unambiguous definition of events of various kind occurring in the boundary layer is the *Quadrant Analysis*, introduced by Willmarth and Lu [51] (for other studies in which velocity correlations have been used see Brodkey *et al* [52], Eckelmann [53], Wallace *et al* [54], Praturi and Brodkey [55] and Kreplin and Eckelmann [56]). In the Quadrant Analysis the local flow behavior is divided into quadrants, depending on the sign of the streamwise and normal fluctuating components of the velocity $u'$ and $v'$. Four quadrants are identified:

- $Q_1$, first quadrant $(u'v')_1$, where $u' > 0$ and $v' > 0$, denoting an event in which high-speed fluid moves toward the center of the flow field;
- $Q_2$, second quadrant $(u'v')_2$, where $u' < 0$ and $v' > 0$, denoting an event in which low-speed fluid moves toward the center of the flow field, away from the wall (ejection);
- $Q_3$, third quadrant $(u'v')_3$, where $u' < 0$ and $v' < 0$, denoting an event in which low-speed fluid moves toward the wall;
- $Q_4$, fourth quadrant $(u'v')_4$, where $u' > 0$ and $v' < 0$, denoting an event in which high-speed fluid moves toward the wall (sweep).

The most relevant events are those of the 2nd and 4th quadrants. Ejections (2nd quadrant) are frequent at a distance from the wall, sweeps (4th quadrant) are frequent near the wall. The ejection and sweep events represent the consequence of the dynamics of vortical structures in the boundary layer, *i.e.* the events mainly responsible for the production of Reynolds stress. Another tool is the *VITA* analysis (Variable-Interval Time-Averaging), introduced by Blackwelder and Kaplan [57]. In performing the *VITA* analysis in a time series of pointwise velocity data, one wants to detect the instants in which the highest velocity fluctuations occur. The notion of local average is introduced, an averaging operation over a time interval of the order of the time scale of the phenomenon under study. The method basically consists in the identification of the instants in which the variance of the velocity data in a significant time interval is greater than the variance of the entire series. For this scope, a localized variance is formulated, defined as (case of the streamwise velocity $u$):

$$var(x_i, t, T) = \langle u^2(x_i, t, T) \rangle - \langle u(x_i, t, T) \rangle^2 \tag{12}$$

(note that also the spatial counterpart of *VITA* exists, the *VISA* analysis, Variable-Interval Space-Averaging). Both Quadrant and *VITA* analysis have been extensively used for the evaluation of pointwise velocity data, in particular as turbulent event-detection techniques. The aforementioned event-detection techniques have been used in a series of studies of a numerical nature. Among others, by Kim *et al* [58] in simulations of turbulent channel flow, by Moin and Kim [59] and Kim and Moin [60] in studies directed to the investigation of the structure of the vorticity field in wall bounded flows, by Adrian [61] and Adrian and Moin [62] (in which the idea of conditional eddy – based on the linear stochastic estimation technique – is introduced), by Chu and Karniadakis [63] and Choi *et al* [64] in studies of the drag reduction effects of riblets in channel flow, in which in particular also third- and fourth-order moments of the velocity fluctuations are used for the detection of sweep and ejection events. Bogard and Tiederman [65] performed a comparative evaluation of *VITA*, Quadrant Analysis and other event-detection methods noting that the validity of these techniques is highly related to the values of the operational parameters used. They found that the Quadrant Analysis has the highest probability of detecting ejections and the lowest of false detection. Subsequently Luchik and Tiederman [66] found that inner variables are the best candidates for proper scaling of the average time between turbulent events. A comparison between different conditional sampling techniques has also been performed in Subramanian *et al* [67]. The following picture of coherent motions in the inner region of a turbulent boundary-layer appears as a result of the works in which the aforementioned investigation techniques have been used. The velocity field in the viscous sublayer and in the buffer layer is organized into alternating streaks of high- and low-speed fluid, persistent, quiescent most of the time and randomly distributed. The most relevant part of the turbulent production process in the whole boundary-layer occurs in the buffer layer during outward ejections of low-speed fluid (2nd quadrant) and inrushes of high-speed fluid toward the wall (4th quadrant). The near-wall turbulence production process appears as an intermittent cyclic sequence of turbulent events. The so-called bursting phenomenon is otherwise identified in different ways, *i)* lift-up, oscillation and breakup of low-speed streaks, *ii)* shear-layer interface between sweeps and ejections, *iii)* single-point event detected with the *VITA* procedure, *iv)* ejection generating from a low-speed streak. This picture represents a synthetic result associated to the first perception of the existence of coherent structures in the boundary-layer, suffering from remarkably poor information about their temporal dynamics. Overall, the bursting process is mainly associated with an intermittent eruption of fluid away from the wall. These limits are strictly related to those of the techniques of investigation that have been used in this phase of the research. Other concepts have emerged later, according to which intermittency is more related to space rather than to time, *i.e.* the coherent structures of the boundary-layer are randomly distributed most in space and are subjected to evolution processes in time.

## 3 Outer region

An important issue in turbulent boundary-layer research involves the phenomena occurring in the outer region and their connection with those of the inner region. Kovasznay *et al* [68] performed a series of observations on the character of the vorticity of the *bulges* that occur in the outer layer. One of their conclusions was that the upstream portion of the turbulent/non-turbulent interface is the most active (see also Cantwell *et al* [69] for different flow cases). Another frequent observation is that the bursting processes observed by Kline *et al* [43] are in some way responsible for the large-scale motions occurring in the outer region. Offen and Kline [70] made an attempt to devise a kinematic relationship between the inner and the outer layer by conjecturing that the bulges in the superlayer are the consequence of vortex pairing between vortices associated with the occurrence of turbulent events. Brown and Thomas [71] observed a line of maximum correlation at an angle of 18° from the wall in the streamwise direction and attributed this fact to the presence of an organized structure. Falco [72], introducing the concept of *typical eddy*, noticed a considerable activity on the trailing interface of the outer bulge and associated this phenomenon to the Reynolds stress production due to small scale eddies in the outer layer. Head and Bandyopadhyay [73] performed a study at a Reynolds number greater than most of the previously published works. For boundary-layer flows with Reynolds number (based on momentum thickness) greater than 1000, they noticed the presence of structures, small in the streamwise direction but rather elongated, in lines at 40° to the wall. In the work of Wygnanski and Champagne [74] the process of transition in a turbulent pipe flow is studied. Transition occurs following instabilities of the boundary-layer flow, long before the flow becomes fully turbulent. Slugs develop at any Reynolds number greater than 3200, occupying all the cross section and growing in length by proceeding downstream. The structure of the flow inside the slugs is the same as in the case of fully developed turbulent flow. Where the mean flow evolves from laminar to turbulent, the velocity profiles exhibit inflections and the maximum value of the Reynolds stress occurs there. A picture of the outer-layer dynamics can be synthetically drawn. Three-dimensional bulges with dimension of the order of the boundary-layer thickness form in the turbulent/non-turbulent interface. Irrotational valleys also form at the edges of the bulges, through which free-stream fluid is entrained toward the turbulent region. Weakly irrotational eddies are observed beneath the bulges and fluid at relatively high speed impacts the upstream sides of the large-scale motions forming shear layers. It seems that the outer layer flow structure has only a moderate influence on the near-wall events and this influence is Reynolds number dependent. Still there is not a clear understanding of the physical relationship between the inner layer, characterized by intense turbulence production, and the less active outer region. Large-scale structures in the outer region appear to be inactive and dissipative, extracting little energy from the mean flow (see Townsend [75] – where the *attached eddy* hypothesis derived from the rapid distortion theory is introduced – and also Perry *et al* [76, 77]. The attached eddy is today essentially interpreted as a headless horseshoe vortex, see Section 4). The mechanism of interaction of inner- and outer layer remains actually unclear. A proposed idea [43, 57] is that the bursting process is the result of an inviscid instability of the instantaneous streamwise velocity profile. Another idea [48, 68] is that the bursting process occurs due to an instability of the sublayer produced by the pressure field and induced by large-scale motions of the outer region. Another view [70] is that sweeping motions in the

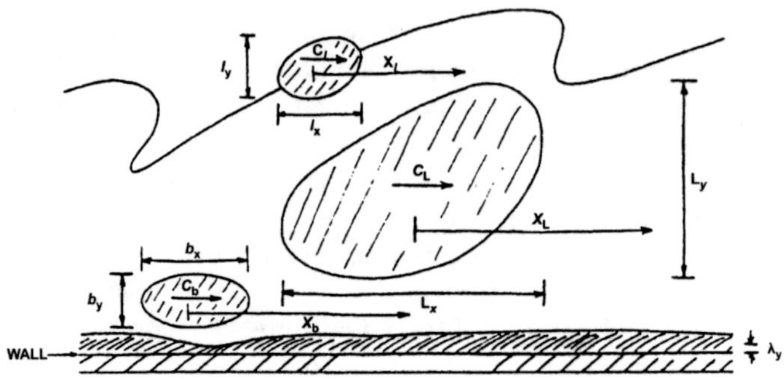


Figure 1: Boundary-layer structure (after Cantwell [35]).

logarithmic region impress on the wall a temporary adverse pressure gradient, which induces the lifting of the streaks. An additional aspect to be considered is the maintenance of the outer-region motions. A widely accepted idea is that the outer flow is a kind of wake formed by a sequence of events that occurred near the wall. In summary (synthesis due to Cantwell [35], fig. 1), the following elements emerge.

Nearest to the wall, streamwise counter-rotating vortices cover most of the wall. Right above the streamwise vortices a layer of fluid is involved by bursts, with intense production of small-scale motions. The outer layer is also affected by small-scale motions, mainly in the upstream portion of the turbulent/non-turbulent interface. The outer small-scale motions are involved in an overall transverse rotation with scale comparable to the thickness of the layer. Vortical structures of various sizes and strengths are present in the boundary-layer and they have a role in the turbulence production cycle and in the transport of momentum between inner- and outer layer.

## 4 Vortical structures

The need of a better understanding of the several phenomena discovered in the inner- and outer layer of a turbulent boundary-layer has brought to consider the dynamics of vortical structures. The concept of vortex is often associated to a coherent structure although, most of the time, the definition of vortex is still intuitive in nature. Following Robinson [34], a vortex can be primarly defined as a "... feature of the flow such as the instantaneous streamlines projected on a plane normal to the vortex core exhibit a roughly circular or spiral pattern...". Traditionally, vortices have been detected by using representations based on vortex lines or vorticity magnitudes. Many efforts in coherent-structures research are devoted to the development of methods for the extraction of structures from the background-, non-coherent vorticity field. Vortical structures have also been identified as elongated advected low-pressure regions (Robinson [34]). One of the first contributions to the issue of the presence of vortices in the boundary layer is due to Theodorsen [78], who introduced the hairpin (horseshoe) vortex. Within a hairpin vortex, a vortex head, neck

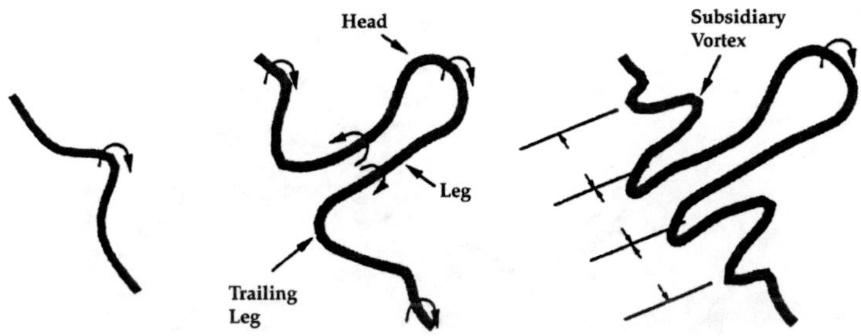


Figure 2: Evolution of a singular symmetric hairpin vortex in a uniform shear flow (after Smith *et al* [81]).

and legs (near the wall), can be distinguished. Robinson [34] confirmed the existence of non-symmetric arches (arch vortices) and quasi-streamwise vortices (rolls), based on the evaluation of *DNS* results. The composition of a quasi-streamwise vortex with an arch vortex may result in a hairpin vortex, complete or, most frequently, one-sided, but this conclusion can strongly depend on the particular technique used for vortex detection. A remarkable group of studies involving the dynamics of the hairpin vortices in the boundary layer has been performed, namely *i)* experimentally by Acarlar and Smith [79, 80], Smith *et al* [81], Haidari and Smith [82] and Perry and Chong [83] and, *ii)* numerically by Singer and Joslin [84]. Mainly based on these studies, a picture of vortex generation and interaction in the boundary layer emerges in which processes of the kind of interaction of existing vortices with wall-layer fluid, viscous-inviscid interaction, generation of new vorticity, redistribution of existing vorticity, vortex stretching near the wall and vortex relaxation in the outer region, are involved. Figure 2 shows the evolution of an inviscid two-dimensional symmetric line vortex with an initial three-dimensional distorsion when placed in a region of uniform shear, as it results from the Biot-Savart kind of simulations of Smith *et al* [81] (note that in particular flow situations the Biot-Savart calculations show failures, cases in which full Navier-Stokes simulations were needed). It can be noticed that subsidiary vortices are generated. Figure 3 shows the evolution of a nonsymmetric vortex in uniform background shear. Subsidiary hairpin vortices also form in this case, with a tendency to become symmetric. In both cases their spanwise spacing mainly depends upon the level of background shear. Figure 4 shows the evolution of a nonsymmetric hairpin vortex when placed in a region of turbulent-flow-type shear profile (Smith *et al* [81]). The legs squeeze together and the head moves away from the wall. A similar process has also been noted by Robinson [34], otherwise described in terms of dynamics of arch vortices.

Overall, individual vortices advected in a shear flow evolve – nonlinearly and mainly inviscidly – into, in most cases, nonsymmetric hairpin-shaped structures, beginning from the portion of the vortex characterized by the highest curvatures. During the development of hairpin vortices, spanwise vorticity is transformed into streamwise vorticity with deformation and birth of subsidiary vortices ([80, 82, 84]). The most important vortex-interaction (inviscid) processes occurring in the bound-

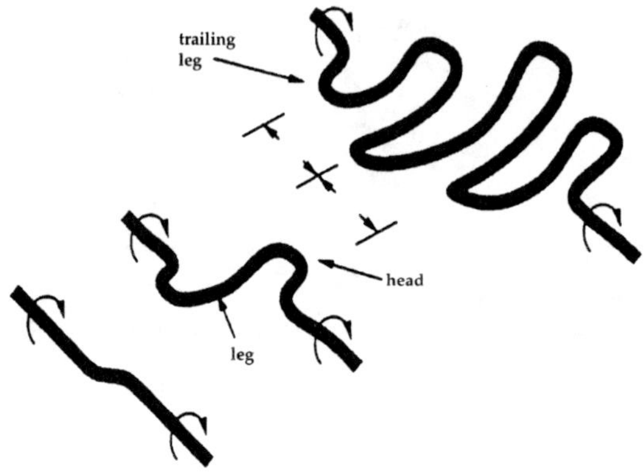


Figure 3: Evolution of singular non-symmetric hairpin vortex in uniform shear flow (after Smith *et al* [81]).

ary layer are: *i)* spanwise vortex compression and stretching in regions of increasing shear, *ii)* spanwise vortex expansion and relaxation in regions of decreasing shear, *iii)* vortex coalescence resulting in larger vortices and, *iv)* vortex reconnection into rings. Note that, in this context, the inner region has to be interpreted as the portion of the boundary layer in which viscous effects dominate and the outer region as the zone in which inviscid effects are prevalent. The evolution process of a hairpin vortex involves the development of vortex legs in regions of increasing shear with intensification of vorticity in the legs themselves. The leg of a vortex – considered in isolation – may appear as a quasi-streamwise vortex near the wall. The head of a vortex instead rises through the shear flow, entering regions of decreasing shear. As a consequence the vorticity in the vortex head diminishes (see also [73, 83]). Processes involving multiple vortex dynamics are more complex. An attempt at a description of this kind of phenomena has been made by Smith *et al* [81] in which

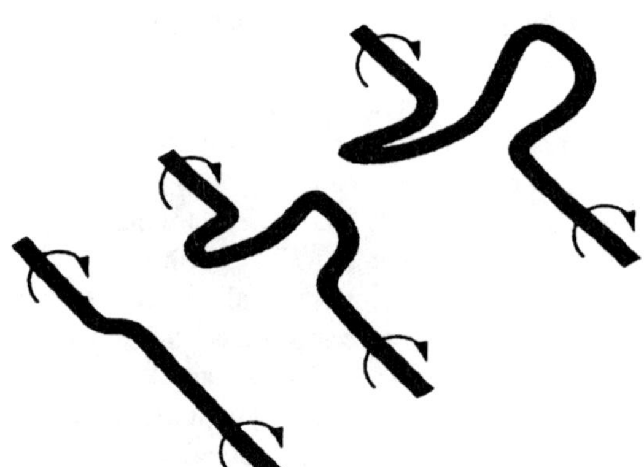

Figure 4: Evolution of a singular non-symmetric hairpin vortex in a turbulent shear flow (after Smith *et al* [81]).

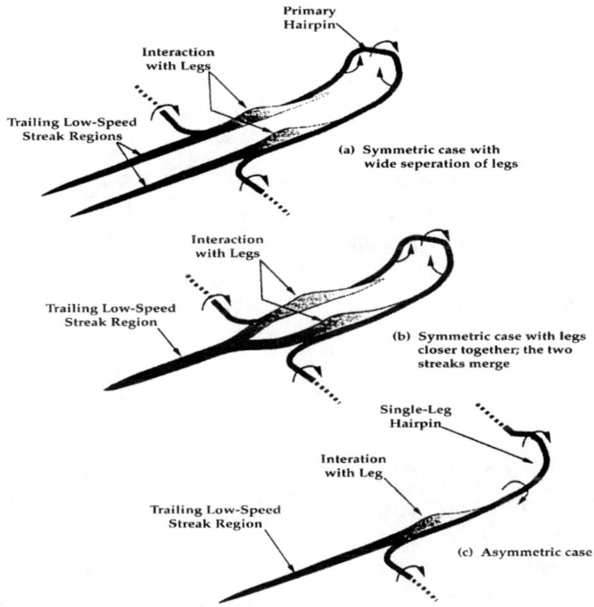


Figure 5: Generation of low-speed streaks caused by hairpin vortices (after Smith *et al* [81]).

the coalescence of small vortices into larger structures is described in terms of intertwining, amalgamation and reinforcement occurring when upward-migrating vortices approach one another. The formation and evolution of multiple vortices grouped into packets is described by Adrian *et al* [85]. The main result of these studies consists in the perception that the migration of small vortices (and vorticity) away from the wall results in the formation of larger vortical structures in the outer region, that are responsible for large-structure dynamics in the outer layer.

## 4.1 Relationship with streaks

A further stage in the process of understanding the role of vortical structures in the boundary layer involves their relationship with the streaks. One basic hypothesis is that the hairpin vortices provide an active mechanism for the formation of the streaks ([74, 80, 83]) in a way as shown in fig. 5. The effect of vortex motion near the wall is that to induce upwelling fluid near the legs of the vortex passing over the wall. The result is that streaks are generated. A streak will initiate only if a vortex is advected sufficiently close to the wall or a vortex leg penetrates through vortices of all other kind and reaches the proximity of the wall. If the streak-generator vortex moves away from the wall, the streak will dissipate due to viscous effects. Meandering of low-speed streaks may be due to vortices overrunning already existing streaks. The main consequence of this process is that the streaks assume the character of inactive motions, essentially being trails of fluid induced by the passage of a hairpin vortex in the vicinity of the wall. The streaks are depicted as transient flow structures, their destiny being exclusively determined by the evolution of the causative hairpin vortex. Not all the streaks are actually of this kind. Low-speed streaks that remain near the wall are inactive streaks. Streaks that are lifted away from the wall become active motions, following events that have to be reconciled with previously introduced concepts, that of the burst in the first place. The need of concept reconciliation mainly lies in the fact that terms like burst, ejection and sweep originated in the framework of pointwise techniques of analysis, mainly Quadrant Analysis and *VITA* analysis, that actually are not the best tools for the description of the evolution processes of complex time-dependent three-dimensional vortical structures.

## 4.2 Relationship with burst

A picture of the bursting phenomenon in terms of evolution of vortical structures will be now drawn. Most generally, the wall layer is subjected to a local breakdown process and erupts into the outer region. Turbulent energy is generated and turbulence itself is perpetuated and sustained. Vorticity previously concentrated near the wall is ejected outward and the eruptive events provide new vorticity to the outer region where shear layers are created with successive roll up in new hairpin vortices. Right after the eruption sweep events take place in terms of high-speed fluid penetrating from upstream close to the wall. After these events quiescent periods occur. The wall-layer breakdown can be interpreted as a viscous reaction of wall-layer fluid to the passage of wall-region vortices (among others, Peridier *et al* [86, 87], Van Dommelen and Cowley [88], Haidari and Smith [82], Singer and Joslin [84] and Doligalski *et al* [89]). A sequence of events occurs, beginning with a discrete eruption of wall-layer fluid and continuing in a strong viscous-inviscid interaction between mainly inviscid vortices and highly viscous outward-erupting fluid, which gives rise to the generation and ejection of a new vortex. Figure 6 shows the reaction of wall-layer fluid to an advecting hairpin vortex. In the regions in which the vortex induces an adverse pressure gradient near to the wall (adjacent to the legs and behind the head of the vortex) eruptions evolving as ridges in the form of tongues develop and surface-layer separation occurs. The ejected tongues penetrate regions of increasing streamwise velocity where the viscous-inviscid interaction process brings the tongue to roll up into a new hairpin-like vortex. At the end of the sequence the tongue completely detaches from the surface layer and gives rise to a new secondary hairpin vortex. The displacement of outward fluid caused by the – eruptive – vortex formation process, is counterbalanced, because of continuity, by an inflow of fast moving – sweeping – fluid toward the wall. These events occur intermittently, with characteristic time scales considerably shorter with respect to other kind of vortex motions (Smith *et al* [81]). A rather complete conceptual model for the evolution processes involving hairpin vortices in the wall region of a boundary layer has been proposed by Acarlar and Smith [80] (fig. 7). The dynamics of hairpin vortices is described, together with that of low-speed streaks, bursts, shear layers, ejections and sweeps, and the bursting of a low-speed streak appears as

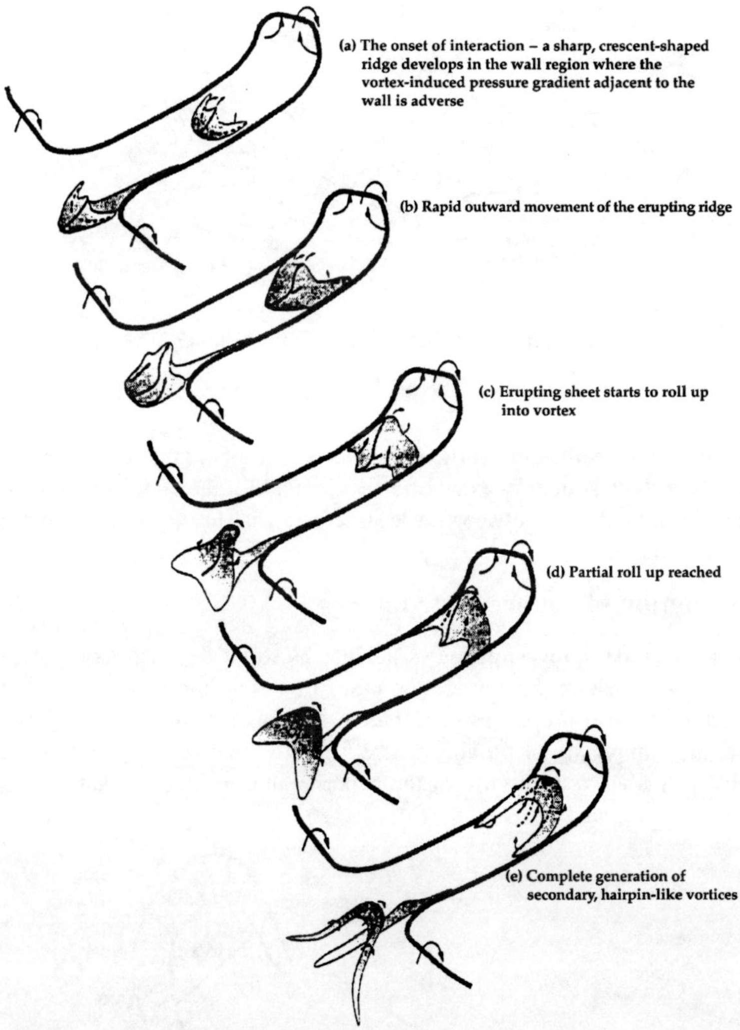


Figure 6: Generation of secondary hairpins from primary symmetric hairpin vortices (after Smith *et al* [81]).

the consequence of vortex roll-up in the unstable shear layer on top and sides of the streak. When formed, a vortex loop moves outward and downstream due to the streamwise velocity gradient. The legs of the vortex remain in the near-wall region, they are stretched and form quasi-streamwise counter-rotating vortices that eject fluid from the wall and accumulate fluid between the legs. Stretched legs of multiple hairpins coalesce, preserving the continuous development of low-speed streaks and outward-growing vortices may agglomerate into large-scale rotational bulges in the outer region. Another model for low-Reynolds-number boundary layers is due to Robinson [34], according to which quasi-streamwise vortices dominate

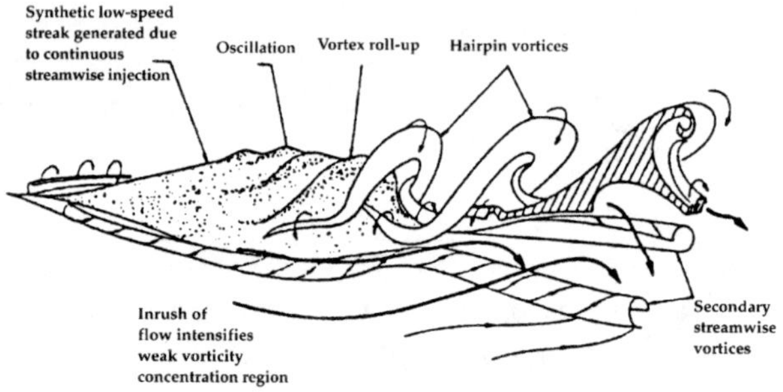


Figure 7: Evolution of hairpin vortices in the boundary-layer (after Acarlar & Smith [80]).

the buffer region, while arch vortices are mainly present in the wake region. In the overlap layer both structures exist, often as elements of the same vortical structure. The mutual relationship between these structures and the ejection/sweep motions, is shown in fig. 8.

## 5 Extraction of coherent structures

The main attitude in investigating turbulence as so far depicted, is of descriptive and/or intuitive nature. Scarce are the cases in which information useful in predicting the physical properties of turbulent flows is actually gained. This is due to both the complexity of turbulence itself and to the kind of information that is possible to gather with the use of the experimental techniques that characterize

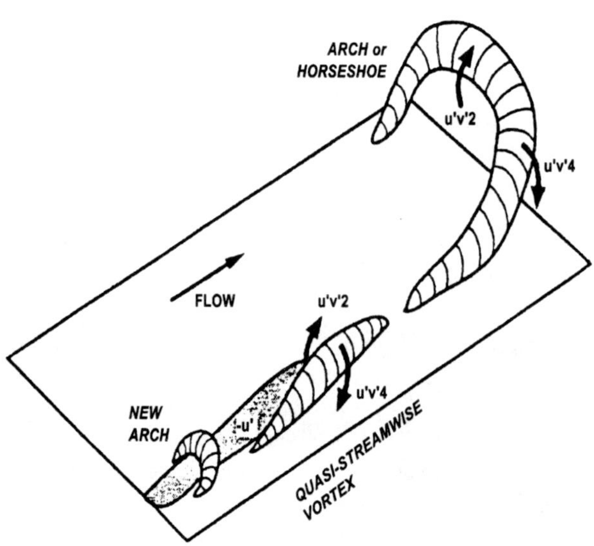


Figure 8: Vortical structures in the boundary-layer (after Robinson [34]).

the great majority of the existing research works in this field, in spite of their advanced nature. Contemporary turbulence research is also characterized by the fact that high-performance computers and computational techniques are extensively used. Advanced computational methods are able to facilitate progresses on some of the leading objectives of turbulence research, *i.e.* the control of turbulence and the production of new predictive models to be incorporated in newly generated high-performance vector and parallel computational Navier-Stokes solvers. An appropriate category to be used for a better scientific understanding of turbulent flows for the aforementioned objectives is that of – three-dimensional in space and evolving in time – coherent structures, where the idea of structures' coherency has to be associated to a formally-expressed definition to be implemented within a procedure of eduction of mathematical nature. In the following subsection the method of the Proper Orthogonal Decomposition for the eduction of the coherent motions in a turbulent flow, is presented. Of the various existing techniques, that of the Proper Orthogonal Decomposition appears to be the most rigorous and, on the basis of the results so far obtained, the most promising.

### 5.1 Proper Orthogonal Decomposition

The Proper Orthogonal Decomposition (*POD*) is an analytically founded statistical technique that can be applied for the extraction of coherent structures from a turbulent flow field. Based on the theory of compact, self-adjoint operators, it allows the selection of a basis for a modal decomposition of an ensemble of signals and its mathematical properties permit to have a clear perception of its capabilities and limits (Berkooz *et al* [90]). The *POD*, also known as Karhunen-Loéve (*KL*) decomposition, was first introduced in turbulence research by Lumley [91] and is extensively presented in Sirovich [92]. By considering an ensemble of temporal realizations of a generally non-homogeneous, square-integrable, three-dimensional, real-valued velocity field $u_i(x_j, t)$ on a finite domain $D$, one wants to find the most similar function to the elements of the ensemble on average, *i.e.* determine the highest mean-square correlated structure with all the members of the ensemble. This problem corresponds to finding a deterministic vector function $\varphi_i(x_j)$ such that $(i, j = 1, 2, 3)$:

$$\max_{\psi} = \frac{\langle |(u_i(x_j, t)\psi_i(x_j))|^2 \rangle}{((\psi_i(x_j)), (\psi_i(x_j)))} = \frac{\langle |(u_i(x_j, t)\varphi_i(x_j))|^2 \rangle}{((\varphi_i(x_j)), (\varphi_i(x_j)))} \quad (13)$$

or, equivalently, find the member of the $\psi_i(x_j)(= \varphi_i(x_j))$ that maximizes the normalized inner product of the candidate structure $\psi_i(x_j)$ with the field $u_i(x_j, t)$. A necessary condition for problem eqn. (13) is that $\varphi_i(x_j)$ is an eigenfunction, solution of the eigenvalue problem and Fredholm integral equation of the second kind:

$$\int_D R_{ij}(x_l, x'_l)\varphi_j(x'_l)dx'_l = \int_D \langle u_i(x_k, t)u_j(x'_k, t)\rangle \varphi_j(x'_k)dx'_k = \lambda\varphi_i(x_k) \quad (14)$$

where $R_{ij} = \langle u_i(x_k, t)u_j(x'_k, t)\rangle$ is the two-point velocity correlation tensor. The maximum of eqn. (13) corresponds to the largest eigenvalue of eqn. (14).

When $D$ is bounded, there exists a denumerable infinity of solutions of eqn. (14) (Hilbert-Schmidt theory) and these solutions are called the empirical eigenfunctions $\varphi_i^{(n)}(x_j)$ (normalized, $\|\varphi_i^{(n)}(x_j)\| = 1$). Orthogonality (orthonormality in this case) implies that structures of different order do not interact with each other in their contribution to second-order statistics. To each eigenfunction is associated a real positive eigenvalue $\lambda^{(n)}$($R_{ij}$ is non-negative by construction) and the eigenfunctions form a complete set. Every member of the ensemble can be reconstructed by means of a modal decomposition in the eigenfunctions themselves:

$$u_i(x_j, t) = \sum_n a_n(t)\varphi_i^{(n)}(x_j) \tag{15}$$

that can be seen as a decomposition of the original random field into deterministic structures with random coefficients. The modal amplitudes are uncorrelated and their mean square values are the eigenvalues themselves:

$$\langle a_n(t)a_m(t)\rangle = \delta_{nm}\lambda^{(n)}. \tag{16}$$

A diagonal decomposition of $R_{ij}$ holds:

$$R_{ij}(x_l, x_l') = \sum_n \lambda^{(n)}\varphi_i^{(n)}(x_l)\varphi_j^{(n)}(x_l') \tag{17}$$

implying that the contribution of each different structure to the turbulent kinetic energy content of the flow, can be separately calculated:

$$E = \int_D \langle u_i(x_j, t)u_i(x_j, t)\rangle = \sum_n \lambda^{(n)} \tag{18}$$

where $E$ is the total turbulent kinetic energy in the domain $D$. Thus, each eigenvalue represents the contribution of each correspondent structure to the total amount of kinetic energy. The *POD* is optimal for modeling or reconstructing a signal $u_i(x_j, t)$ in the sense that, for a given number of modes, the decay of the tail of the empirical eigen-spectrum is always faster (or at most as fast) than the tail of the spectrum based on any other possible basis, Fourier spectrum included.

The Proper Orthogonal Decomposition has been used in Rayleigh-Bénard turbulent convection problems (Sirovich and Park [93], Park and Sirovich [94], Deane and Sirovich [95] and Sirovich and Deane [96]), in studies of free shear flows (Sirovich *et al* [97] and Kirby *et al* [98]) and in the analysis of wall-bounded turbulent flow (Alfonsi *et al* [99–101]). In the field of wall-bounded flows Aubry *et al* [102] used the *POD* in studying the turbulent boundary-layer problem starting from experimental pipe flow data. They introduced the so-called bi-orthogonal decomposition that can be otherwise reduced to a particular case of the general POD formulation. Moin and Moser [103], Sirovich *et al* [104] and Ball *et al* [105], applied the method of the Proper Orthogonal Decomposition to the turbulent channel flow. The two homogeneous directions (streamwise and spanwise) are treated by means of Fourier decomposition and $R_{ij}$ has to be evaluated only along the direction orthogonal to the solid walls. Webber *et al* [106] analyzed with the method of the *KL* decomposition a database obtained by using the minimal channel flow domain. They showed that the most energetic modes of the flow are streamwise rollers followed by outward tilted quasi-streamwise vortices, very similar to structures already observed in physical experiments. This work actually follows another work of Sirovich *et al* [107] in which *DNS* data of turbulent channel flow are analyzed with the method of the *POD*. The analysis reveals the presence of propagating plane waves in the turbulent boundary layer and the interaction of these waves appears to be essential in the process of turbulence production through bursting or sweeping events, with the further suggestion that the fast-acting plane waves *actually trigger* the turbulence-production events. Handler *et al* [108] presented results of direct numerical simulations of turbulent channel flow in which a forcing is introduced as derived from the randomization of selected Fourier modes. An increase of 30% in the maximum mass flux with respect to normal turbulent condition is declared, corresponding to a drag reduction of 58%. The authors claim that numerical drag reduction by phase randomization is due to the destruction of coherency in the turbulence-producing structures near the wall – the plane waves of [107] – actually inhibiting the bursting mechanism. Here the viewpoint is emphasized that turbulence results from coherent triad interactions of plane waves and roll modes of the flow, so that, in order to control turbulence (with the aim of obtaining skin friction reduction) this coherency has to be destroyed. Levich *et al* [109] showed that the energy-transfer process to small scales of turbulence requires a specific phase coherency of helicity-associated fluctuations. Levich [110], in discussing classical and modern concepts in turbulence and in particular the insufficiency of the classical semi-empirical approaches to turbulence closures, argues that intermittency in physical space is in correspondence with certain phase coherency of turbulence in an appropriate dual space and analyzes phase coherency and intermittency for turbulence control. As a physical counterpart, Sirovich and Karlsson [111] performed a laboratory experiment in which randomized arrays of appropriately designed protrusions on the wall of a channel resulted in a measured drag reduction of the 10% with respect to the smooth-wall case.

## 6 Concluding remarks

The issue of coherent motions in turbulent shear flows has been reviewed. The rapid evolution of research methods and approaches in both experimental and numerical fields is supported by the advent of new concepts in describing and interpreting turbulence. One of these concepts is that of coherent structures. Coherent structures of turbulence represent a promising category for the physical description of turbulent flows, particularly as regards the leading objectives in modern fluid technology. Of greatest interest is the control of turbulence and the development of new predictive models for the numerical calculation of high-Reynolds-number flows of relevance to applications.

This work was supported by the Italian Ministry of Scientific Research, project PRIN 2002 "Influence of vorticity and turbulence in interactions of water bodies with their boundary elements and effect on hydraulic design".

## References


[1] Reynolds, O., On the dynamical theory of incompressible viscous fluids and the determination of the criterion, *Phil. Trans. R. Soc. London, A* **186**, pp. 123–161, 1894.

[2] Prandtl, L., Bericht über untersuchungen zur ausgebildeten turbulenz, *Z. Angew. Math. Mech. (ZAMM),* **5**, pp. 136–139, 1925.

[3] Taylor, G.I., The statistical theory of turbulence, Parts I-IV, *Proc. R. Soc. London, A* **151**, pp. 421–511, 1935.

[4] Kolmogorov, A.N., The local structure of turbulence in an incompressible fluid for very large Reynolds numbers, *Dokl. Akad. Nauk. SSSR,* **30**, pp. 299–303, 1941.

[5] Heisenberg, W., On the theory of statistical and isotropic turbulence, *Proc. R. Soc. London, A* **195**, pp. 402–406, 1948.

[6] Batchelor, G.K., *The theory of homogeneous turbulence*, Cambridge Univ. Press, 1953.

[7] Sreenivasan, K.R. & Antonia, R.A., The phenomenology of small-scale turbulence, *Ann. Rev. Fluid, Mech.,* **29**, pp. 435–472, 1997.

[8] Shen, X. & Warhaft, Z., The anisotropy of the small scale structure in high Reynolds number ($R_\lambda \approx 1000$) turbulent shear flow, *Phys. Fluids,* **12**, pp. 2976–2989, 2000.

[9] Townsend, A.A., *The structure of turbulent shear flow*, Cambridge Univ. Press, 1st ed., 1956.

[10] Corssin, S. & Kistler, A., The free stream boundaries of turbulent flows, *NACA Tech.* Note 3133, 1954.

[11] Klebanoff, P.S., Characteristics of turbulence in a boundary layer with zero pressure gradient, *NACA Tech.* Note 3178, 1954.

[12] Comte-Bellot, G., Hot-Wire Anemometry, *Ann. Rev. Fluid, Mech.,* **8**, pp. 209–231, 1976.

[13] Buchhave, P. & George, W.K., The measurement of turbulence with the Laser-Doppler anemometer, *Ann. Rev. Fluid, Mech.,* **11**, pp. 443–503, 1979.

[14] Alfonsi, G., Analysis of streamwise velocity fluctuations in turbulent pipe flow with the use of an ultrasonic Doppler flowmeter, *Flow Turb. Combust.,* **67**, pp. 137–142, 2001.

[15] Adrian, R.J., Particle-imaging techniques for experimental fluid mechanics, *Ann. Rev. Fluid, Mech.,* **23**, pp. 261–304, 1991.

[16] Canuto, C., Hussaini, M.Y., Quarteroni, A. & Zang, T.A., *Spectral methods in fluid dynamics,* Springer Verlag, 1987.

[17] Glowinski, R., Finite element methods for Navier-Stokes equations, *Ann. Rev. Fluid, Mech.,* **24**, pp. 167–204, 1992.

[18] Karniadakis, E.M. & Sherwin, S.J., *Spectral/hp element methods for CFD,* Oxford Univ. Press, 1999.

[19] Alfonsi, G., Giorgini, A., Nonlinear perturbation of the vortex shedding from a circular cylinder, *J. Fluid, Mech.,* **222**, pp. 267–291, 1991.

[20] Alfonsi, G., Passoni, G., Pancaldo, L. & Zampaglione, D., A spectral-finite difference solution of the Navier-Stokes equations in three dimensions, *Int. J. Num. Meth. Fluids,* **28**, pp. 129–142, 1998.

[21] Passoni, G., Alfonsi, G. & Galbiati, M., Analysis of hybrid algorithms for the Navier-Stokes equations with respect to hydrodynamic stability theory, *Int. J. Num. Meth. Fluids,* **38**, pp. 1069–1089, 2002.

[22] Speziale, C.G., Analytical methods for the development of Reynolds-stress closures in turbulence, *Ann. Rev. Fluid, Mech.,* **23**, pp. 107–157, 1991.

[23] Lesieur, M. & Métais, O., New trends in large-eddy simulation of turbulence, *Ann. Rev. Fluid, Mech.,* **28**, pp. 45–82, 1996.

[24] Moin, P. & Mahesh, K., Direct numerical simulation: a tool in turbulence research, *Ann. Rev. Fluid, Mech.,* **30**, pp. 539–578, 1998.

[25] Smagorinsky, J., General circulation experiments with the primitive equations, *Monthly Weather, Rev.,* **91**, pp. 99–164, 1963.

[26] Bardina, J., Ferziger, J.H. & Reynolds, W.C., Improved subgrid models for large-eddy simulation, *AIAA Pap.* pp. 80–1357, 1980.

[27] Kraichnan, R.H., Eddy viscosity in two and three dimensions, *J. Atmos. Sci.,* **33**, pp. 1521–1536, 1976.

[28] Métais, O. & Lesieur, M., Statistical predictability of decaying turbulence, *J. Atmos. Sci.,* **43**, pp. 857–870, 1986.

[29] Yakhot, A., Orszag, S.A., Yakhot, V. & Israeli, M., Renormalization group formulation of large-eddy simulation, *J. Sci. Comput.,* **4**, pp. 139–158, 1989.

[30] Germano, M., Turbulence, the filtering approach, *J. Fluid, Mech.,* **238**, pp. 325–336, 1992.

[31] Grötzbach, G., Spatial resolution requirements for Direct Numerical Simulation of the Rayleigh-Bénard convection, *J. Comput. Phys.,* **49**, pp. 241–264, 1983.

[32] Passoni, G., Alfonsi, G., Tula, G. & Cardu, U., A wavenumber parallel computational code for the numerical integration of the Navier-Stokes equations, *Parall. Comp.,* **25**, pp. 593–611, 1999.

[33] Passoni, G., Cremonesi, P. & Alfonsi, G., Analysis and implementation of a parallelization strategy on a Navier-Stokes solver for shear flow simulation, *Parall. Comp.,* **27**, pp. 1665–1685, 2001.

[34] Robinson, S.K., Coherent motions in the turbulent boundary layer, *Ann. Rev. Fluid, Mech.,* **23**, pp. 601–639, 1991.

[35] Cantwell, B.J., Organized motion in turbulent flow, *Ann. Rev. Fluid, Mech.,* **13**, pp. 457–515, 1981.

[36] Panton, R.L., Overview of the self-sustaining mechanisms of wall turbulence, *Prog. Aerosp. Sci.,* **37**, pp. 341–383, 2001.

[37] Coles, D.E., The law of the wake in the turbulent boundary layer, *J. Fluid, Mech.,* **1**, pp. 191–226, 1956.

[38] Barenblatt, G.I., Scaling laws for fully developed turbulent shear layer flows. Part 1. Basic hypotesis and analysis, *J. Fluid, Mech.,* **248**, pp. 513–520, 1993.

[39] Barenblatt, G.I. & Prostokishin, V.M., Scaling laws for fully developed turbulent shear flows. Part 2. Processing of experimental data, *J. Fluid Mech.*, **248**, pp. 521–529, 1993.

[40] Nikuradse, J., Gesetzmässigkeiten der turbulenten strömung in glatten rohren, *NACA* TT F-10, 1932.

[41] Zagarola, M.V., Perry, A.C. & Smits, A.J., Log laws or power laws: the scaling in the overlap region, *Phys. Fluids,* **9**, pp. 2094–2100, 1997.

[42] Alfonsi, G., Brambilla, S. & Chiuch, D., On the use of an ultrasonic Doppler velocimeter in turbulent pipe flow, *Exp. Fluids,* **35**, pp. 553–559, 2003.

[43] Kline, S.J., Reynolds, W.C., Schraub, F.A. & Rundstadler, P.W., The structure of turbulent boundary layers, *J. Fluid, Mech.,* **30**, pp. 741–773, 1967.

[44] Guezennec, Y.G., Piomelli, U. & Kim, J., On the shape and dynamics of wall structures in turbulent channel flow, *Phys. Fluids, A* **1**, pp. 764–766, 1989.

[45] Corino, E.R. & Brodkey, R.S., A visual investigation of the wall region in turbulent flow, *J. Fluid, Mech.,* **37**, pp. 1–30, 1969.

[46] Kim, H.T., Kline, S.J. & Reynolds, W.C., The production of turbulence near a smooth wall, *J. Fluid, Mech.,* **50**, pp. 133–160, 1971.

[47] Grass, A.J., Structural features of turbulent flow over smooth and rough boundaries, *J. Fluid, Mech.,* **50**, pp. 233–255, 1971.

[48] Rao, K.N., Narasimha, R. & Narayanan, M.A.B., Bursting in a turbulent boundary layer, *J. Fluid, Mech.,* **48**, pp. 339–352, 1971.

[49] Antonia, R.A., Conditional sampling in turbulence measurement, *Ann. Rev. Fluid, Mech.,* **13**, pp. 131–156, 1981.

[50] Willmarth, W.W. & Wooldridge, C.E., Measurements of the fluctuating pressure at the wall beneath a thick turbulent boundary layer, *J. Fluid, Mech.,* **14**, pp. 187–210, 1962.

[51] Willmarth, W.W., Lu, S.S., Structure of the Reynolds stress near the wall, *J. Fluid, Mech.,* **55**, pp. 65–92, 1972.

[52] Brodkey, R.S., Wallace, J.M. & Eckelmann, H., Some properties of truncated turbulence signals in bounded shear flows, *J. Fluid, Mech.,* **63**, pp. 209–224, 1974.

[53] Eckelmann, H., The structure of the viscous sublayer and the adjacent wall region in a turbulent channel flow, *J. Fluid, Mech.,* **65**, pp. 439–459, 1974.

[54] Wallace, J.M., Brodkey, R.S. & Eckelmann, H., Pattern-recognized structures in bounded turbulent shear flows, *J. Fluid, Mech.,* **83**, pp. 673–693, 1977.

[55] Praturi, A.K. & Brodkey, R.S., A stereoscopic visual study of coherent structures in turbulent shear flow, *J. Fluid, Mech.,* **89**, pp. 251–272, 1978.

[56] Kreplin, H.P. & Eckelmann, H., Propagation of perturbations in the viscous sublayer and adjacent wall region, *J. Fluid, Mech.,* **95**, pp. 305–322, 1979.

[57] Blackwelder, R.F. & Kaplan, R.E., On the wall structure of the turbulent boundary layer, *J. Fluid, Mech.,* **76**, pp. 89–112, 1976.

[58] Kim, J., Moin, P. & Moser, R., Turbulence statistics in fully developed channel flow at low Reynolds number, *J. Fluid, Mech.,* **177**, pp. 133–166, 1987.

[59] Moin, P. & Kim, J., The structure of the vorticity field in turbulent channel flow. Part 1. Analysis of instantaneous fields and statistical correlations, *J. Fluid, Mech.,* **155**, pp. 441–464, 1985.

[60] Kim, J. & Moin, P., The structure of the vorticity field in turbulent channel flow. Part 2. Study of ensemble-averaged fields, *J. Fluid, Mech.,* **162**, pp. 339–363, 1986.

[61] Adrian, R.J., Conditional eddies in isotropic turbulence, *Phys. Fluids,* **22**, pp. 2065–2070, 1979.

[62] Adrian, R.J. & Moin, P., Stochastic estimation of organized turbulent structure: homogeneous shear flow, *J. Fluid, Mech.,* **190**, pp. 531–559, 1988.

[63] Chu, C.D. & Karniadakis, E.G., A direct numerical simulation of laminar and turbulent flow over riblet-mounted surfaces, *J. Fluid, Mech.,* **250**, pp. 213–240, 1993.

[64] Choi, H., Moin, P. & Kim, J., Direct numerical simulation of turbulent flow over riblets, *J. Fluid, Mech.,* **255**, pp. 503–539, 1993.

[65] Bogard, D.G. & Tiederman, W.G., Burst detection with single-point velocity measurements, *J. Fluid, Mech.,* **162**, pp. 389–413, 1986.

[66] Luchik, T.S. & Tiederman, W.G., Timescale and structure of ejections and bursts in turbulent channel flows, *J. Fluid, Mech.,* **174**, pp. 529–552, 1987.

[67] Subramanian, C.S., Rajagopalan, S., Antonia, R.A. & Chambers, A.J., Comparison of conditional sampling and averaging techniques in a turbulent boundary layer, *J. Fluid, Mech.,* **123**, pp. 335–362, 1982.

[68] Kovasznay, L.S.G., Kibens, V. & Blackwelder, R., Large scale motion in the intermittent region of a turbulent boundary layer, *J. Fluid, Mech.,* **41**, pp. 283–325, 1970.

[69] Cantwell, B.J., Coles, D.E. & Dimotakis, P.E., Structure and entrainment in the plane of symmetry of a turbulent spot, *J. Fluid, Mech.,* **87**, pp. 641–672, 1978.

[70] Offen, G.R. & Kline, S.J., Combined dye-streak and hydrogen-bubble visual observations of a turbulent boundary layer, *J. Fluid, Mech.,* **62**, pp. 223–239, 1974.

[71] Brown, G.L. & Thomas, A.S.W., Large structure in a turbulent boundary layer, *Phys. Fluids,* **20**, S 243–252, 1977.

[72] Falco, R.E., Coherent motions in the outer region of turbulent boundary layer, *Phys. Fluids,* **20**, S 124–132, 1977.

[73] Head, M.R. & Bandyopadhyay, P., New aspects of turbulent boundary-layer structure, *J. Fluid, Mech.,* **107**, pp. 297–338, 1981.

[74] Wygnanski, I.J. & Champagne, F.H., On transition in a pipe. Part I. The origin of puffs and slugs and the flow in a turbulent slug, *J. Fluid, Mech.,* **59**, pp. 281–335, 1973.

[75] Townsend, A.A., *The structure of turbulent shear flow,* Cambridge Univ. Press, 2nd ed., 1976.

[76] Perry, A.E., Henbest, S. & Chong, M.S., A theoretical and experimental study of wall turbulence, *J. Fluid, Mech.,* **165**, pp. 163–199, 1986.

[77] Perry, A.E., Li, J.D. & Marusich, I., Towards a closure scheme for turbulent boundary layers using the attached eddy hypothesis, *Phil. Trans. R. Soc. London,* **336**, pp. 67–79, 1991.

[78] Theodorsen, T., Mechanism of turbulence, *Proc. 2nd Midwest. Fluid Mech. Conf.,* Columbus, Ohio, pp. 1–18, 1952.

[79] Acarlar, M.S. & Smith, C.R., A study of hairpin vortices in a laminar boundary layer. Part 1. Hairpin vortices generated by a hemisphere protuberance, *J. Fluid, Mech.,* **175**, pp. 1–41, 1987.

[80] Acarlar, M.S. & Smith, C.R., A study of hairpin vortices in a laminar boundary layer. Part 2. Hairpin vortices generated by fluid injection, *J. Fluid, Mech.,* **175**, pp. 43–83, 1987.

[81] Smith, C.R., Walker, J.D.A., Haidari, A.H. & Soburn, U., On the dynamics of near-wall turbulence, *Phil. Trans. R. Soc. London,* **336**, pp. 131–175, 1991.

[82] Haidari, A.H. & Smith, C.R., The generation and regeneration of single hairpin vortices, *J. Fluid, Mech.,* **277**, pp. 135–162, 1994.

[83] Perry, A.E. & Chong, M.S., On the mechanism of wall turbulence, *J. Fluid, Mech.,* **119**, pp. 173–217, 1982.

[84] Singer, B.A. & Joslin, R.D., Metamorphosis of a hairpin vortex into a young turbulent spot, *Phys. Fluids,* **6**, pp. 3724–3736, 1994.

[85] Adrian, R.J., Meinhart, C.D. & Tomkins, C.D., Vortex organization in the outer region of the turbulent boundary layer, *J. Fluid, Mech.,* **422**, pp. 1–54, 2000.

[86] Peridier, V.J., Smith, F.T. & Walker, J.D.A., Vortex-induced boundary-layer separation. Part 1. The unsteady limit problem $Re \to \infty$, *J. Fluid, Mech.,* **232**, pp. 99–131, 1991.

[87] Peridier, V.J., Smith, F.T. & Walker, J.D.A., Vortex-induced boundary-layer separation. Part 2. Unsteady interacting boundary-layer theory, *J. Fluid Mech.,* **232**, pp. 133–165, 1991.

[88] Van Dommelen, L.L. & Cowley, S.J., One the Lagrangian description of unsteady boundary-layer separation. Part 1. General theory, *J. Fluid Mech.,* **210**, pp. 593–626, 1990.

[89] Doligalski, T.L. & Smith, C.R., Walker, J.D.A., Vortex interactions with walls, *Annual Rev. Fluid, Mech.,* **26**, pp. 573–616, 1994.

[90] Berkooz, G., Holmes, P. & Lumley, J.L., The proper orthogonal decomposition in the analysis of turbulent flows, *Ann. Rev. Fluid, Mech.,* **25**, pp. 539–575, 1993.

[91] Lumley, J.L., *Stochastic tools in turbulence,* Academic Press, 1971.

[92] Sirovich, L., Turbulence and the dynamics of coherent structures Part I: coherent structures. Part II: symmetries and transformations. Part III: dynamics and scaling, *Quart. Appl. Math.,* **45**, pp. 561–590, 1987.

[93] Sirovich, L. & Park, H., Turbulent thermal convection in a finite domain. Part I. Theory, *Phys. Fluids, A* **2**, pp. 1649–1658, 1990.

[94] Park, H. & Sirovich, L., Turbulent thermal convection in a finite domain. Part II. Numerical results, *Phys. Fluids, A* **2**, pp. 1659–1668, 1990.

[95] Deane, A.E. & Sirovich, L., A computational study of Rayleigh-Bénard convection. Part I. Rayleigh-number scaling, *J. Fluid, Mech.,* **222**, pp. 231–250, 1991.

[96] Sirovich, L. & Deane, A.E., A computational study of Rayleigh-Bénard convection. Part II. Dimension considerations, *J. Fluid, Mech.,* **222**, pp. 251–265, 1991.

[97] Sirovich, L., Kirby, M. & Winter, M., An eigenfunction approach to large scale transitional structures in jet flow, *Phys. Fluids, A* **2**, pp. 127–136, 1990.

[98] Kirby, M., Boris, J. & Sirovich, L., An eigenfunction analysis for axisymmetric jet flow, *J. Comput. Phys.,* **90**, pp. 98–122, 1990.

[99] Alfonsi, G. & Primavera, L., Coherent structure dynamics in turbulent channel flow, *J. Flow Visual. Im. Proc.,* **9**, pp. 89–98, 2002.

[100] Alfonsi, G., Primavera, L. & Felisari, R., On the behavior of POD modes of the flow past a perforated plate, *J. Flow Visual. Im. Proc.,* **10**, pp. 105–117, 2003.

[101] Alfonsi, G., Restano, C. & Primavera, L., Coherent structures of the flow around a surface-mounted cubic obstacle in turbulent channel flow, *J. Wind Eng. Ind. Aerodyn.,* **91**, pp. 495–511, 2003.

[102] Aubry, N., Holmes, P., Lumley, J.L. & Stone, E., The dynamics of coherent structures in the wall region of a turbulent boundary layer, *J. Fluid Mech.,* **192**, pp. 115–173, 1988.

[103] Moin, P. & Moser, R.D., Characteristic-eddy decomposition of turbulence in a channel, *J. Fluid, Mech.,* **200**, pp. 471–509, 1989.

[104] Sirovich, L., Ball, K.S. & Keefe, L.R., Plane waves and structures in turbulent channel flow, *Phys. Fluids, A* **2**, pp. 2217–2226, 1990.

[105] Ball, K.S., Sirovich, L. & Keefe, L.R., Dynamical eigenfunction decomposition of turbulent channel flow, *Int. J. Num. Meth. Fluids,* **12**, pp. 585–604, 1991.

[106] Webber, G.A., Handler, R.A. & Sirovich, L., The Karhunen-Loéve decomposition of minimal channel flow, *Phys. Fluids,* **9**, pp. 1054–1066, 1997.

[107] Sirovich, L., Plane waves and structures in turbulent channel flow, *Phys. Fluids, A* **2**, pp. 2217–2226, 1990.

[108] Handler, R.A., Levich, E. & Sirovich, L., Drag reduction in turbulent channel flow by phase randomization, *Phys. Fluids, A* **5**, pp. 686–694, 1993.

[109] Levich, E., Shtilman, L. & Tur, A.V., The origin of coherence in hydrodynamical turbulence, *Physica A,* **176**, pp. 241–296, 1991.

[110] Levich, E., New developments and classical theories of turbulence, *Int. J. Mod. Phys.,* **10**, pp. 2325–2392, 1996.

[111] Sirovich, L. & Karlsson, S., Turbulent drag reduction by passive mechanisms, *Nature,* **388**, pp. 753–755, 1997.

# CHAPTER 2

# Results on large eddy simulations of some environmental flows

V. Armenio[1] & S. Salon[2]
[1]*Dipartimento di Ingegneria Civile e Ambientale, Universitá di Trieste, Italy.*
[2]*Istituto Nazionale di Oceanografia e Geofisica Sperimentale - OGS, Sgonico (TS), Italy.*

## Abstract

In the present paper a synthesis of the scientific activity of the group of Environmental Hydraulics of the Universitá di Trieste is given. Numerical simulations of flow fields of interest in Environmental engineering are presented and discussed. The simulations were carried out using Large-eddy simulation that is nowadays considered as a formidable tool for the detailed investigation of turbulence. The model employed in the simulation is briefly described. The results are shown for the following classes of problems: turbulent field over a topography; turbulent stratified wall-bounded flows; unsteady wall-bounded flows; particle dispersion in turbulent flows.

## 1 Introduction

Numerical simulations represent nowadays a well-established tool for investigating turbulence. Turbulence research spans from civil engineering to geophysics, from industrial applications to fundamental studies, and the need to model turbulent flows is by now a meeting point for many researchers coming from different backgrounds.

Numerical techniques can be generally divided in three kinds of approaches: Reynolds-averaged Navier-Stokes equations (RANS), large-eddy simulation (LES) and direct numerical simulation (DNS). The differences among these methods are substantially due to the averaging treatment adopted in the resolution of the primitive equations: RANS gives an estimate of the mean fields (velocity and density) and parameterizes the contribution of the fluctuating ones by closures of the Reynolds stresses and density flux tensors; LES directly solves the large scales of turbulence and statistically models the small, isotropic and dissipative scales with a subgrid scale (SGS) model; DNS does not use any kind of closure or filtering and solves all the scales of turbulence directly. However, none of these techniques is superior to the others: each of them has a well-defined field of application which is roughly related to the Reynolds number relative to the problem studied. In particular, due to the computational requirements typical of each method, RANS is commonly used for high Reynolds numbers; DNS is suited for low Reynolds number problems. LES stays in the middle: the computational power is not as so burdensome as that required by DNS, but anyhow higher than RANS.

The problem related to the use of RANS is that, since all the turbulent scales must be parameterized, including the large and not isotropic ones, the performance of the model can be very sensitive to the physical characteristics of the turbulent field. From a practical point of view, this means that the constants of the model, finely tuned based on the results coming from a typical experiment, can be off-designed for a turbulent field that does not fit the original conditions. In other words, the performance of the model is often strongly dependent on the empiric tuning of the constants.

On the other hand, in large eddy simulation, since only the small and more isotropic scales of turbulence are parametrized, the quality of the numerical results is nearly independent on the characteristics of the turbulent field. In particular, as will be discussed in the next section, since in dynamic models the constant is calculated *a posteriori* during the calculation, based on the local characteristics of the turbulent field, the model is basically free of empiricism in the choice of the constant.

Large eddy simulation gives results at values of Reynolds number that, although far from being of applicative use, are large enough to minimize Reynolds-number effects on the qualitative characteristics of the flow fields. It follows that databases of LES may be employed for testing and developing RANS-like models. This may be more difficult for DNS: specifically, so far DNS has been applied to very low Reynolds number flow field, and the extrapolation of the results to the cases of high Reynolds number flows is not straightforward.

In the present paper a review of LES of flows relevant in environmental applications, performed by the Group of Environmental Hydraulics of the Universitá di Trieste (particles dispersion, topography effect, stratified flows, unsteadily driven flows) is illustrated, showing the most significant results.

## 2 The mathematical formulation

As previously hinted, in large-eddy simulation a filtering operation is performed to separate the large scales of motion from the small ones. A direct consequence of filtering the Navier-Stokes equations is the presence of second-order terms determining the energetic contribution associated to the small scales which needs to be modeled.

A number of SGS models have been developed and used over the last decades. For a detailed discussion the reader is referred to [1]. Moreover, as far as wall-bounded applications are concerned, two kinds of LES are commonly carried out: large-scale simulations where the near-wall viscous layer is parameterized through the use of wall-layer models (see for details [2, 3]), and simulations where the viscous layer is directly resolved, called *resolved* LES. The former are able to give results at applicative values of the Reynolds number, although the effectiveness of the near-wall parameterization in situations characterized by complex geometry and physics (rotation, stratification, local re-laminarization) is still under analysis. The latter give very accurate results in a large class of problems but remain applicable at small-to-moderate values of the Reynolds number. In spite of this limitation, resolved LES still remains a formidable tool for understanding new physical mechanisms occurring in turbulence.

In this paper we focus on resolved LES carried out using dynamic mixed models that have been proved to be able to accurately simulate equilibrium as well as non-equilibrium turbulent flows [4, 5]. Moreover, dynamic models have also been demonstrated to simulate correctly flow fields characterized by sharp transition to turbulence and local re-laminarization (see for example [6, 7]). As regards unsteady turbulent flows subjected to periodic forcing, LES has been successfully employed by Scotti & Piomelli [8] to study the pulsating flow in a channel and by Salon *et al* [9] for the investigation of the turbulent Stokes boundary layer.

In the past, most cases studied *via* LES with dynamic models were characterized by simple geometry. SGS models have been extensively validated within the Cartesian framework, and only recently their response under coordinate transformation was extensively evaluated. On the basis of the findings of Jordan [10], recently Armenio & Piomelli [11] have reformulated the dynamic mixed model of Wu & Squires [12] in a novel contravariant formulation, with the aim to investigate turbulent flow developing over topography or, in general, over complex geometry, that is the typical case of environmental applications.

The model, developed and employed in a wide variety of problems by the group of Environmental Hydraulics of the Universitá di Trieste, is described in detail in [11]. The performance of the model was evaluated using as test case the canonical turbulent channel flow. The results of the simulations showed the response of the model to be insensitive to grid deformation. Small differences in the turbulent intensities were detected between the response of the dynamic eddy-viscosity model and the dynamic mixed model, whereas it was observed that the mixed model is able to give a more accurate velocity profile. Overall, the curvilinear dynamic-mixed model proved to be able to simulate correctly turbulent flow field even in cases where highly distorted grids were required.

In spite of the number of algebraic operations required for going back and forth from the computational to the physical space, the present formulation has proved to be not appreciably more expensive than the direct one. This is due to the fact that filtering in the physical space, when the computational domain is not regular, requires the use of a special filter function that increases the computational cost of the simulation (for details see [10]).

# 3 Analysis of turbulent flows over topography

A typical feature of flows of interest in environmental fluid mechanics is the presence of a topography. As an example, the flow over wavy walls is relevant for the comprehension of mixing properties in the low atmosphere when the wind flows over two-dimensional hills, or for the analysis of vertical mixing in the oceans in the presence of large-scale topography. Another example worth mentioning is when a turbulent flow develops along a canyon-like geometry characterized by the presence of large-scale longitudinal topography. In this case, vertical mixing is associated to the presence of large-scale, cross-sectional secondary recirculations that are able to drive fluid from the bottom toward the top and *vice versa*.

In this section some results of LES applications to flows which involve domains with topography are described. In particular, we present here: (i) a turbulent flow over a two-dimensional wavy wall, and (ii) a turbulent flow in a channel with longitudinal ridges over the walls.

## 3.1 Channel with a wavy wall

As regards the turbulent flow over a wavy wall, it is known that if the wave slope, defined as the wave amplitude ($2a$) to the wavelength ($\lambda$) ratio, is large enough ($2a/\lambda > 0.05$) the flow exhibits a large recirculation zone downstream of the wave crest [13, 14]. In this case the flow field can be considered as composed of four regions: an outer region, above the wave crest, for which the waviness constitutes a sort of wall roughness, an intense shear layer downstream the wave crest where turbulent kinetic energy is produced, the mean recirculation region below the shear layer, and a thin boundary layer developing downstream the recirculation, in the upslope part of the wall. The re-attachment region is characterized by the presence of large space-time fluctuations of the shear stress and of the pressure that can be responsible of phenomena of erosion and sediment transport. Many experimental and numerical studies have been devoted to the analysis of the turbulent flow over wavy walls. For a review, the reader is referred to [11].

Large-eddy simulations were carried out by Armenio & Piomelli [11] considering two different cases, both characterized by a wave height such to yield the development of a recirculation region. In both cases, a channel flow was considered, having a bottom wavy wall and a flat wall at the top. One of the cases aimed at reproducing the experiment of Hudson *et al* [14], whereas the second one matched the experiment of Buckles [15]. In both cases, following [16] and [17], a periodic domain containing two waves was considered. The Lagrangian technique of Meneveau *et al* [18] was employed to localize the dynamic mixed model employed for the simulations. The experiments of Hudson *et al* [14] were carried out considering a moderate-amplitude wave slope ($2a/\lambda = 0.1$) and a value of the Reynolds number $Re_b = 6600$, referred to the bulk velocity $u_b$ and the domain's height. The results of the LES simulations were compared with the DNS data of Maas & Schumann [19] and the experimental data of Hudson *et al* [14]. The present LES predicted a zone of reversed flow, characterized by the recirculating region, larger than that measured experimentally. However, the streamwise location of the separation and

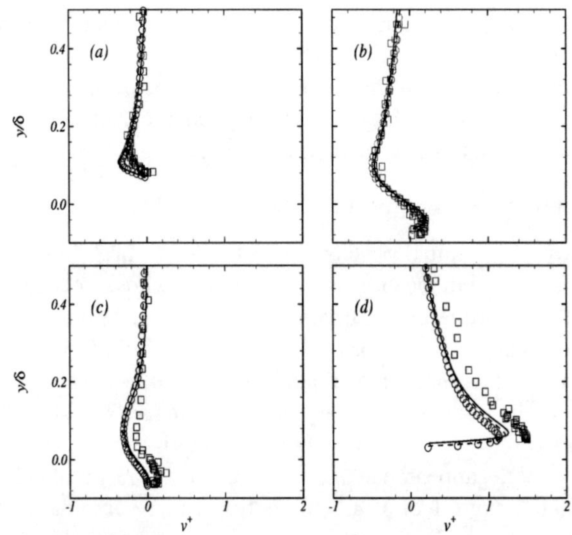


Figure 1: Wavy channel with $2a/\lambda = 0.1$: Mean vertical velocity made non dimensional with the friction velocity at four different streamwise sections (a) $x/\lambda = 0.1$; (b) $x/\lambda = 0.4$; (c) $x/\lambda = 0.6$; (d) $x\lambda = 0.8$. Dashed line, coarse LES; solid line, fine LES; circles, DNS of [19]; squares, experiment of [14]. From [11].

reattachment points is in agreement with the numerical results of Maas & Schumann [19] and the LES of Calhoun [16]. The averaged velocity profiles and the turbulence intensities were compared with the reference ones at four locations: downstream of the wave crest ($x/\lambda = 0.1$), in the middle of the recirculation zone ($x/\lambda = 0.4$), near the reattachment point ($x/\lambda = 0.6$) and beyond it ($x/\lambda = 0.8$). The streamwise and wall-normal velocity profiles predicted by the use of LES were in good agreement with both DNS and experimental data, except at $x/\lambda = 0.6$ where all the numerical simulations evaluated the flow on the verge of reattachment while the experiment measured it in the recovery region (see for example fig. 1). Similarly, also the turbulence intensities and the Reynolds shear stresses gave a satisfactory agreement with reference results with small discrepancies near the recirculation zone, at $x/\lambda = 0.4$ and 0.8. In general, the agreement between LES and DNS data was systematically better than that with the experimental ones, probably due to experimental uncertainty and small differences in the boundary conditions and in the geometry.

The experiments of Buckles [15] and the simulations of Henn & Sykes [17] were characterized by $Re_b = 12000$ and a large wave amplitude ($2a/\lambda = 0.2$). Also in this case, the numerical simulations gave a recirculation region larger than that measured in the experiments, although in this case the difference was more evident at the separation point. Similarly to the previous case, peaks of $\langle u'v' \rangle$, $q^2$ and of $\nu_T/\nu$ were shown to be present in the shear layer, but in this case another maximum of $q^2$ was detected near the wall beyond the reattachment point. Such intense value was shown to be related to a growth of the spanwise turbulence intensity, which, differently from the previous case, is even larger than the maximum streamwise component, due to the splash phenomena occurring in the up-slope part of the topography, beyond the reattachment region.

### 3.2 Turbulent flow over longitudinally ridged walls

The presence of longitudinal obstacles in a boundary layer produces the rise of secondary recirculation in the cross-sectional plane, whose intensity is of the order of some percentage of the mean streamwise velocity. In spite of their own weakness, such large-scale recirculations are able to drive fluid from the bottom to the top and *vice versa*, thus supplying an additional mean for vertical mixing of mass and momentum. The study of secondary recirculation has been shown to be relevant for the analysis of the sand bars developing in the river beds, since, as addressed by Colombini [20] the appearance of a sand ridge is the result of a delicate balance between the stabilizing effect of the gravitational force and the destabilizing one given by the secondary wall shear stress. Falcomer & Armenio [5] performed large-eddy simulations reproducing case "I" of the experiments of Nezu & Nakagawa [21], who considered a rectangular duct (with aspect ratio $b/h = 2.25$ with $b$ and $h$ the width and the height of the duct) with the horizontal walls equipped with longitudinal ridges. The friction Reynolds number of the experiments (based on the half height of the duct and the mean friction velocity) was set equal to $580$. The trapezoidal ridge, in wall units was $70$ high, $290$ wide at the bottom and $145$ wide at the top. In the case reproduced in the simulations, the ridge spacing along the spanwise direction was equal to $590$ wall units. Periodic conditions were used in the streamwise and spanwise directions, thus considering an infinite array of ridges. The dynamic mixed model with the constant averaged along the streamwise direction of homogeneity was used. Details on the computational parameters are in [5]. The comparison between the results of the simulation and the experimental data was in general satisfactory, although differences were detected both in the spanwise distribution of the streamwise velocity component (fig. 2) and in the position of the center of the main secondary recirculation. Such differences where attributed to the difference between the numerical and the experimental set up. The numerical simulations performed considering a channel unbounded in the spanwise direction; the experiments were carried out in a low aspect-ratio duct.

The analysis of the LES results showed that, at the value of the Reynolds number investigated, a large-scale recirculation that spans the whole half height of the domain was detected together with two small cellular circulations, respectively at the bottom corner and at the top of the ridge (fig. 3). The presence of such additional recirculations was explained by analysis of the source terms of the transport equation of the mean streamwise vorticity. Such small cellular near-wall circulations were shown to have a dramatic effect on the spanwise distribution and magnitude of the secondary wall stress.

Finally, the analysis of the coherent structures showed that the main effect of the ridge is a reorganization and alignment of the structures compared to an

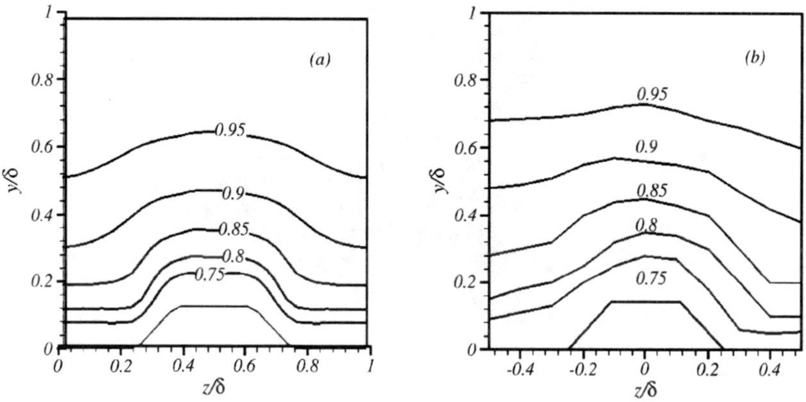


Figure 2: Mean streamwise velocity (made non-dimensional with $U_{max}$) in the cross-stream section ($Re_\tau = 580$): (a) LES data; (b) data from [21]. From [5].

equivalent plane-channel flow. The analysis also proved that the structures evolving along the crest of the ridge are nearly independent on those evolving in the trough. The distribution of the wall stresses reflected the evolution of the coherent structures. In particular, the spanwise vorticity at the wall (and thus the primary wall stress) appeared well correlated over a length of about 850 wall units, whereas, the streamwise vorticity at the wall (and consequently the secondary wall stress) was observed to have a spot-like structure.

## 4 Stably stratified flows

One of the main characteristics of environmental flows is the fact that, in general, their temperature and/or the concentration of a dissolved species is a function of the depth within the fluid column. This makes the column stratified. As is well known, when the mean density increases with the depth, the fluid is stably stratified and vertical mixing of mass and momentum is inhibited. Due to the many practical applications, stably stratified flows are of great importance in environmental fluid mechanics. As an example, thermal inversion in the low atmosphere causes the stagnation of pollutants and particulates that degrades the quality of air, whereas in the oceans, stable stratification suppresses vertical mixing of nutrients. Generally, stable stratification can strongly influence the dynamic and the anisotropic characters of turbulence, leading to qualitative and quantitative changes in the small-scale mixing not only of momentum and mass, but also of salinity, pollutant and nutrients in the oceans.

Most numerical and experimental studies dealing with the interaction between a mean shear, that is the source of turbulent mixing, and a mean, stable, density gradient acting toward the suppression of turbulence, have been carried out in the very simple case of homogeneous turbulence (see for example [22–26]. In this

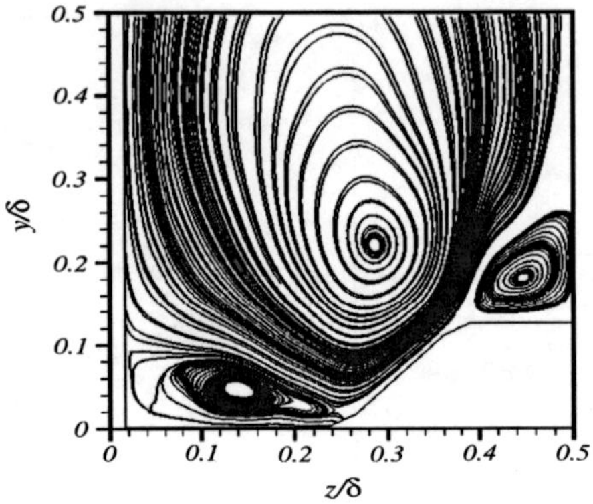


Figure 3: Mean stream-tracers in the cross-stream plane at $Re_\tau = 580$. From [5].

case a constant mean shear and a mean linear density profile are considered, and thus, the gradient Richardson number, $Ri_g = N^2/S^2$ (where $N^2 = g/\rho_0 \; d\rho/dz$ is the square of the Brunt-Väisälä frequency, with $g$ the gravitational acceleration and $\rho_0$ the reference density, and $S$ is the mean shear rate) is constant in space and in time. The fate of vertical mixing in wall-bounded stably stratified turbulence has been investigated by few authors (for example [27–29]), and among them by Armenio and co-authors [7, 30, 31]. These investigations were carried out using large-eddy simulation and the numerical method discussed in the previous sections. The governing equations were written under the assumption that the variations in the fluid density are very small compared to the reference density $\rho_0$, with $\rho_{tot} = \rho_0 + \rho$, and that inertial effects related to the variation of the perturbation density $\rho$ are negligible compared to those related to the gravitational field (Boussinesq approximation). Under these assumptions the flow field can be considered to be solenoidal, the effect of density variation on the velocity field comes from the gravitational term and the energy equation can be re-written as a transport equation for the perturbation density. In the LES framework, the filtered equations have to be considered and SGS density fluxes need to be modeled. A dynamic eddy diffusivity model was used for the closure of the SGS fluxes. Details are in [7].

Three problems were investigated:

- the first one is the turbulent flow that develops between two parallel, horizontal and infinite solid plates, with imposed temperature at the solid walls. In this case the shear and the density gradient are aligned, and they are functions of the vertical position in the channel;
- the second one is the turbulent flow between two parallel and vertical walls, in the presence of stable stratification. This problem is characterized by the fact that the mean shear and the density gradient are orthogonal to each other;
- the third one is characterized by the presence of a topography. Specifically, the turbulent free-surface flow evolving over longitudinally ridged walls was investigated under stable stratification.

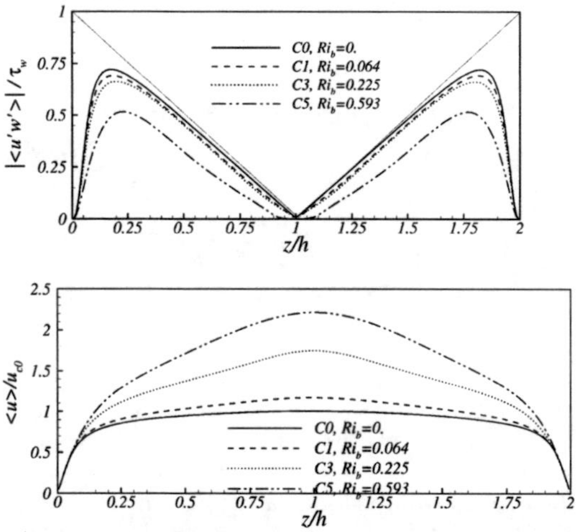


Figure 4: The influence of stratification on: (a) total turbulent momentum flux (sum of resolved and SGS quantities); (b) mean velocity profile scaled with the centerline velocity. From [7].

### 4.1 Stably stratified turbulent channel flow

The LES study of Armenio & Sarkar [7] investigated turbulent mixing in a channel flow with stable stratification, sustained by constant values of temperature at the solid walls. A main topic of the research was to investigate the role of the gradient Richardson number $Ri_g$ in inhomogeneous turbulence. The simulations were carried out with an imposed driving pressure gradient that gave a constant wall stress and hence a constant value of the friction Reynolds number $Re_\tau$, that in this case was set equal to 180. The Prandtl number was set equal to 0.71, corresponding to thermally stratified air. Six cases were studied spanning a wide range of stratification levels, from the case of passive scalar ($Ri_\tau = 0$) to the case of strong stratification ($Ri_\tau = 480$). Note that the friction Richardson number is defined as $Ri_\tau = (g\Delta\rho\ h)/(\rho_0 u_\tau^2)$ where $\Delta\rho$ is a reference density gap and $u_\tau$ the friction velocity.

The simulations showed that the the increase of stratification causes a general suppression of turbulent activity. The Reynolds shear stress appeared reduced (fig. 4a) and, thus, the integral balance required a corresponding increase of the molecular shear stress; this produced the increase of the mean velocity in the channel (fig. 4b) and, consequently of the bulk Reynolds number. Since the friction coefficient is $c_f = 2Re_\tau^2/Re_b^2$, and the friction Reynolds number does not change with increasing $Ri_\tau$, then the increase of stable stratification causes a reduction of the friction coefficient. Specifically, the coefficient dropped by a factor 3.4 from the case of passive scalar ($c_f = 8.18 \times 10^{-3}$) to that of strong stratification ($c_f = 2.4 \times 10^{-3}$ for $Ri_\tau = 480$).

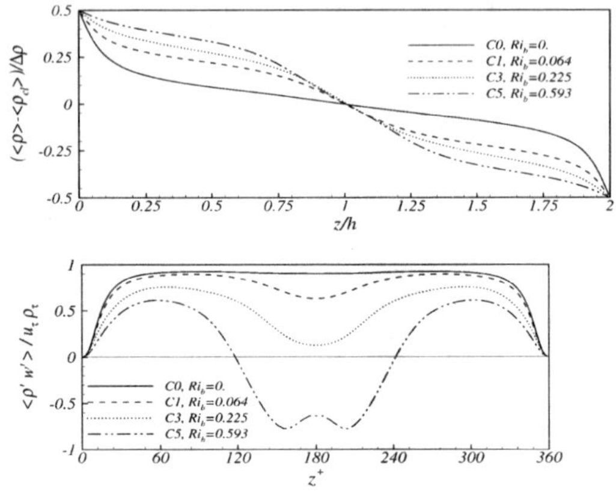

Figure 5: The influence of stratification on: (a) mean density profile made non dimensional with the density gap $\Delta\rho$; (b) turbulent buoyancy flux (sum of resolved and SGS quantities). The abscissa is the vertical coordinate made non-dimensional with $u_\tau/\nu$. From [7].

The mean density profile and the density fluctuations also appeared affected by stratification. The density gradient at the wall appeared decreased with increased stratification (fig. 5) and consequently the Nusselt number, that quantifies the turbulent heat flux at the wall, compared to the molecular one, appeared reduced. In agreement with the findings of Moore & Long [32] and Crapper & Linden [33], the mean density profile sharpened in the central region of the channel showing also, in case C5 ($Ri_\tau = 480$) and in agreement with Komori *et al* [28], the presence of countergradient density flux, together with the tendency to create a density interface (fig. 5a and b). On the other hand, even in the case of strong stratification, turbulence activity and well-mixed regions were still observed in the near-wall region.

The analysis of the map of the invariants of the anisotropy tensor of the Reynolds stresses (Lumley triangle) proved that the log-zone ($30 < z^+ < 110$) was the one most affected by the stable stratification: as $Ri_b$ increased, the fluctuating motion became more horizontal, and the vertical component appeared more and more reduced compared to the horizontal ones. As a result, near the wall ($z^+ < 30$) the standard cigar-like shape was observed whereas in the log-zone, while the low-$Ri$ case persisted on the cigar-like turbulence, the high-$Ri$ case tended to pancake-like shape.

The complete picture of the role of stratification appeared very clear when the local value of the gradient Richardson number $Ri_g = N^2/S^2$ was considered. The parameter $Ri_g(z^+)$ is linear from the wall up to the height where $Ri_g \simeq 0.2$ (according to the theory as shown in [7]), and after this point the slope abruptly increases. Moreover, the point where the slope changes moves toward the wall as stratification increases. The region where $Ri_g > 0.2$ roughly corresponded to the layer interested by the strong suppression of turbulence. An interesting result was that the correlation coefficients of turbulent momentum fluxes and buoyancy flux plotted against the *local* parameter $Ri_g$ for different values of the *overall* parameter $Ri_\tau$ tended to collapse over a very thin range of values. Moreover, they were nearly constant for $Ri_g < 0.15 - 0.20$ and they sharply decayed for $Ri_g > 0.2$, indicating a rapid change from unstratified turbulence with classical mixing characteristics to stably stratified turbulence with inhibited vertical mixing. The study has thus highlighted the presence of two separate regions in wall bounded stably stratified turbulence: a near wall one (*buoyancy affected region*), where $Ri_g < 0.2$, characterized by boundary-layer turbulence; a zone with $Ri_g > 0.2$, or (*buoyancy dominated region*), where classical turbulence was observed.

A side result of the research was that a dynamic model was able to predict by itself the rapid increase of the turbulent Prandtl number that occurs in the buoyancy dominated regime, without the need of any *ad-hoc* adjustment. This finding may be expected to have a strong impact as regards the simulation of strongly stratified turbulent flows.

## 4.2 Stably stratified turbulent channel flow with vertical walls

In environmental applications, there are situations in which the mean shear is not aligned with stratification. For instance, this is the case of canyon-like flows. The presence of vertical walls in a stably stratified channel flow induces horizontal gradient of mean velocity which competes with the vertical density gradient in changing the character of turbulence. The work by Armenio & Sarkar [30] deals with such application, considering a flow through a channel with infinite vertical walls separated by a distance $2h$ and with a uniform vertical stratification. The friction Reynolds number was set equal to $Re_\tau = 390$, the Prandtl number was $Pr = 5$, that corresponded to thermally stratified water.

Different cases of horizontal shear (HS) with stratification were simulated; moreover, in order to highlight the differences between the HS case and the most investigated vertical shear (VS), two cases characterized by the presence of VS were also run (see table 1). The LES adopted is the same as used in the previous section, coupled with a dynamic mixed SGS model.

The profiles of the gradient Richardson number for HS and VS along the wall-normal distance (fig. 6) showed that, for the same level of stratification, $Ri_g$ was generally larger in the HS case, thus producing narrower buoyancy-affected regions when compared to the VS cases. A main result of the study was that, in the case where the mean shear is orthogonal to the direction of stratification, turbulent mixing was much less affected by stratification than in an equivalent VS case. The analysis of the correlation coefficients plotted against the local gradient Richardson number showed that in the HS case, active turbulence is present for values of $Ri_g < 2$, one order of magnitude larger than that found in the VS case.

The study showed that horizontal mean shear is able to promote vertical transport even in case of stable stratification. Fluctuating horizontal vorticity was found to be well correlated with overturning events, in particular in the near-wall region. With increased stratification, the magnitude of the fluctuating horizontal vorticity remained nearly unaltered, but the barrier of potential energy in the flow field

Table 1: Parameters of the simulations and bulk quantities of the flow field for the cases studied in [30]. Cases *C0–C3* correspond to horizontal shear channel flow while cases *CV1* and *CV2* correspond to vertical shear channel flow.

| Case | $Re_\tau$ | $Re_b$ | $Ri_\tau$ | $Ri_b$ | $c_f \times 10^2$ |
|---|---|---|---|---|---|
| $C0$ | 390 | 7320 | 0 | 0 | 0.64 |
| $C1$ | 390 | 7530 | 15 | 0.041 | 0.55 |
| $C2$ | 390 | 9320 | 100 | 0.200 | 0.41 |
| $C3$ | 390 | 11470 | 500 | 0.590 | 0.237 |
| $CV1$ | 390 | 8700 | 100 | 0.210 | 0.42 |
| $CV2$ | 390 | 9380 | 200 | 0.360 | 0.36 |

increased due to the large mean density gradient; density overturning was thus strongly inhibited (fig. 7) and, as a consequence, vertical mixing appeared to be suppressed.

### 4.3 Stably stratified flow over longitudinal ridges

Many studies have been devoted to problems characterized by simple flow field and geometry, while only a few others have tackled with the presence of topography. The work by Armenio *et al* [31] dealt with along-ridge (or canyon-like) topography, relevant both in atmospheric and in oceanographic applications. It is known that secondary, large-scale recirculations rise when a fluid flows over a bottom wall that

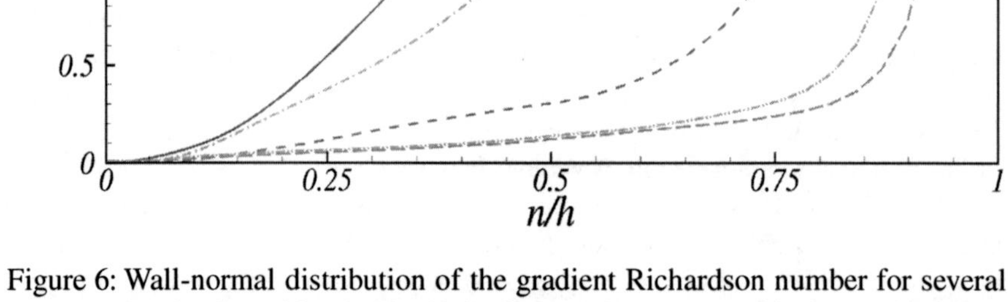


Figure 6: Wall-normal distribution of the gradient Richardson number for several levels of stratification in the horizontal shear case and in the vertical shear case ($n$ denotes the wall-normal direction). From [30].

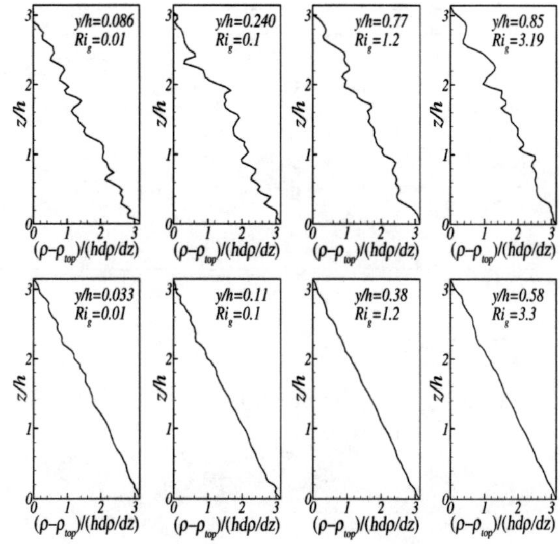


Figure 7: Vertical profiles of instantaneous density at a particular time and different distances from the vertical wall: (a) case C1 with $Ri_b = 0.041$, (b) case C3 with $Ri_b = 0.59$. From [30].

has finite amplitude spanwise perturbations. As discussed in a previous section, the recirculations develop because of the unbalance of the normal Reynolds stress in the cross-stream plate that are able to supply additional vertical mixing of mass and momentum [5]. The research of Armenio *et al* [31] aimed at understanding the role played by stable stratification in the large-scale secondary motion that, in a neutrally stratified flow, is responsible for vertical transport of fluid from the bottom to the top.

A free surface channel flow was considered, evolving over a bottom wall constituted by an infinite array of trapezoidal, finite-amplitude ridges ($1/8$ of the channel height) longitudinally placed over the wall. The Prandtl number was set equal to 5, that corresponds to thermally stratified water whereas the friction Reynolds number was set equal to 400. Several levels of stratification were considered, from the passive scalar case to the case of strong stratification. The numerical method and the SGS model of Armenio & Piomelli [7] were employed.

The presence of a secondary recirculation in the cross-stream plane due to the ridges was well reproduced by the computations. Figure 8 shows the iso lines of the three velocity components in the cross stream plane for three levels of stratification. Similarly to the case of stably stratified flow over a plane wall, stratification reduced the shear stress $\langle u'w' \rangle$ and, hence, causes the increase of mean vertical shear $\partial U/\partial z$ (as observable in fig. 8a) and of the bulk velocity $U_b$. The iso-contours of fig. 8b,c clearly show the presence of a large secondary circulation and of a small cellular motion at the trough of the ridge. Stratification has a double effect on the secondary recirculations: it reduces its vertical extension; it increases its

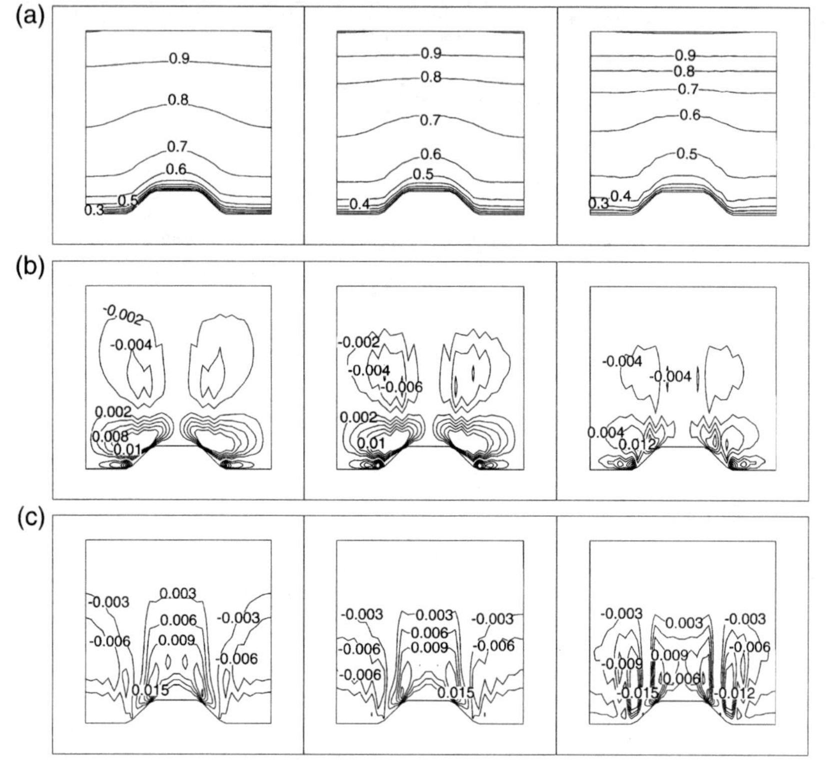


Figure 8: Cross-stream distribution of the mean velocity components for three levels of stratification, C1,C2,C4 from the left to the right: (a) $U/U_{max}$; (b) $V/U_{max}$; (c) $W/U_{max}$. From [31].

intensity. As explained in [7], the first effect is due to the presence of the strong thermocline that develops in the free surface region, that strongly suppresses both the fluctuating and the mean vertical motion (fig. 9). It represents a barrier of potential energy for the secondary recirculations and consequently it suppresses the vertical motion which remains confined near the wall as the level of stratification increases; the second effect is due to the increase, with stratification, of the mean streamwise component of the vorticity $\Omega_x$. Since the transport equation for $\Omega_x$ is unchanged from the neutral case, most of its production is related to the variation of the terms of anisotropy of the normal and the cross-stream Reynolds stress, respectively $\langle v'v' \rangle - \langle w'w' \rangle$ and $\langle v'w' \rangle$, in the core region and at the trough of the ridge. Anisotropy in the cross-stream plane increases with stratification, and consequently the production term of $\Omega_z$ is enhanced by stratification, thus producing more intense recirculation regions. The role of stratification is also reflected in the evolution of the tangential stresses: the primary wall stress ($\tau_x$), related to $\langle u'w' \rangle$, decreases in the trough and increases at the crest of the ridge, due to the redistribution of the vertical shear stress; the secondary stress ($\tau_y$) is coupled

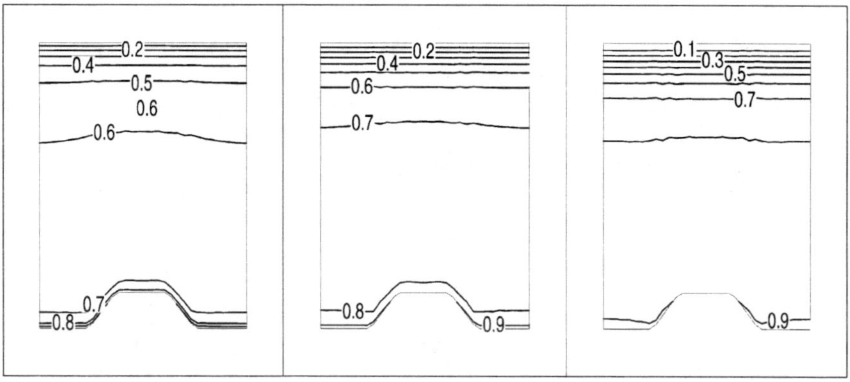


Figure 9: Cross-stream distribution of $\langle\rho\rangle/\Delta\rho$ for three levels of stratification, C1,C2,C4 from the left to the right. From [31].

with the small-scale motions near the bottom corner (fig. 8b), which intensify with stratification.

# 5 Unsteadily driven flows

Most research in wall-bounded turbulence has been concerned with steady flows, leaving not clear many characteristics of wall turbulence in unsteady flows. However, unsteady turbulent flows occur very often in environmental fluid mechanics, for instance in coastal and offshore engineering due to the tidal or wave forcing, and in bio-fluid mechanics. The oscillating boundary layer is prototypical in the frame of the full comprehension of the unsteadily driven turbulent flows. In this section we report results of a research program aimed at the study of turbulent mixing in a tidal rotating flow. The understanding of this class of flow field is relevant in geophysical applications, as far as mixing associated to the interaction between a tide and the sea-bottom is concerned. First we discuss the purely oscillating boundary layer (also known as Stokes boundary layer) in the turbulent regime; afterward, outcomes of the oscillating-rotating boundary layer will be given. It is worth noting that, to the best of our knowledge, these two studies represent the first numerical investigations that describe:

- the details of the turbulent field in a purely oscillating flow at a value of the Reynolds number such that most of the cycle of oscillation is characterized by the presence of fully developed turbulence;
- the turbulent rotating-oscillating flow (Stokes-Ekman layer).

## 5.1 Stokes boundary layer in the turbulent regime

In the Stokes boundary layer, a zero-mean, harmonic velocity field drives the flow, and the Reynolds number $Re_\delta = U_0\delta_S/\nu$ is commonly defined by means of the maximum amplitude of the outer velocity $U_0$ and the Stokes-layer thickness

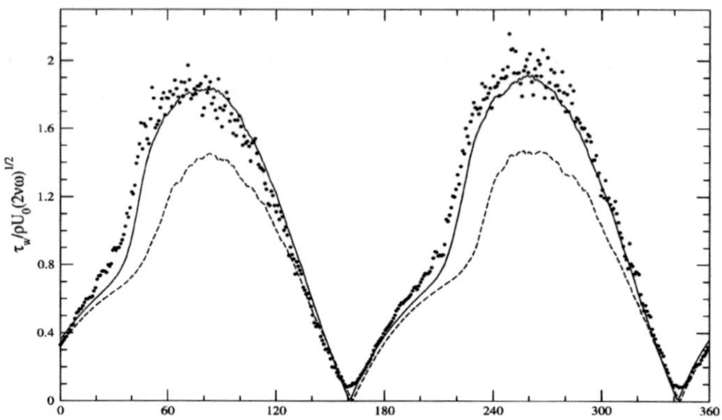


Figure 10: Non-dimensional mean wall shear stress phase-averaged over the 14 cycles: coarse grid (dashed line), fine grid (solid line), experimental data of Jensen *et al* [37] (dots). From [9].

$\delta_S = \sqrt{2\nu/\omega}$, where, as usual, $\nu$ is the kinematic viscosity of the fluid and $\omega$ is the angular frequency of the oscillations.

Previous investigations [34–40] have shown that the Stokes boundary layers present four different flow regimes, depending on the value of $Re_\delta$: the laminar regime, the disturbed laminar regime, the intermittent turbulent regime and the fully developed turbulent regime. At small values of $Re_\delta$ turbulence first appears at the beginning of the decelerating phase, associated with the presence of explosive near-wall bursts [36] as $Re_\delta$ increases turbulence involves earlier and earlier phases of the cycle. Turbulence is present everywhere during the cycle only in the fully developed turbulent regime: according to the experimental analysis by Hino *et al* [36] such regime exists for $Re_\delta > 800$. Moreover, the experiments of Jensen *et al* [37] showed that turbulence is present throughout the cycle at $Re_\delta \sim 3500$, and Sarpkaya [38] reported that at $Re_\delta \sim 1800$ turbulence is already present in most of the cycle of oscillation.

In the work by Salon *et al* [9], large-eddy simulations of a Stokes boundary layer in the turbulent regime ($Re_\delta = 1790$) were performed. The dynamic-mixed SGS model of Armenio & Piomelli [11] was used with the constant averaged over the planes of homogeneity.

The turbulent statistics accumulated after 28 half-periods of simulation were analyzed and compared with the experimental data of Jensen *et al* [37]. The results corroborated and extended the findings of the relevant literature studies: the alternating phases of acceleration and deceleration were correctly reproduced, as like as the beginning of the turbulence activity, occurring at $\omega t \approx 45°$, and its maximum between 90° and 105°. Two grids of different resolution were used, the difference being in the spanwise resolution: the coarse one had a grid spacing, in wall units, $\Delta z^+ = 63$ while the fine one used $\Delta z^+ = 31$. The fine grid correctly reproduced the experimental data, whereas the coarse one underpredicted the wall shear stress by more than 25% (fig. 10). This was basically due to the fact that the coarse grid

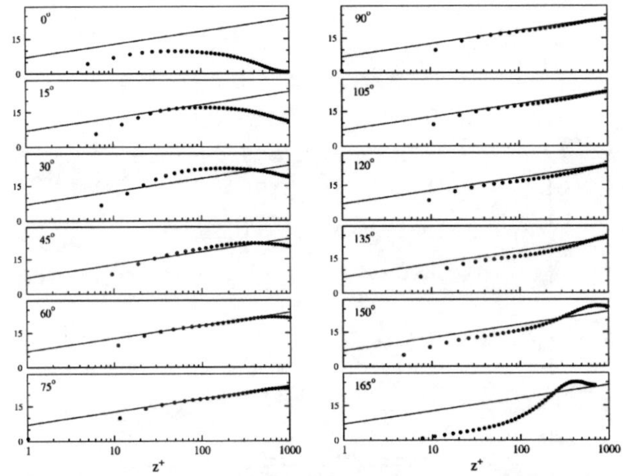

Figure 11: Mean streamwise velocity profiles in semi-log plot. Solid line, log-law with $\kappa = 0.41$ and $A = 7$. From [9].

simulation was not able to solve adequately the low-speed streaks that are generated near the wall during the acceleration phase of the cycle, thus resulting in a deficit in the energy transfer.

The mean streamwise velocity and the second-order statistics were in good agreement with the measures of Jensen *et al* [37]. The study also focused on the structure of turbulence, whose knowledge is key to understand the characteristics of horizontal and vertical mixing during the tidal oscillation. The analysis of turbulent energy spectra, of the Lumley invariant map, of the instantaneous near-wall

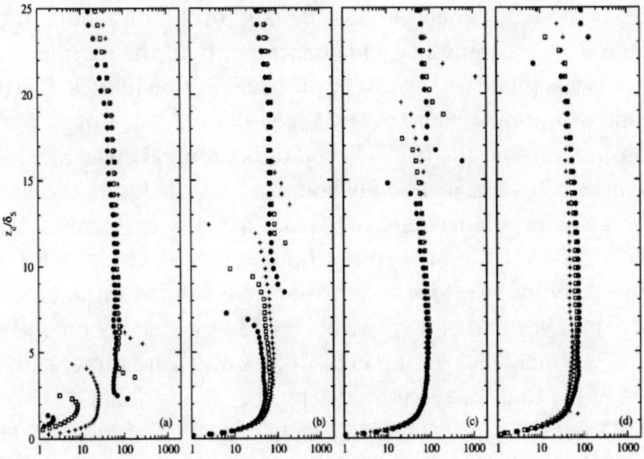

Figure 12: Non-dimensional eddy viscosity $\nu_T/\nu$: (a) 15° (•), 30° (□), 45° (+); (b) 60° (•), 75° (□), 90° (+); (c) 105° (•), 120° (□), 135° (+); (d) 150° (•), 165° (□), 180° (+).

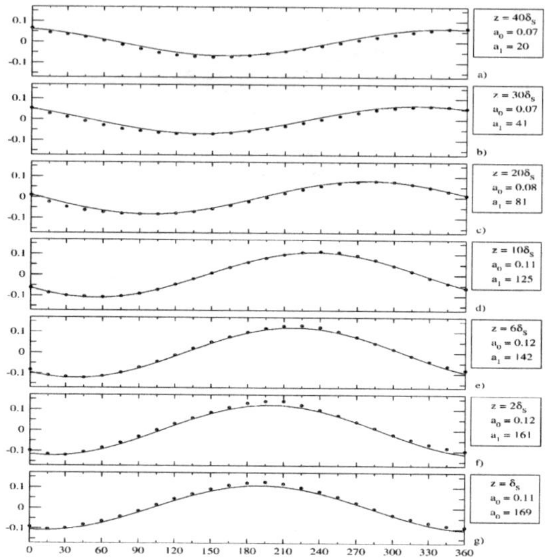


Figure 13: Non-dimensional mean spanwise velocity $\langle v\rangle/U_0$ at different planes (dots) and fit with cosinusoidal function $y = a_0 cos(x + a_1)$. See inlets for details. From [42].

turbulent structures showed the following scenario: Elongated structures, aligned with the direction of the flow, are generated during the early phases of oscillation. Such structures tend to coalesce and successively to break into smaller structures during the acceleration phases. This event corresponds to a sharp increase of the wall shear stress. The presence of such a mechanism has also been found at low values of Reynolds number in the intermittent regime (see for example [38] and [41]), hence it appears to be an inherent characteristic of the flow field, independent on the value of the Reynolds number and consequently of the regime of motion.

Turbulence is sustained up to most of the deceleration phases. During the phases of fully turbulent motion (from $60°$ to $150°$) the presence of a log-layer is well detected in the velocity profile (fig. 11) and the canonical cigar-like shape of turbulence is recognized, just like in a steady boundary layer. During the late deceleration phases of the cycle, turbulence tends to decay in a very anisotropic manner. In particular, in the near-wall region vertical fluctuations decay much faster than the horizontal ones, giving rise to a sort of pancake-shaped turbulence. Conversely, in the outer region, vertical and spanwise fluctuations decay much faster than the streamwise fluctuations, thus giving cigar-like toward one-dimensional turbulence.

Inspection of the total eddy viscosity (fig. 12) showed that it is about two order of magnitude larger than the molecular one along most of the fluid column (up to about $20\delta_s$), well above the height of the nominal thickness of the boundary layer (defined as the point where the vertical shear vanishes at the phase of $90°$). This has been attributed to the fact that in the unsteady BL a mean shear and a mean turbulent shear stress are present above the point of zero shear stress.

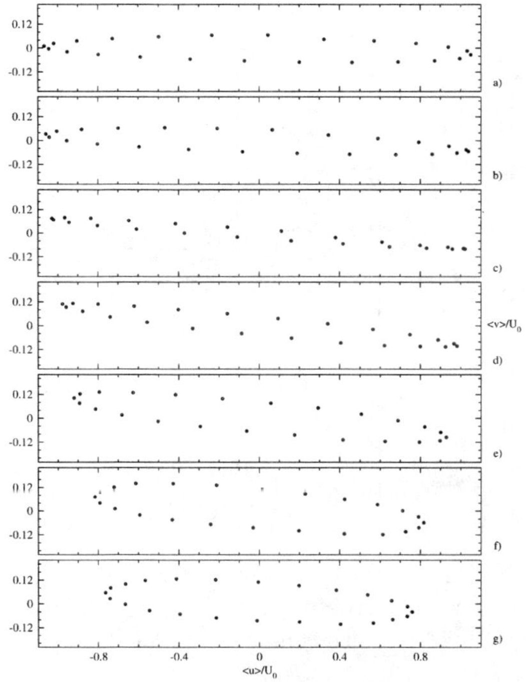


Figure 14: Elliptic paths: mean spanwise *vs* mean streamwise velocity at different planes: (a) $z_d = 40\delta_S$; (b) $z_d = 30\delta_S$; (c) $z_d = 20\delta_S$; (d) $z_d = 10\delta_S$; (e) $z_d = 6\delta_S$; (f) $z_d = 2\delta_S$; (g) $z_d = \delta_S$. From [42].

Finally, the research demonstrated that the dynamic SGS model properly adjusts to the level of turbulent activity during the cycle of oscillation, and therefore is able to supply the contribution to turbulence coming from the small scales.

## 5.2 Rotating Stokes boundary layer in the turbulent regime

The turbulent oscillating flow discussed in the previous sections was recast in a rotational frame in order to describe the effects due to the Earth rotation on a tidal flow at mid-latitudes (details are in [42]). Consistently with Coleman *et al* [43], both the vertical and the horizontal components of the rotation vector were considered in the equations governing the flow.

The first significant result was that the rotation of the frame of reference breaks the symmetry between the two half cycles of the oscillation period. As known, the Coriolis force gives rise to a cross-stream pressure gradient, and therefore to a non-zero mean cross-stream velocity. The mean profile of the spanwise velocity $\langle v \rangle$ was observed to oscillate during the cycle, and its amplitude to be about one order of magnitude smaller than that of the streamwise component $\langle u \rangle$, in agreement with the DNS results of Coleman *et al* [43] for the steady Ekman layer, but phase-shifted due to the oscillating motion.

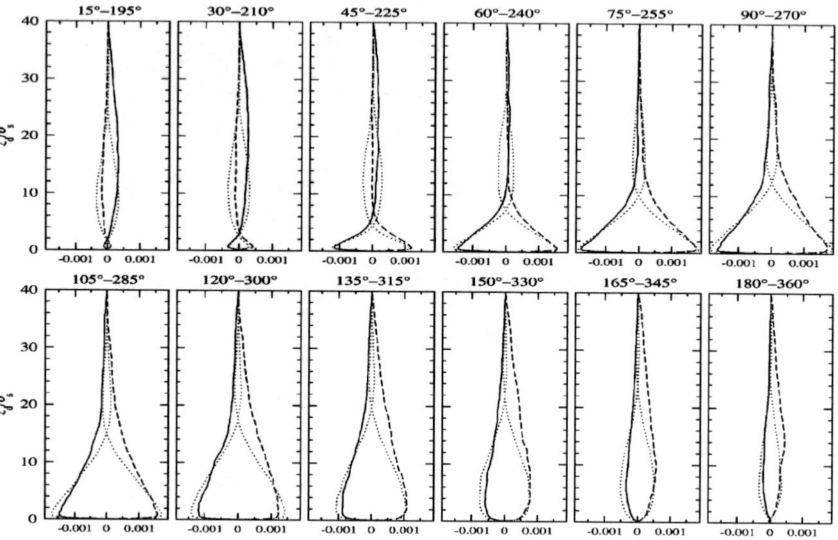


Figure 15: Non-dimensional mean Reynolds shear stress $\langle u'w' \rangle / U_0^2$ for the rotating-oscillating flow from 15° to 180° (–), from 195° to 360° (– –) and for the pure oscillating flow (· · ·). From [42].

Figure 13 shows, for seven different horizontal planes, the evolution of $\langle v \rangle / U_0$, and the relative fits with a cosinusoidal function. Larger amplitudes occur between

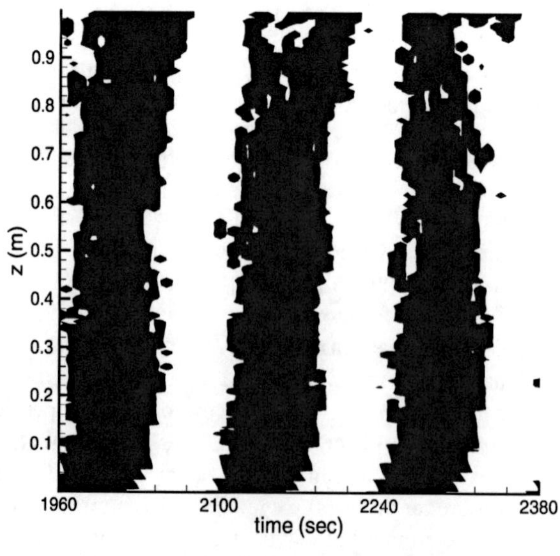


Figure 16: Bradshaw number $B$ for three cycles of simulation (period is 140 s): shading represents positive values of $B$. From [42].

$z = 2\delta_S$ and $z = 6\delta_S$ (the height of the water column is $40\delta_S$), and the phase-lag decreases going further from the wall. As a result, narrow elliptic paths characterize the water column, with the major axis decreasing in amplitude and rotating going from the surface toward the bottom (fig. 14). Because of the combined effect of oscillation and rotation, and the phase-lag between the two horizontal velocity components, the "Ekman spiral" does not develop.

Rotation was shown to have a twofold effect on the system dynamics: in the first half cycle, corresponding to forcing from SW to NE, the mean vorticity (related to the mean vertical shear) is parallel to the background vorticity and consequently turbulence tends to be reduced; conversely, in the second half period, where forcing goes from NE to SW, the mean vorticity is opposite to the background one and thus turbulence tends to be enhanced.

Therefore, as shown in fig. 15, turbulence activity increases when compared to the pure oscillating case, in particular in the bottom half of the water column and during the decelerating phases of the cycle (from 90° to 165° and from 270° to 360°), being always more intense in the second half period. Such a stabilizing/destabilizing effect agrees with theory, as described in [44] and [43], and emphasizes the importance played by the horizontal component of the Earth rotation vector in simulations of turbulent Ekman layers.

Our results showed non-zero correlations between horizontal velocity fluctuations ($\tau_{12}$) and between spanwise and vertical components ($\tau_{23}$), and also an increase of the vertical and cross-stream turbulence intensities when compared to the pure oscillating flow. This picture describes thus a highly three-dimensional character of turbulence, affecting all the three spatial directions.

The Bradshaw number was defined by Tritton [45] as $B = R(R+1)$, where $R$ is the ratio between background and mean vorticity, and rules the effect of the rotation over the flow: destabilizing when $B < 0$, stabilizing when $B > 0$. The evolution of $B$ throughout the cycle was also correctly reproduced (fig. 16), together with the enhancement trend of turbulence observed with inertial forcing coming from eastern quarters of the compass, as also addressed by Coleman *et al* [43].

## 6 Particle-laden flows

Finally, we very briefly show some results of large-eddy simulations of flow fields characterized by the presence of a dispersed phase. This class of problem is relevant in environmental applications. Typical problems characterized by the presence of particles evolving in a flow field are the dispersion of pollutants in air or in water, or transport of contaminants in industrial processes. Phenomena of dispersion of a particulate (*i.e.* organic matter) in seawater constitute also a challenge for the analysis of biological species in water reservoirs.

In the present section we deal with the Lagrangian-Eulerian approach, in which the particles are treated in a Lagrangian way, and evolve within an Eulerian field. This approach has been successfully employed in the past for the comprehension of the mechanisms of interaction between a cloud of particles and a surrounding turbulent field (see for instance [46–49]). It is well known that very light particles

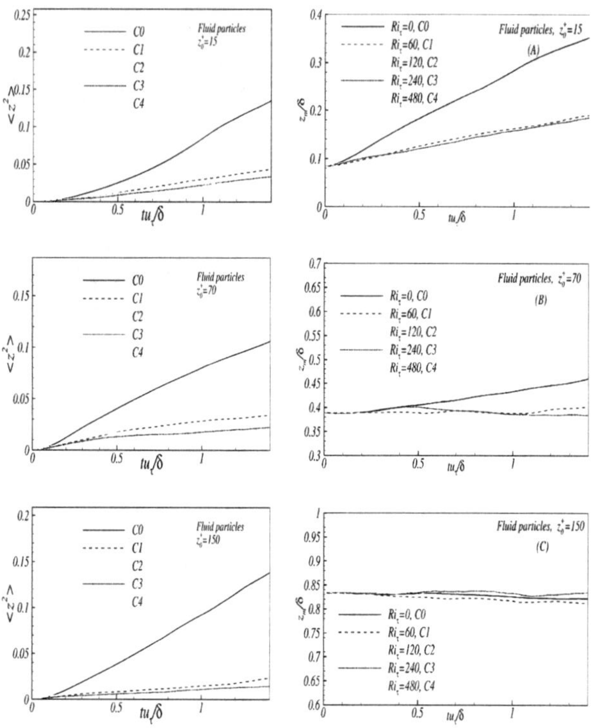


Figure 17: Left: vertical dispersion of tracers versus time for different levels of stratification, from the passive scalar case (C0) to strongly stratified case $Ri_\tau = 480$ (C4): (a1) particles released at $z^+ = 15$; (b1) particles released at $z^+ = 70$; (c1) particles released at $z^+ = 150$. Right: mean vertical displacement of tracers versus time for different levels of stratification, from the passive scalar case (C0) to strongly stratified case $Ri_\tau = 480$ (C4): (a2) particles released at $z^+ = 15$; (b2) particles released at $z^+ = 70$; (c2) particles released at $z^+ = 150$.

behave like tracers. On the opposite side, heavy particles are likely to be sensitive to the largest scales of the motion. For heavy particles, when the concentration is large enough [50], two-way coupling has to be considered, that means that the cloud of particles is such to affect the characteristics of the turbulent field. The cases discussed in the present section are concerned with tracers or particles concentrations small enough to be regarded to belong to the one-way coupling regime (particles are driven by the Eulerian field and do not have a feed-back effect on it).

As known, in LES the small and dissipative scales are filtered out, and, as a consequence, in a Lagrangian-Eulerian approach the particles are driven by a filtered field that contains the large, energy-carrying scales of the motion. The Lagrangian-Eulerian technique, used in conjunction with LES, has been shown to be able to give a fairly good estimate of the characteristics of dispersion of a particulate [51]. The study of Armenio *et al* [52] was devoted to quantify the error associated to the use of a Eulerian filtered (LES) field. To this scope, a turbulent channel flow at a friction Reynolds number equal to 180 was considered, and particles were initially located at different distances from the walls. In order to discern the effect of pure filtering from the modeling one, computations were carried out moving the particles with a field obtained by filtering step-by-step a Eulerian field obtained using a direct simulation (DNS), and then, comparing the particle statistics with those obtained moving the particle with an actual LES field. The results of the study showed that filtering out the small scales, always produces an underestimation of the dispersion coefficients. This effect appeared more pronounced in the near-wall region. The maximum underestimation is of the order of 8–10% when a filter width typical of that used in *resolved* LES is used. The effect of modeling the SGS scales was shown to be negligible if compared to the filtering itself, when dynamic models were used. Conversely, the use of the Smagorinsky model gave a poor prediction of the particle statistics. A companion study (see [53]) was devoted to the effect of LES modeling on the evaluation of the deposition characteristics of an ensemble of heavy particles. In this case also, the use of a *resolved* LES in conjunction with a dynamic model was proved to be able to give good predictions of the deposition rate of the particles at the wall.

Having shown and quantified the ability of Lagrangian-Eulerian techniques in simulating the evolution of a dispersed phase in a turbulent flow field, even in the case in which the Eulerian field is evaluated by means of *resolved* LES, the analysis of particle dispersion in a stratified wall-bounded flow was investigated. As base flow, the turbulent stratified channel flow of Armenio & Sakar [7] was considered. The particles were placed over horizontal planes located at different vertical positions. Tracers and heavy particles were considered for several levels of stratification, and their characteristics of dispersion were evaluated. Here we show and discuss the results for the case of tracers evolving in flow fields characterized by different levels of stratification. Vertical dispersion appeared strongly inhibited by stratification (fig. 17, left), independently on the distance of the particles from the walls. This result is consistent with those obtained by Kimura & Herring [54] for the case of stratified homogeneous turbulence. It is noteworthy that even in the case of very weak stratification, the vertical dispersion dropped by a factor 2 in the near-wall region and even more in the core region. Further increases of stratification did not produce an effect as large as that already observed. This is due to the fact that the turbulent eddy diffusivity $k_T = \langle w'\rho' \rangle / d\rho/dz$ that is directly related to the vertical dispersion, strongly decreases with increased stratification for two reasons: increases the mean density gradient and decreases the vertical buoyancy flux. The mean height of the cloud of particle also appeared to be affected by stratification. In particular, the study showed that particles released in the core region, on average, tend to maintain the vertical position of their center of mass, independently on the level of stratification. Conversely, for particles released in the near-wall region, the vertical position of their center of mass was strongly affected by stratification. Specifically, in the case of stratified flow, the particles on average tend to remain confined at a certain height (fig. 17, right).

## 7 Conclusions

Large eddy simulation is nowadays considered as a robust tool for the investigation of turbulent flows fields. Recent extension of this methodology to cases characterized by complex geometry (presence of topography) and physics (inclusion of effects of stratification, rotation, dispersed phase *etc.*) makes the methodology well suited in environmental engineering.

A main advantage of LES over the direct numerical simulation, is in its own ability to deal with values of the Reynolds number about one order of magnitude larger than those of a typical DNS. This makes the results of the simulation more meaningful, since Reynolds number effects (scale effects) on the characteristics of the physical problem are expected to be of less significance than in the case of DNS.

As a matter of fact, most results shown in the present paper are for values of the Reynolds number such that an inertial subrange is detectable, and, for such reason, Reynolds number effects on the qualitative response of the flow field are unimportant. Such results nowadays, can be easily obtained by the use of desktop computers in a few hours, whereas DNS at comparable values of Reynolds number still require the use of massive computations on parallel supercomputers. As an example of the versatility of LES, in the present paper we have shown LES results for the Stokes boundary layer in an actual turbulent regime, that is nowadays beyond the capabilities of direct numerical simulations.

A main limitation of resolved LES (where the viscous sub-layer is completely solved) is in its own capability in dealing with very large values of the Reynolds number, typical of practical applications. The scientific community is now working to overcome this problem, by coupling an actual LES that solves the turbulent field from the log-region up to the core one, with a RANS-like wall model that is designed to give a parameterization of the physics of the near-wall structures. Such very promising models are expected to be routinely used in CFD labs in the next few years.

The authors wish to acknowledge the anonymous referees who have contributed to improve the manuscript. This study has received financial support by the Italian Ministry of Scientific and Technology Research, project PRIN 2002 Influence of vorticity and turbulence in interactions of water bodies with their boundary elements and effect on hydraulic design.

## References

[1] Piomelli, U., Large-eddy simulation: achievements and challenges. *Progress in Aerospace Sciences*, **35**, pp. 335–362, 1999.

[2] Piomelli, U. & Balaras, E., Wall-layer models for large-eddy simulations. *Ann. Rev. Fluid Mech.*, **34**, pp. 349–374, 2002.

[3] Squires, K.D., Detached-eddy simulation: current status and perspectives. *Invited lecture at Direct and Large-Eddy Simulation V: Proceedings of the Fifth International ERCOFTAC Workshop on Direct and Large-Eddy*

*Simulation*, eds. R. Friedrich, B.J. Geurts & O. Métais, Kluwer: Dordrecht, pp. 465–480, 2003.

[4] Sarghini, F., Piomelli, U. & Balaras, E., Scale similar models for Large-eddy simulations. *Phys. Fluids*, **11**, pp. 1596–1607, 1999.

[5] Falcomer, L. & Armenio, V., Large-eddy simulation of secondary flow over longitudinally ridged walls. *J. Turbulence*, **3**, pp. 1–17, 2002.

[6] Germano, M., Piomelli, U., Moin, P. & Cabot, W.H., A dynamic subgrid-scale eddy viscosity model. *Phys. Fluids A*, **3**, pp. 1760–1765, 1991.

[7] Armenio, V. & Sarkar, S., An investigation of stably-stratified turbulent channel flow using large eddy simulation. *J. Fluid Mech.*, **459**, pp. 1–42, 2002.

[8] Scotti, A. & Piomelli, U., Numerical simulation of pulsating turbulent channel flow. *Phys. Fluids*, **13**, pp. 1367–1384, 2001.

[9] Salon, S., Armenio, V. & Crise, A., Studio numerico dello strato limite di Stokes nel regime turbolento. *Proc. of the XXIX Convegno di Idraulica e Costruzioni Idrauliche, Trento*, **1**, Editoriale Bios: Cosenza, pp. 291–298, 2004.

[10] Jordan, S.A., A large-eddy simulation methodology in generalized curvilinear coordinates. *J. Comput. Phys.*, **148**, pp. 322–340, 1999.

[11] Armenio, V. & Piomelli, U., A Lagrangian mixed subgrid-scale model in generalized coordinates. *Flow Turb. and Comb.*, **65**, pp. 51–81, 2000.

[12] Wu, X., & Squires, K.D., Prediction of the three-dimensional turbulent boundary layer over a swept bump. *AIAA J.*, **36**, pp. 505–514, 1998.

[13] Buckles, J., Hanratty, T.J. & Adrian, R.J., Turbulent flow over large amplitude wavy surfaces. *J. Fluid Mech.*, **140**, pp. 27–44, 1984.

[14] Hudson, J.D., Dykhno, L. & Hanratty, T.J., Turbulence production in flow over a wavy wall. *Exp. Fluids*, **20**, pp. 257–265, 1996.

[15] Buckles, J., Turbulent separated flow over wavy surfaces. Ph.D. Dissertation, University of Illinois, 1983.

[16] Calhoun, R.J., Numerical investigation of turbulent flow over complex terrain. Ph.D. Dissertation, Stanford University, 1998.

[17] Henn, D.S. & Sykes, R.I., Large eddy simulation of flow over wavy surfaces. *J. Fluid Mech.*, **383**, pp. 75–112, 1999.

[18] Meneveau, C., Lund, T.S., & Cabot, W.H., A Lagrangian dynamic subgrid-scale model of turbulence. *J. Fluid Mech.*, **319**, pp. 353–385, 1996.

[19] Maas, C. & Schumann, U., Direct numerical simulation of separated turbulent flow over a wavy boundary. In: Hirschel, E.H. (ed.), *Flow Simulation with High Performance Computers, II*. Notes on Numerical Fluid Dynamics, Vol. 52, pp. 227–241, 1996.

[20] Colombini, M., Turbulence-driven secondary flows and formation of sand ridges. *J. Fluid Mech.*, **254**, pp. 701–719, 1993.

[21] Nezu, I. & Nakagawa, H., Cellular secondary currents in straight conduit. *J. Hydr. Eng.*, **110**, pp. 173–193, 1984.

[22] Gerz, T., Schumann, U. & Elghobashi, S.E., Direct numerical simulation of stratified homogeneous turbulent shear flows. *J. Fluid Mech.*, **200**, pp. 563–594, 1989.

[23] Holt, S.E., Koseff, J.R. & Ferziger, J.H., A numerical study of the evolution and structure of homogeneous stably stratified sheared turbulence. *J. Fluid Mech.*, **237**, pp. 499–539, 1992.

[24] Kaltenbach, H.-J., Gerz, T. & Schumann, U., Large-eddy simulation of homogeneous turbulence and diffusion in stably stratified shear flow. *J. Fluid Mech.*, **280**, pp. 1–40, 1994.

[25] Piccirillo, P.S. & Van Atta, C.W., The evolution of a uniformly sheared thermally stratified turbulent flow. *J. Fluid Mech.*, **334**, pp. 61–86, 1995.

[26] Jacobitz, F.G., Sarkar, S. & Van Atta, C.W., Direct numerical simulations of the turbulence evolution in a uniformly sheared and stably stratified flow. *J. Fluid Mech.* **342**, pp. 231–261, 1998.

[27] Arya, S.P.S., Buoyancy effects in an horizontal flat-plane boundary layer. *J. Fluid Mech.*, **68**, pp. 321–343, 1975.

[28] Komori, S., Ueda, H., Ogino, F. & Mizushina, T., Turbulent structure in stably stratified open-channel flow. *J. Fluid Mech.*, **130**, pp. 13–26, 1983.

[29] Garg, R.P., Ferziger, J.H., Monismith, S.G. & Koseff, J.R., Stably stratified turbulent channel flows. I. Stratification regimes and turbulence suppression mechanism. *Phys. Fluids*, **12**, pp. 2569–2594, 2000.

[30] Armenio, V. & Sarkar, S., Mixing in a stably-stratified medium by horizontal shear near vertical walls. *Theor. Comput. Fluid Dyn.*, **17**, pp. 331–349, 2004.

[31] Armenio, V., Falcomer, L. & Carnevale, G.C., Large-eddy simulation of a stably stratified flow over longitudinally ridged walls. *Direct and Large-Eddy Simulation V: Proceedings of the Fifth International ERCOFTAC Workshop on Direct and Large-eddy simulation*, eds. R. Friedrich, B.J. Geurts & O. Métais, Kluwer: Dordrecht, pp. 299–306, 2003.

[32] Moore, M.J. & Long, R.R., An experimental investigation of turbulent stratified shearing flow. *J. Fluid Mech.*, **49**, pp. 635–655, 1971.

[33] Crapper, P.F. & Linden, P.F., The structure of turbulent density interfaces. *J. Fluid Mech.*, **65**, pp. 45–63, 1974.

[34] Hino, M., Sawamoto, M. & Takasu, S., Experiments on transition to turbulence in an oscillatory pipe flow. *J. Fluid Mech.*, **75**, pp. 193–207, 1976.

[35] Blondeaux, P. & Seminara, G., Transizione incipiente al fondo di un'onda di gravitá. *Rendiconti Accad. Naz. Lincei*, **67**, pp. 407–417, 1979.

[36] Hino, M., Kashiwayanagi, M., Nakayama, A. & Hara, T., Experiments on the turbulence statistics and the structure of a reciprocating oscillatory flow. *J. Fluid Mech.*, **131**, pp. 363–400, 1983.

[37] Jensen, B.L., Sumer, B.M. & Fredsøe, J., Turbulent oscillatory boundary layers at high Reynolds numbers. *J. Fluid Mech.*, **206**, pp. 265–297, 1989.

[38] Sarpkaya, T., Coherent structures in oscillatory boundary layers. *J. Fluid Mech.*, **253**, pp. 105–140, 1993.

[39] Blondeaux, P. & Vittori, G., Wall imperfection as a triggering mechanism for Stokes-layer transition. *J. Fluid Mech.*, **264**, pp. 107–135, 1994.

[40] Vittori, G. & Verzicco, R., Direct simulation of transition in an oscillatory boundary layer. *J. Fluid Mech.*, **371**, pp. 207–232, 1998.

[41] Costamagna, P., Vittori, G. & Blondeaux, P., Coherent structures in oscillatory boundary layers. *J. Fluid Mech.*, **474**, pp. 1–33, 2003.

[42] Salon, S., Turbulent mixing in the Gulf of Trieste under critical conditions. Tesi di Dottorato di Ricerca in Geofisica Applicata e Idraulica, XVI ciclo. Universitá degli Studi di Trieste, Trieste, 2004.

[43] Coleman, G.N., Ferziger, J.H. & Spalart, P.R., A numerical study of the turbulent Ekman layer. *J. Fluid Mech.*, **213**, pp. 313–348, 1990.

[44] Hopfinger, E.J. & Linden, P.F., The effect of background rotation on fluid motions: a report on Euromech 245. *J. Fluid Mech.*, **211**, pp. 417–435, 1990.

[45] Tritton, D.J., Stabilization and destabilization of turbulent shear flow in a rotating fluid. *J. Fluid Mech.*, **241**, pp. 503–523, 1992.

[46] McLaughlin, J.B., Aerosol particle deposition in numerically simulated channel flow. *Phys. Fluids A*, **1**, pp. 1211–1224, 1989.

[47] Elghobashi, S. & Truesdell, G.C., Direct simulation of particle dispersion in a decaying isotropic turbulence. *J. Fluid Mech.*, **242**, pp. 655–700, 1992.

[48] Pedinotti, S., Mariotti, G. & Banerjee, S., Direct numerical simulation of particle behavior in the wall region of turbulent flows in horizontal channels. *Int. J. Multiphase Flow*, **18**, pp. 927–941, 1990.

[49] Kulick, J.D., Fessler, J.D. & Eaton, J.K., Particle response and turbulent modification in fully turbulent channel flow. *J. Fluid Mech.*, **277**, pp. 109–134, 1994.

[50] Elghobashi, S., On predicting particle-laden turbulent flows. *Appl. Sci. Res.*, **52**, pp. 309–329, 1994.

[51] Wang, Q. & Squires, K.D., Large eddy simulation of particle-laden turbulent channel flow. *Phys. Fluids*, **8**, pp. 1207–1223, 1996.

[52] Armenio, V., Piomelli, U. & Fiorotto, V., Effect of the subgrid scales on particle motion. *Phys. Fluids*, **11**, pp. 3030–3042, 1999.

[53] Armenio, V., Piomelli, U. & Fiorotto, V., On the application of large-eddy simulation to particle-laden flows. *Proc. TSFP I*, S. Banerjee and J.K. Eaton Editors, Begell House, NY., pp. 139–144, 1999.

[54] Kimura, Y. & Herring, J.R., Diffusion in stably stratified turbulence. *J. Fluid Mech.*, **328**, pp. 253–269, 1996.

# CHAPTER 3

# Nearshore mixing and macrovortices

M. Brocchini[1], A. Piattella[2], L. Soldini[2] & A. Mancinelli[2]
[1]*DIAM, Genoa University, Italy.*
[2]*Polytechnic University of Marche, Italy.*

## Abstract

Horizontal mixing of shallow coastal flows is studied with a specific focus on the role played by large-scale horizontal eddies (macrovortices). Within the classic depth-averaged Nonlinear Shallow Water Equations (NSWE) framework, generation of such macrovortices can be described through one single mechanism for which lateral gradients of shock-type solutions introduce vorticity in the flow. This mechanism is intensely activated when waves break over discontinuous topographic features like natural longshore sand bars or man-made submerged breakwaters. Description of macrovortex-induced mixing is given on the basis of numerical solutions of the NSWE and interpreting the results of specifically-designed laboratory experiments. Deterministic results concerning the generation/evolution of macrovortices are obtained and statistics of passive tracers are used to interpret the overall dynamics in terms of 2D turbulence theory. Preliminary results indicate differences in the mixing features of flows induced by isolated and arrays of submerged breakwaters. A discussion is also proposed on possible approaches for improving our knowledge/modeling of such type of mixing.

## 1 Introduction

The object of this paper is the analysis of large-scale features of shallow-water turbulence which characterizes the flows of nearshore waters. The latter evolves as shallow-water flows in which the horizontal scale is much larger than the vertical scale and are most often analysed in terms of depth-averaged properties like in the case of the classic Nonlinear Shallow Water Equations (NSWE) on which we base our subsequent, quantitative analysis. The importance for shallow coastal flows of horizontal, large-scale eddies (macrovortices hereinafter) has been widely reported [1–6]. Large-scale, horizontal mixing of coastal flows is greatly promoted by macrovortices which are generated because of a spatially-nonuniform breaking of the incoming waves [2, 7]. Although such differential breaking may be induced by various reasons (irregularity of the incoming wave field, wave-wave interaction, *etc.*) the major cause of breaking unevenness is due to bottom topography. This is often characterized by longshore, isolated (natural bumps or manmade submerged breakwaters) or almost-continuous features (bars or arrays of submerged breakwaters) over which uniform wave fronts break with large lateral gradients. Hence, macrovortices can be shed which alter the hydro-morphodynamics of either the region between the isolated topographic feature and the shore [5, 7] or the area seaside of the structures while propagating toward the offshore in conjunction with rip-currents [3, 8]. In coastal areas dams and harbour breakwaters are loci of generation of macrovortices which have an important impact on the morphology of large regions [4, 9].

Field observations of nearshore vortices are rare [10]. To our knowledge there have been very few reported laboratory observations specifically focussed on breaking wave-generated macrovortices, in part because they are difficult to measure using fixed current meters. Vortex shedding at the lee-side of a topographic obstacle is a well-documented phenomenon both in marine [11] and atmospheric environments [12, 13]. Once transition to vortex shedding occurs a number of parameters like the vortex size and the shedding period are analyzed, as useful for both hindcasting and forecasting purposes, in relation to steady current conditions [12], to an oscillatory shallow-water flow around an island [14] and to breaking wave conditions [7]. Two typical situations are characteristic of the generation and transport of startup macrovortices for topographically-controlled wave breaking: wave breaking at breakwater heads and on rip current topographies [7]. Vortices generated on opposite ends of a breakwater are widely separated, typically have little mutual interaction and travel towards the shoreline mainly because of the wave field and self-advection, *i.e.* the contribution to the vortex motion due to the presence of a sloping bed which forces the vortex to move along isobaths. In contrast, oppositely-signed vortices in a rip current topography are extremely close together, have significant interactions, and may travel offshore as a pair. The problem is highly complex as many processes of similar strength operate simultaneously [7, 8].

The role of macrovortices is also fundamental to any water quality analyses of coastal areas. In most cases the evolution of passive tracers, like non-reacting pollutants, is predicted by means of a depth-averaged advection-diffusion equation for the mean tracers concentration $C$ as advected by the depth-averaged velocity vector $\mathbf{v} = (u,v)$ and in dependence of a depth-averaged diffusivity tensor $\mathbf{K}$ [15]. This equation can only be solved for $C$ if, beyond the flow field, the diffusivity is known through a constitutive relationship of Fickian-type. Such a closure is largely dominated by the presence of large-scale coherent features like macrovortices and is typical of the flow conditions at hand. Examples of closures for coastal flows can be found, among others, in Fischer *et al* [16], in Larson & Kraus [17] and in Takewaka *et al* [18].

In the attempt of placing solid foundations to the study of generation and evolution of macrovortices in shallow flows, Jirka [19] proposed a classification of macrovortices in dependence of their generation mechanism. The three types of generation mechanisms are due to topographic forcing (the most important one), to internal transverse shear instabilities and to secondary instabilities of the base flow. We note that, at times, it is very hard to distinguish between topographic forcings and shear instabilities. For example, the background shear flow, whose instability gives rise to the 2D vertical structures, is often regarded as the primary agent for macrovortex generation; however, it is evident that such flow is very often only the result of a topographic forcing. In this perspective Soldini *et al* [20], analysing in detail compound channel flows, show that within the pseudoinviscid NSWE framework most of the macrovortices, usually regarded as generated via an instability, can, alternatively, be regarded as generated by the lateral gradients of shock-type solutions. Hence, an approach analysing, both numerically and analytically, the propagation of NSWE shocks can be used to investigate generation of macrovortices alternative to the classic stability analyses used for coastal flows [21, 22].

However, NSWE can be used as a simple and effective modelling tool for the flows at hand provided it is clear that shallow flows are quasi-2D rather than exactly 2D (for example 2D flows are divergenceless while shallow-water flows are horizontally divergent). This means that phenomena due to vertical flow disuniformities are only approximately accounted for in the NSWE framework. For example, as partially mentioned above, dispersive mixing is often parameterized through a Fickian-type closure. Then diffusive terms are introduced in which eddy viscosities (momentum mixing) or eddy diffusivities (mass mixing) are used as "catch all" parameters (hence the name of "effective eddy viscosity") *i.e.* such that various 3D effects (secondary flows, small-scale turbulence, *etc.*) are all represented through one single function. This is the approach used in many ocean and atmosphere circulation models in which only the most energetic part of the horizontal flow is resolved while the small-scale part of horizontal mixing and vertical mixing are modelled as subgrid processes. This Horizontal Large-Eddy Simulation (HLES) approach is currently being taken up also for coastal circulation computations [3]. Much of the work, then, goes in the representation of the effective parameters which can either be prescribed through an algebraic closure [3] or a one-equation closure. Obviously the type of closure influences the accuracy of the modelling of the small-scale flows.

The HLES approach seems well justified also in view of the results coming from recent experimental studies of shallow-water turbulence. Virtually all of them suggest that shallow-water turbulence of barotropic fluids, like that generated in wakes [23], shallow jets [24] and mixing layers [25], is characterized by spectral properties typical of 2D turbulence. In other words both a direct enstrophy cascade at large wavenumbers and an inverse energy cascade at small wavenumbers are well evident with decay rates typical of 2D turbulence *i.e.* "-5/3 law" and "-3 law" respectively for energy and enstrophy cascading [26, 27]. In this respect it is also auspicable to model the transport properties of shallow-water macrovortices in analogy to those of coherent barotropic vortices of 2D turbulence [28].

In a 2D turbulent flow characterized by large-scale coherent structures the evolution of tracers and the flow dynamics are so intimately connected that knowledge of the former may give a predictive key for the latter, and, obviously, *vice versa*. This approach, which has been usefully employed to investigate atmospheric [29, 30] and oceanic [31, 32] flows, is now becoming of interest also for nearshore flows [33, 34]. This is also connected with the recent developments made in the monitoring of coastal waters by means of video techniques [35]. With such equipment floats/dye released near the shore can be monitored for times/areas large enough to provide the fundamental data for any dispersion analysis. For example, the recent work of Takewaka *et al* [18] shows how it is possible to apply the mentioned approach to compute dispersive parameters of dye patches released near the breaking region. In this perspective, and with the aim of using information coming from this prototype-scale experiment, Piattella [36, 37] attempted at creating a theoretical framework useful for the interpretation of statistics of tracers released in coastal areas.

In the rest of the paper we give a broad overview of the generation mechanism for macrovortices in the presence of topographic gradients (section 2) and of their evolution as described from numerical solutions of the NSWE for practically-important coastal flows (section 3). Nearshore mixing is then analysed by means of statistical properties of passive tracers and interpreted in terms of 2D turbulence theory (section 4). Some conclusions and a description of ongoing research are given in section 5.

## 2 Topographic-induced generation of macrovortices in coastal environments

Various types of models/approaches can be used to analyse vorticity and circulation arising from differential wave breaking induced by topographic features [7].

In the "bore evolution" approach, the amount of vorticity or circulation generated by the bore is directly related to the depth jump across a bore and no modelling is required other than suitably following the breaker. This is naturally embodied in the wave-resolving depth-averaged (2DH) NSWE. The two greatest advantages of studies based on this approach are that they have some fundamental validity for finite wave heights, and that dissipation follows directly from the bore geometry. This was exploited by Peregrine [2], who showed that the instantaneous rate of change of circulation around a closed material curve passing once only through a single bore is equal to the rate of energy loss through that bore.

Other wave-resolving approaches can be used, like Boussinesq-type models, in which the energy dissipation process and the induced circulation is accounted for by a parametric formulation which models the dynamics occurring at the breaker's front. The best examples of such models, without bore dissipation built into the numerical discretisation, are the various types of wave breaking schemes for Boussinesq wave models [3, 38–41, and numerous others]. These schemes differ considerably from each other but generally have two commonalities, which tend to give them similar performance: dissipation is localised on the front face of the breaking wave and dissipative forces are momentum-conserving. Localising breaking forces on the front face of waves mimics bore dissipation, as bores do not form naturally in these nonlinear-dispersive models. Momentum-conserving breaking is essential for the proper representation of breaking wave-induced processes such as setup; without it, all predictions are completely wrong.

A third approach stems from scale-separating fluid motion into short waves and currents, and then averaging over the short time scale. Phase-averaged equations for wave evolution vary considerably. To a reasonable level of approximation, wave action relations based on geometrical optics show many relevant features. Looking at the initial stage of vorticity generation, when mean currents and departures from still water levels are small, Brocchini *et al* [7] proposed a phase-averaged model, which in some ways resembles the analysis of Bühler & Jacobson [42]; in the latter a Reynolds-type decomposition of the flow was used to separate wave and current components and the seabed friction effects were neglected. In additions to previous similar analyses, Brocchini *et al*'s approach [7] allows for frequency dispersion and gives this method a somewhat broader area of application.

## 2.1 Vorticity generation in the classic 2DH NSWE

In this section we illustrate a model for which vorticity or, better, potential vorticity (PV hereinafter) can be generated by shock-type solutions of the classic, wave-resolving and depth-averaged NSWE. In absence of dissipative body forces these can be written as

$$d_{,t} + (ud)_{,x} + (vd)_{,y} = 0, \tag{1}$$

$$u_{,t} + uu_{,x} + vu_{,y} + gd_{,x} = gh_{,x} - \frac{c_f|\mathbf{v}|u}{d}, \tag{2}$$

$$v_{,t} + uv_{,x} + vv_{,y} + gd_{,y} = gh_{,y} - \frac{c_f|\mathbf{v}|v}{d} \tag{3}$$

in which the symbol $(\cdot)_{,i}$ represents partial differentiation with respect to the generic variable *i*, $d = h + \eta$ is the total water depth, *h* the undisturbed water depth, $\mathbf{v} = (u,v)$ the vector of the depth-averaged velocity and $c_f$ the Chezy bed resistance coefficient. Since the flow evolves in the $(x, y)$-plane, with $x$ as the main flow direction (*i.e.* that of waves or currents depending on the flow at hand), the only non-zero component of the vorticity vector is

$$\omega \equiv v_{,x} - u_{,y} \tag{4}$$

which measures the local flow rotation around a vertical axis. Then, assuming $c_f = 0$ and operating the combination $(3_{,x} - 2_{,y})$ the following equivalent equations for $\omega$ and $\Omega \equiv \omega/d$ are found:

$$\frac{D\omega}{Dt} \equiv \frac{\omega}{d}\frac{Dd}{Dt} \tag{5}$$

or

$$\frac{D\Omega}{Dt} = 0 \tag{6}$$

where $\frac{D}{Dt} \equiv \frac{\partial}{\partial t} + \mathbf{v} \cdot \nabla$ and $\nabla \equiv (\frac{\partial}{\partial x}, \frac{\partial}{\partial y})$.

Note that the equation for $\omega$, eqn. (5), does not contain neither sources nor sinks *i.e.* according to such equation $\omega$ can only be transported or locally intensified/reduced if $d$ increases/decreases when following a "water column" which represents a coherent body of water of constant volume. Water columns are in the 2DH NSWE scheme the equivalent of water particles in a general 3D scheme. The equation for $\Omega$, eqn. (6), states that following the water columns the quantity $\Omega$, *i.e.* the PV is conserved. From the above it is evident that no generation of either $\omega$ or $\Omega$ is present in the pseudoinviscid NSWE framework in the absence of shock-type solutions. However, if shocks are present in the domain, jump conditions, also known as Rankine-Hugoniot conditions, hold across the discontinuity. These conditions introduce a generation mechanism of vorticity/PV not accounted for by eqns. (5) and (6). In particular if dissipative body forces, typically due to bores or hydraulic jumps (*i.e.* shocks), are accounted for, eqn. (6) is modified so that the curl of such force appears at the right hand side [42], hence stating that PV generated by shocks moves inside the fluid body with the water columns. Following the approach of Pratt [43] we assume, for simplicity, that a shock of straight, finite front propagates at velocity $V$ in the $x$-direction (see fig. 1, a simple rotation allows to generalize the following to any shock incidence). If points of coordinates $x_A$ and $x_B$ lay, respectively, upstream and downstream of the shock there is a jump of $\Omega$ across the shock which reads:

$$[\Omega]_{x_A}^{x_B} = -\left[\frac{2}{g[d(x_A)+d(x_B)]d(x_A)d(x_B)}\right]^{1/2} \frac{\partial E_D}{\partial y} \tag{7}$$

with

$$E_D = \frac{[d(x_B)-d(x_A)]^3}{4d(x_A)d(x_B)} \quad \text{and} \quad [\Omega]_{x_A}^{x_B} \equiv \Omega(x_B) - \Omega(x_A). \tag{8}$$

$E_D$ is the specific (per unit weight) energy dissipation rate occurring at a steady (hydraulic jump) or moving (bore) flow discontinuity. Hence, PV is generated at locations where there is a cross-flow variation of $E_D$. This is maximum where there is an abrupt cross-flow change of $[d]_{x_A}^{x_B}$. Note that the sign of the vorticity generated is opposite to the sign of $\partial[d(x_B) - d(x_A)]/\partial y$.

### 2.1.1 Vorticity generation by breaking waves

The above generation mechanism can be applied to a number of nearshore flow conditions in which breakers of finite longshore length are present, originated under various circumstances [2, 5, 7, 8]. Two examples are reported in fig. 1 which are of considerable importance for nearshore circulation. The former (left panel of fig. 1) illustrates the case in which two uniform wave fronts propagate towards

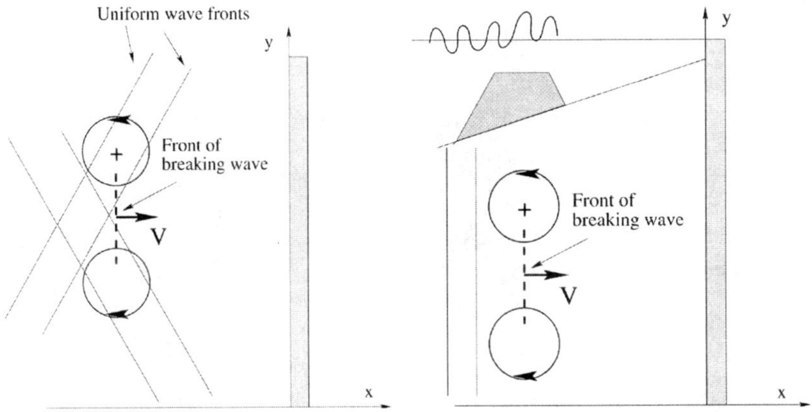


Figure 1: Generation of breakers of finite length: uniform waves crossing at an angle over a uniform beach (left panel) and uniform waves surmounting a submerged breakwater and breaking locally (right panel). Circles and arrows give a schematic representation of the flow rotation at the edges of the breaker. (Adapted from [5]).

the shoreline over a uniformly-sloping beach from different directions. Their interaction can lead to local steepening and breaking so that a breaker of finite longshore length is generated. At the edges of such breaker PV is generated due to the large value of $\partial[d(x_B) - d(x_A)]/\partial y$. For a more exhaustive analysis of generation of vorticity by breakers of finite length we refer the reader to the work of Peregrine [2]. The latter case (see right panel of fig. 1) is of greater practical importance as models the flow conditions generated by waves approaching the shore and locally breaking over a submerged breakwater. Vorticity generation at the edges of a submerged breakwater and evolution of macrovortices in the nearshore is currently being investigated in great detail [5, 7, 8].

## 3 Deterministic results on macrovortex evolution

By means of the WAF NSWE solver described in [44], many numerical tests have been run in order to understand and model the mechanism of formation of macrovortices when steep waves overpass a submerged breakwater. The numerical domain represents real-life topographic conditions in which coastal defences are placed or sand bars evolve (*e.g.* [5]). An 80m-long submerged breakwater with seaward slope of $1 : 2.5$ and a shoreward slope of $1 : 1.5$ was placed at a water depth of 3.0m with a submergence of $h_c = 0.5$m. The beach slope was made to vary over the range $s = 0.005 \div 0.1$. A radiating boundary condition has been enforced at the seaward boundary of the domain (left end in fig. 2) while open conditions have been used at the lateral boundaries. Sinusoidal waves were generated with periods $T_{in} = 5$s, 10s and 5 different wave heights in the range $H_{in} = (0.5 \div 2.5)$m were used. The dependence of the bottom friction was accounted for by varying $c_f$ over two values

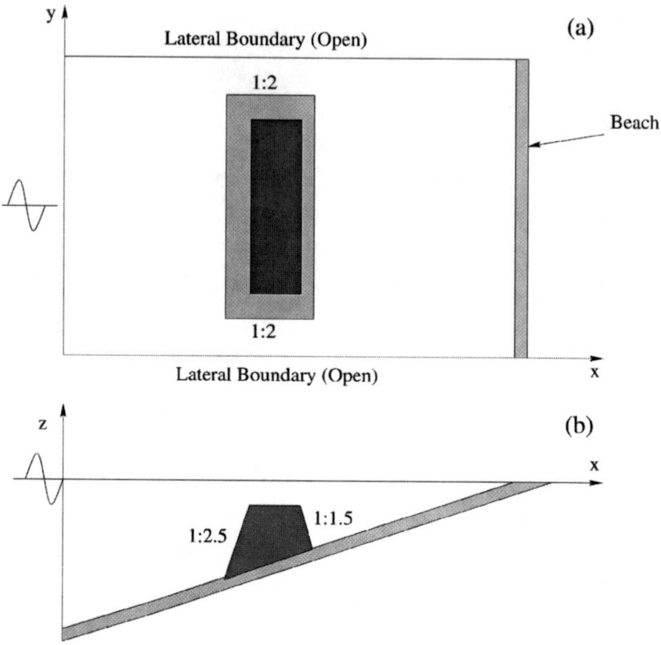


Figure 2: Sketch of domain used in the numerical tests: (a) top view, (b) side view. (Adapted from [5]).

only: $c_f = 0 - 0.01$. The discretization used, which allowed for both accurate and feasible numerical experiments, was such that $\triangle x = 1$m, $\triangle y = 2$m. In particular, the most intense macrovortices have a diameter of the order of $(10 \div 15)$m and, hence, are adequately resolved in our computations.

### 3.1 Macrovortices at submerged breakwaters

The analysis of numerical tests, aiming at an evaluation of the impact of macrovortices on the nearshore dynamics, reveals the complexity of vorticity generation and re-organization. As an example we show the patterns of PV for the test case characterized by $H_{in} = 0.5$m and $T_{in} = 5$s (see figs. 3 and 4). It is clear that vorticity is generated by the breaking waves at the edges of the submerged breakwater, increases in intensity while the first waves pass over the structure, re-organizes in the shape of coherent vortices which move towards the shoreline undergoing a complex deformation. These two examples well characterize the flow evolution of all the other cases and, hence, are taken as representative of the different vortex motion over a $1:30$ and $1:10$ beach. Analysis of numerical simulations reveals that the beach slope seems to largely control the vortex motion.

We can see that for the milder slope conditions macrovortices move along a rather complex route. In the following description the generation instant is taken as the zero time datum. During the initial stages of their motion ($t < 60$s) vortices, due

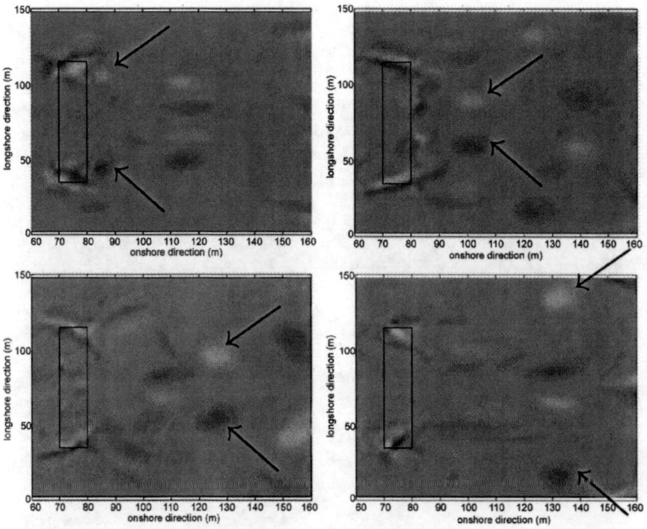

Figure 3: Maps of PV at various stages of evolution for flow conditions (from left to right and from top to bottom for times $t = 45\text{s}, 130\text{s}, 195\text{s}, 250\text{s}$) of $s = 1:30$, $T_{in} = 5\text{s}$, $H_{in} = 0.5\text{m}$ and $c_f = 0$. The vorticity intensity increases from black (negative) to white (positive).

to the strong interaction with the steep slopes of the breakwater, self-advect around the corner from the side slope, and propagate parallel to the breakwater itself (top left panel of fig. 3). They then migrate towards the shoreline along a route which until $t \approx 190\text{s}$ is almost orthogonal to the shoreline and almost coinciding with the breakwater mid-line ($2^{\text{nd}}$ and $3^{\text{rd}}$ panel of fig. 3). This shoreward migration is due to the coupling with the opposite-signed vortex shed from the opposite edge of the breakwater, hence forming a vortex pair. Nearer the shoreline, because of the very shallow-water, self-advection becomes dominant and stronger than mutual advection so that vortices moves along isobathes hence the pair splits and for the last 50s of motion vortices move diagonally *i.e.* still towards the shore but away from the breakwater mid-line (last panel of fig. 3). This shoreline motion is qualitatively similar to that reported in [45] and interested readers should consult this reference for a detailed experimental investigation of vortex couples near shorelines.

For the $1:10$ steeper slope (fig. 4) macrovortices are shed from the breakwater side slope but their route to the shore is less complex. After re-organisation, vortices migrate along a diagonal track which bends away from the breakwater. The overall effect of the steep breakwater slopes, which controls the vortex motion for the gentler $1:30$ beach slope, appears here much reduced. In other words the vortices, being of considerable size (comparable with the breakwater berm), seem to interact more strongly with the beach than with the breakwater. For the $1:10$ beach slope the breakwater slope does not influence much the vortex path, while for the $1:30$ beach slope it only affects the motion prior to detachment. In fig. 5 we

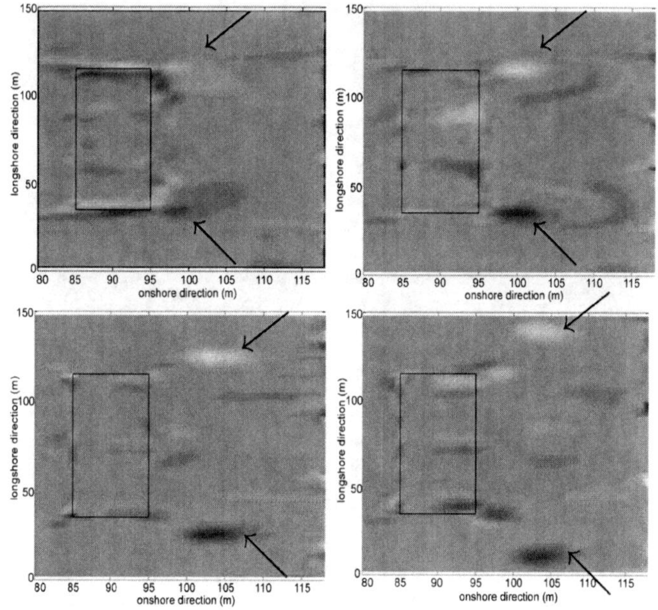


Figure 4: Maps of PV at various stages of evolution for flow conditions (from left to right and from top to bottom for times $t = 5\text{s}, 25\text{s}, 45\text{s}, 65\text{s}$) of $s = 1:10$, $T_{in} = 5\text{s}$, $H_{in} = 0.5\text{m}$ and $c_f = 0$. The vorticity intensity increases from black (negative) to white (positive).

summarize the information on the vortex trajectories for the two cases illustrated in figs. 3 and 4. With similar graphs it is quite easy to analyze many important features of macrovortex evolution. In summary it seems that two distinct phases characterize the life of vorticity/PV. A first phase includes generation, re-organization into coherent vortices and, eventually, minor migration around the breakwater (see case $s = 1:30$). During this phase the vorticity patch, which is becoming a vortex, increases its rotational speed and may or may not have significant migration. A second phase then begins in which the vortices may either dissipate or migrate away from the breakwater possibly undergoing deformation.

### 3.1.1 Detachment period

For widely-spaced breakwaters, we examine the detachment period $T_d$ of the vortices and give a theoretical estimate for $T_d$ using simple dimensional arguments [7]. We define $T_d$ as the time for which the vortex reaches, under the action of the breaking waves and of self-advection, a distance from the breakwater equal to its own size $R$. Then, if we designate by $A_d$ the onshore acceleration at which the vortex speeds away from the breakwater, we have

$$R \approx \frac{A_d T_d^2}{2} \implies T_d \approx \sqrt{\frac{2R}{A_d}}. \tag{9}$$

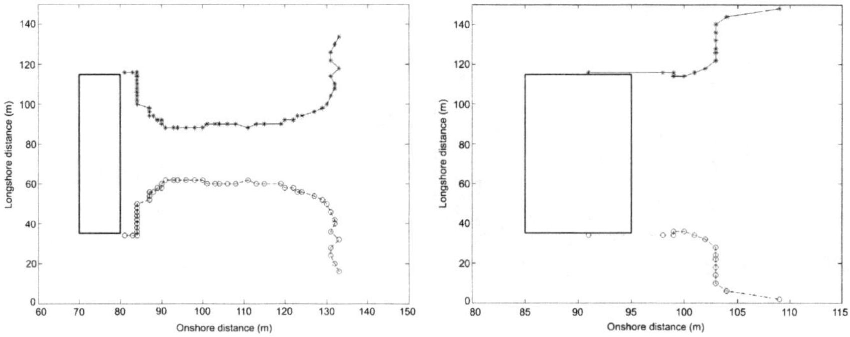


Figure 5: Typical cases of macrovortex trajectories for different beach slopes. The thick black line represents the breakwater berm, while trajectories of positive vortices are given in continuous lines and those of negative vortices in dotted lines. Left panel: trajectories of vortices emitted for flow conditions of $s = 1 : 30$, $T_{in} = 5$s, $H_{in} = 0.5$m and $c_f = 0$, shoreline at $x = 150$m. Right panel: trajectories of vortices emitted for flow conditions of $s = 1 : 10$, $T_{in} = 5$s, $H_{in} = 0.5$m and $c_f = 0$, shoreline at $x = 115$m. (Adapted from [7]).

We compute the vortex onshore velocity and acceleration using the energy dissipated through the bore $E_D$

$$E_D = \frac{gH_B^3}{4d^2 - H_B^2} \tag{10}$$

in which $H_B$ is the instantaneous bore height and $d$ is the mean water level across the breakwater. Then, taking the reference local depth as the computed mean depth at detachment $d_d$, we get the following estimate:

$$T_d \approx \sqrt{\frac{2R}{E_D}\left\{\frac{1}{\frac{1}{2R} + \frac{s_l}{d_d}\frac{1}{4\pi}\left[\log\left(\frac{8d_d}{s_l R}\right) - \frac{1}{4}\right]}\right\}} \tag{11}$$

in which $s_l$ is the side slope of the breakwater, here $s_l = 1 : 2$ and accounting for the self-advection in the onshore direction. Note that according to eqn. (11) $T_d$ is evaluated on the basis of local flow properties rather than on global properties as with other available descriptions [12]. This is obviously due to the much more complicated flow here investigated in comparison to that evaluated in the available literature.

A comparison between estimated detachment period, $T_d$, evaluated using eqns. (10) and (11), and results from the numerical simulations is given in fig. 6. It is clear that the dimensionless evaluated detachment period $T_d^e/(\sqrt{h_c/g})$ of eqn. (11), reported along the ordinate axis, slightly overestimates the dimensionless measured value $T_d^m/(\sqrt{h_c/g})$ reported along the abscissa axis. In fact most of

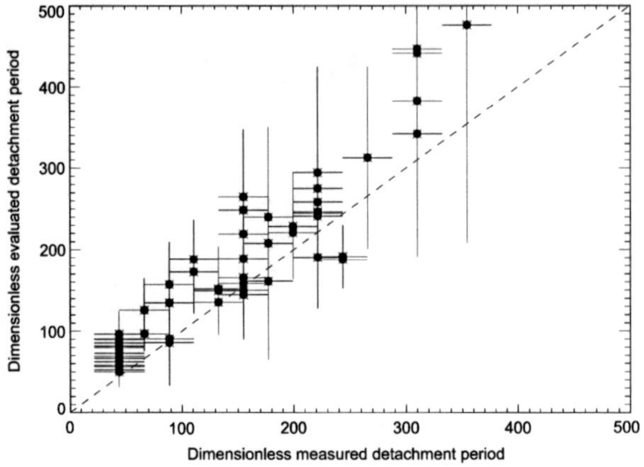


Figure 6: Comparison of the dimensionless detachment period predicted by eqns. (10) and (11) with that measured from the numerical experiments described in the previous section. Detachment periods have been made dimensionless with the time scale $\sqrt{h_c/g}$. (Adapted from [7]).

the solid circles lie above the dashed line which represents a perfect agreement. Error bars have been superposed which correspond to the sampling time for $T_d^m$ and the confidence range of 95% for $T_d^e$. The discrepancy between $T_d^e$ and $T_d^m$ can, alternatively, be measured also by the relative error

$$\Delta T_d = \frac{\sum_{i=1}^{N} |T_{d_i}^m - T_{d_i}^e|}{\sum_{i=1}^{N} T_{d_i}^m} \tag{12}$$

which is found to be of about 28%.

Note that a natural lower limit of one detachment period per wave is visible for conditions of large waves and/or small vortices. We believe that, notwithstanding the number of estimated parameters (like $E_D$ and $R$) influencing eqn. (10), such equation provides a quite accurate and useful means for predicting $T_d$.

## 4 Statistical results on flow mixing

The large-scale horizontal eddies strongly determine the mixing properties of a 2D turbulent flow. The knowledge of the flow dynamics can give important informations on the evolution of passive tracers trajectories. *Vice versa* mixing properties, analysed statistically in terms of absolute/relative diffusity, can give informations on flow hydrodynamics. Thus, by means of the statistical analysis of both velocity

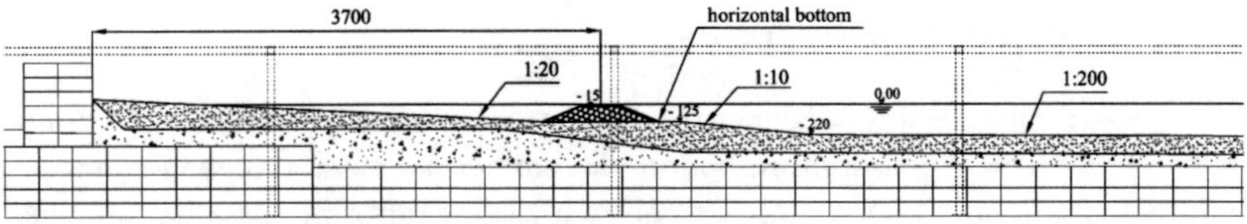

Figure 7: Cross-shore section of the Bari experimental set-up, in correspondence of the submerged breakwater.

data and floaters trajectories, we try to give a theoretical background for the determination of the hydrodynamics and of the mixing features of the flow due to wave overpassing submerged breakwaters.

In this section, we describe the main results of a large-scale laboratory experiment [46, 47] finalized to the analysis of the above-mentioned issues. In particular we performed a spectral analysis of the ADV velocities and a statistical analysis of the trajectories of passive tracers released both for the "single breakwater configuration" and for the "rip current configuration".

## 4.1 Laboratory experiments

The experiments were carried out at the large-scale wave basin of the Polytechnic of Bari (90m long, 50m wide). Model breakwaters (4m long and with berm width of about 0.3m) were placed over a sandy beach ($d_{50} = 0.2$mm) of almost uniform slope both offshore of the breakwaters ($s = 1 : 200$) and inshore of them ($s = 1 : 20$), as shown in fig. 7.

Two distinct configurations were analysed (see fig. 8, in which the locations of the Acoustic Doppler Velocimeters (ADVs) are also shown). In the "single breakwater" case the considered structure is far from any other structures while in the "rip current" case the submerged breakwaters were separated by narrow gaps. At the offshore boundary of the domain, with still water depth of 0.79m, both regular and irregular waves were generated with periods in the range $(0.91 \div 1.82)$s and heights in the range $(1.67 \div 6.67)$cm. Not only flow measurements were made (*i.e.* velocities around the breakwaters and water level over them), but also floaters $[10 \div 25$ wooden spheres with diameter of $(25 \div 42)$mm$]$ were released around the breakwaters and their meandering tracked-down with a fixed videocamera.

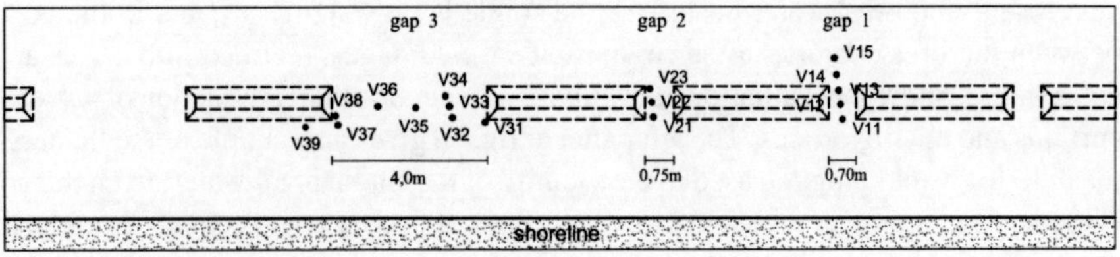


Figure 8: Planimetric layout of the Bari experimental set-up and locations of the ADVs.

## 4.2 Hydrodynamics characteristics

Previous experiences [48–50] underline the fact that wave propagation, even normal to the beach, over a non-uniform bottom determines a general circulation characterized by both longshore and rip currents, which constitute the so-called "circulation cells". This "primary circulation" is determined by the presence of breaking-induced mean water level gradients both in the longshore direction (for the presence of the rip channels) and in the crosshore direction. Waves break over the submerged breakwaters and produce a crosshore setup of the water surface; the latter is less pronounced in the rip channel, in which the interaction with the seaward-flowing rip current modifies the approaching waves. Waves flowing to shore directly through the rip channel induce a finite-length breaker very close to the shore and, consequently, a pair of macrovortices which rotate in opposition to the vortices of the "primary circulation" ("secondary circulation"). Rip currents are often unstable and the velocities in the rip channel are greater in the middle of the gap, being one of the most important causes of localized erosion and offshore sand transport, and become lower seaward for the onshore waves propagation.

This behaviour is confirmed by the experimental velocity data collected during the Bari experiments. In particular we analyse the vertical distribution of crosshore, time-averaged velocities $\overline{u}$ (see fig. 9), measured in the gaps 1 and 2 of fig. 8; these refer to the test characterized by a regular waves of height $H_{in} = 5$cm and period $T_{in} = 1.8$s. The velocities are measured by ADVs once the flow pattern reaches a quasi-steady state for an interval 30s and with a sampling frequency equal to 20Hz. In particular velocities reach the maximum values of about $\overline{u} = 0.19$m/s at the inshore middle of the gap, both for gap 1 and gap 2, and become almost vanishing within a crosshore distance of about $4 \div 5$ gap widths. This seems to confirm the numerical results of Mancinelli *et al* [51], in which the rip currents are locally intensified near the breakwater but made spatially unstable by the presence of macrovortices generated by the depth gradients at the ends of the submerged breakwaters ("local circulation"). On the contrary, numerical simulations show that in the case of isolated breakwaters macrovortices propagating towards the shoreline become one of the most important forcings, together with the waves, of the general circulation. Experimental data collected near the isolated breakwater of fig. 8 are being analysed to confirm such numerical evidence.

The "primary", the "secondary" and "local" circulations are also investigated by means of passive tracers trajectories. The initial locations of floaters were random but close to the breakwater heads, for the single breakwater cases, and in the gap between the breakwaters, in the rip current cases. Images, rectified into cartesian coordinates, allow for evaluation of the floaters dispersion under the action of waves, currents and macrovortices. The left panel of fig. 10 gives an example of the floaters meandering after being released in the vicinity of the single breakwater; from this it is evident the presence of a drift current of few centimetres per second, seemingly due to density effects. The right panel of fig. 10 shows the particles trajectories in the case of an array of breakwaters. The presence of the rip current and its effects on the particles dispersion are rather evident. In particular it is clear that the rip

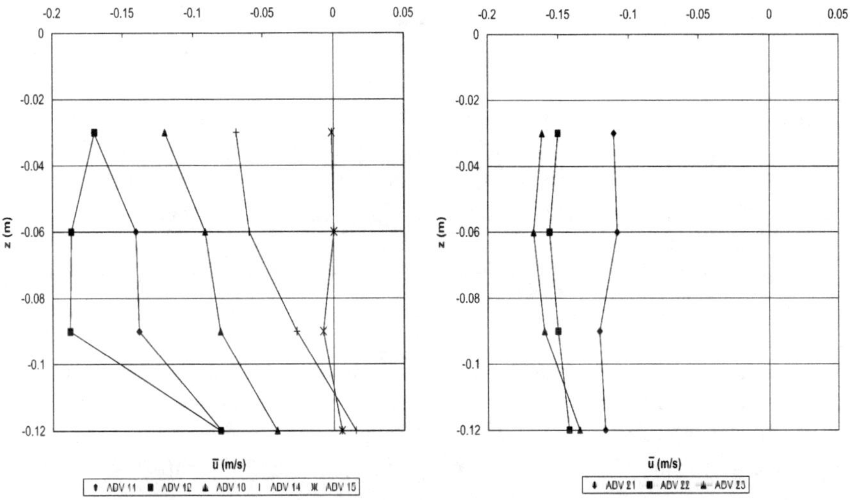


Figure 9: Vertical distribution of the time-averaged crosshore velocity $\overline{u}$. Left panel: gap 1; right panel: gap 2.

current determines a strongly-anisotropic field, with a crosshore dispersion greater than the longshore one, above all in the rip neck. It is also possible to observe that the circulation is characterized by a size comparable with the breakwater length ($\sim$ 4m).

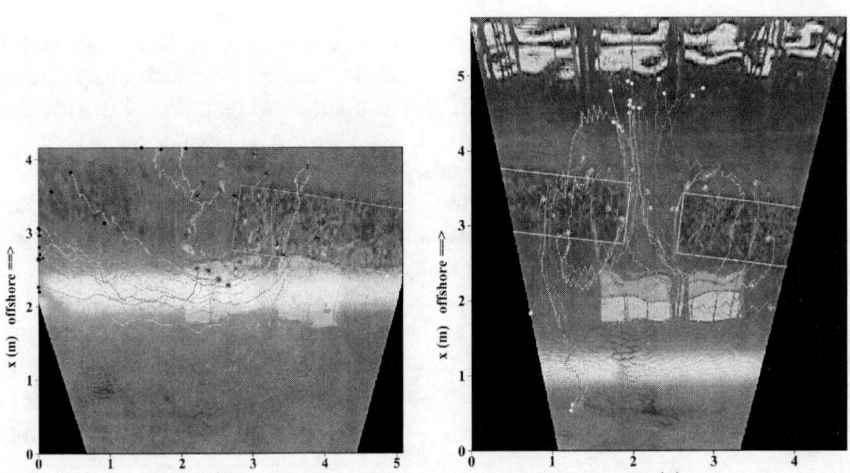


Figure 10: Typical particles trajectories for the "Bari experiments". Left panel: "single breakwater configuration". Right panel: "rip current configuration". White straight lines give the breakwaters location.

### 4.2.1 Spectral analysis of ADV velocities

Signatures of circulation features are currently being analysed with the aid of a spectral analysis of ADV velocities.

The velocities used for this analysis are those measured in the rip channel (gap 1 of fig. 8) and near the head of the single breakwater (gap 3 of fig. 8), at a depth of $-3$cm below the still water level. We are first using surface data for a better comparison with the statistical analysis of the passive tracers trajectories of the following subsection.

Preliminary results seem to indicate that different decay rates characterize the spectra for $f < f_i$ and for $f > f_i$, $f_i$ being the forcing (*i.e.* wave) frequency. In more detail it seems that $E(f) \propto f^{(-5/3)}$ for $f < f_i$ while $E(f) \propto f^{(-3)}$ for $f > f_i$. Moreover, for the "rip current configuration" much of the energy is stored in the $f^{(-5/3)}$–branch while the opposite occurs for the "single breakwater configuration". This would indicate a possible dominance of a "shearing regime" [52] for the "rip current" and a dominance of enstrophy cascade [26] for the "single breakwater".

These different behaviours between the two configurations are also supported by the statistical analysis of the passive tracers trajectories [37] summarized in the following section.

## 4.3 Statistical analysis of passive tracers trajectories

To illustrate the evolution of passive tracers we have analysed the trajectories of water particles released close to the breakwaters. Such trajectories are described by the simple equation:

$$\frac{d}{dt}(x, y) = (u, v) \tag{13}$$

in which $(u, v)$ is the vector of the local, depth-averaged velocity. These trajectories can only be suitably used to model the motion of tracers for which many effects important for the motion of solid particles or floaters (weight, drag, lift, *etc.*) can be neglected. This analysis, specifically designed to evaluate features of 2D turbulence induced by both "single breakwater" and "rip current" configurations, is, in spirit, very similar to that performed to study the mixing features of oceanic eddies [32]. The advantage of the present analysis, based on laboratory data, is the possibility of strictly controlling, and eventually repeating, the input flow conditions.

Flow mixing is here quantified in terms of statistics of the floaters motion. We start our analysis by computing the absolute dispersion $\langle X^2 \rangle$ and the diffusivity $K^{(1)}$, the latter defined as the time derivative of the absolute dispersion, *i.e.*

$$K^{(1)} \equiv \frac{1}{2}\frac{d}{dt}\langle X^2 \rangle. \tag{14}$$

Both for the "single breakwater" and the "rip current" configuration the total absolute dispersion exhibits typical "small times" and "large times" behaviours: an initial quadratic growth is followed by an intermediate regime and, later, by a linear growth, with transitions respectively occurring around the time at which the waves

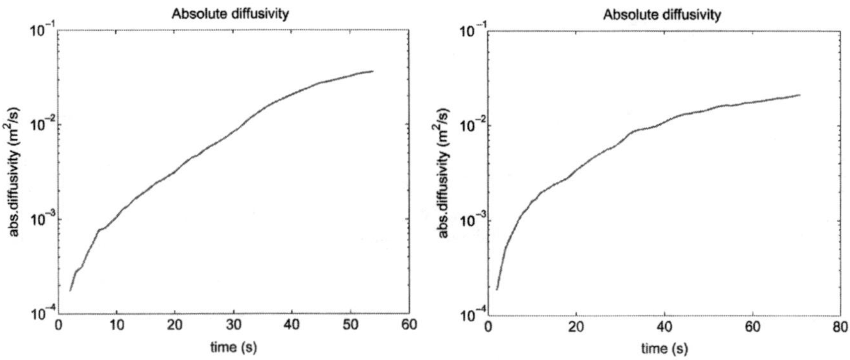


Figure 11: The absolute diffusivity $K^{(1)}$. Left panel: "single breakwater configuration". Right panel: "rip current configuration".

reach the breakwaters $t \approx (10 \div 15)$s and around few Lagrangian decorrelation times ($t \approx (30 \div 40)$s, *i.e.* $t = (3 \div 4) \times T_L$). The Lagrangian decorrelation time $T_L$ is defined as:

$$T_L \equiv \int_0^\infty R(t)dt \tag{15}$$

in which $R(t)$ is the correlation coefficient (or "lagrangian autocorrelation coefficient"):

$$R(t - t_0) = \frac{\langle \mathbf{V}_i(t_0 + t) \cdot \mathbf{V}_i(t_0) \rangle}{(\langle \mathbf{V}_i(t_0 + t) \rangle^2 \langle \mathbf{V}_i(t_0) \rangle^2)^{1/2}} \tag{16}$$

$\mathbf{V}_i(t) = (U_i, V_i)$ being the Lagrangian velocity of the $i$th particle and the angular brackets indicating an average over the whole set of tracers.

We then compute the absolute diffusivity as the time derivative of the absolute dispersion; accordingly with the previous results, as shown in fig. 11, for all cases we observe a linear growth in time for $t < (10 \div 15)$s (ballistic regime) and an almost constant value, for large times $t > (30 \div 40)$s (asymptotic or brownian regime).

For the determination of both relative dispersion $\langle D^2 \rangle$ and diffusivity $K^{(2)}$, three different initial separations are considered, $D_0 = (0.60 \div 0.80, 1.50, 2.86)$m; the lowest and the highest values of initial separation are respectively representative of the breakwater berm width and of the breakwater distance from the shoreline while the intermediate value is representative of the dimension of the largest observed vortices. In fig. 12 we plot the total relative diffusivities for the three initial separations, against the distance $D$ and we can observe that, for the scales of the intermediate regime, the diffusivities exhibit a different power law dependence for the "single breakwater" (left panel of fig. 12) and the "rip current case" (right panel of fig. 12). The case of "single breakwater" exhibits a $K^{(2)} \propto D^2$ law which seems to suggest an enstrophy cascade rather than a shear-dominated flow

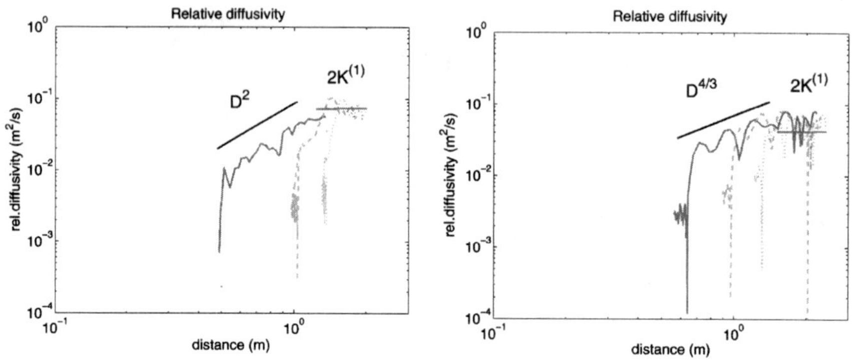


Figure 12: Relative diffusivity $K^{(2)}$ for initial separations: $D_0 \leq (0.60 \div 0.80)$m (solid), $D_0 \leq 1.5$m (dashed) and $D_0 \leq 2.86$m (dotted). Left panel: "single breakwater configuration". Right panel: "rip current configuration".

like for the "rip current case", represented by the $K^{(2)} \propto D^{4/3}$ fit. Bennett [52] found that $\langle D^2 \rangle \propto t^3$, and thus $K^{(2)} \propto D^{4/3}$, in the case of particles pairs taking independent random walks in the $y$-direction in the presence of a shear flow in the $x$-direction. Hence, the strong anisotropy due to the horizontal shear, here represented by the rip current, seems to cause the $D^{4/3}$ law dependence for the relative diffusivity. This anisotropy is confirmed by the fact that the relative diffusivity in the crosshore direction is greater than that in the longshore direction, $K_x^{(2)} \gg K_y^{(2)}$ for all the "rip current cases" analysed. Johnson & Pattiaratchi [34], on the basis of field data collected in the presence of transient rip currents, have found a similar $D^{4/3}$ behaviour (see their fig. 17).

In all cases an asymptotic constant value of $K^{(2)}$ is reached which is slightly smaller than twice the absolute diffusivity $K^{(1)}$. Quantitative results, both in terms of $K^{(1)}$ and $K^{(2)}$ seem important in view of a synthetic description of the mixing properties due to waves incident on either a single or an array of breakwaters and for use in practical computations of mixing made with a convective-diffusive equation.

We can also determine the same statistical features of dispersion using some numerical simulations performed by means of the NSWE solver. The numerical results are reasonably similar to the experimental ones in terms of the absolute and relative statistics, of the growth rates and of the asymptotic values, for both the single breakwater and the rip current configurations. We here give (see fig. 13) an example of comparison between the experimental (left panel) and the numerical results (right panel) for the "single breakwater configuration" ($H_{in} = 0.05$m and $T_{in} = 0.9$s) only.

The vorticity pattern, relative to the same numerical solution, is plotted in fig. 14 and can give some qualitative information about the mixing properties of macrovortices in shallow waters. The shape of the macrovortices generated at the lee side of the breakwater is shown; we can note that the shearing field due to macrovortices is so strong that intense stretching of the vortex sheets placed between the large-scale structures occurs. In these conditions, as described by Kraichnan [26], the

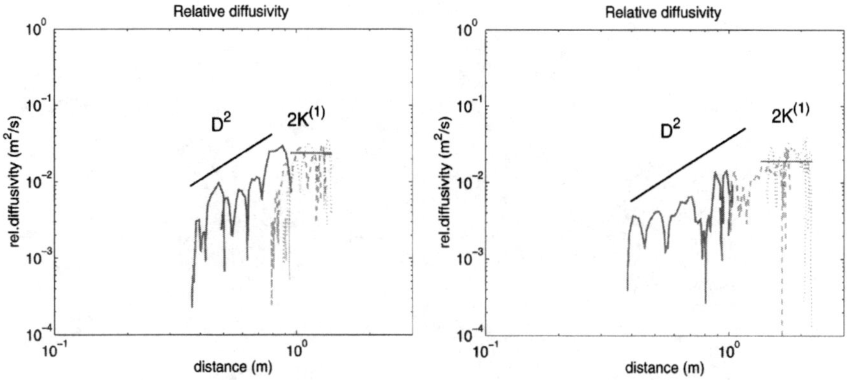


Figure 13: Relative diffusivity $K^{(2)}$ for the "single breakwater configuration". Initial separations: $D_0 \leq 0.60$m (solid), $D_0 \leq 1.5$m (dashed) and $D_0 \leq 2.86$m (dotted). Left panel: experimental results. Right panel: numerical results.

small-scale vorticity is intensified representing the mechanism typical for a direct enstrophy cascade.

## 5 Conclusions and description of ongoing research

An analysis of macrovortices, based on the depth-averaged classic NSWE, has been proposed and large-scale features of shallow-water turbulence (or quasi-2D turbulence) which characterize the flows of nearshore waters have been studied. The latters evolve as shallow-water flows in which the horizontal scale is much larger than the vertical scale. The generation of macrovortices and their role on horizontal mass and momentum mixing of the flows at hand has been clarified. In particular it has been shown that within the NSWE framework most of the macrovortices, usually regarded as generated via an instability can, alternatively, be regarded as generated by the lateral gradients of shock-type solutions. Hence, an approach analysing, both numerically and analytically, the propagation of NSWE shocks can be used to investigate generation of macrovortices alternative to classic stability analyses. Moreover, sample computations have been used to describe in detail the evolution of macrovortices generated by topographic forcings in coastal waters (*e.g.* at submerged breakwaters). Coastal macrovortices are seen to propagate under the action of three main forcings which are the self-advection (*i.e.* interaction with the topography), the mutual-advection (*i.e.* the interaction with other vortical structures) and the interaction with the background flows due to waves and currents. Depending on the relative strength of these mechanisms the macrovortices may either propagate towards the shoreline or towards the offshore. Since in a 2D turbulent flow characterized by large-scale coherent structures, the evolution of tracers and the flow dynamics are so intimately connected that knowledge of the former may give a predictive key for the latter, and, obviously, *vice versa*, mixing properties

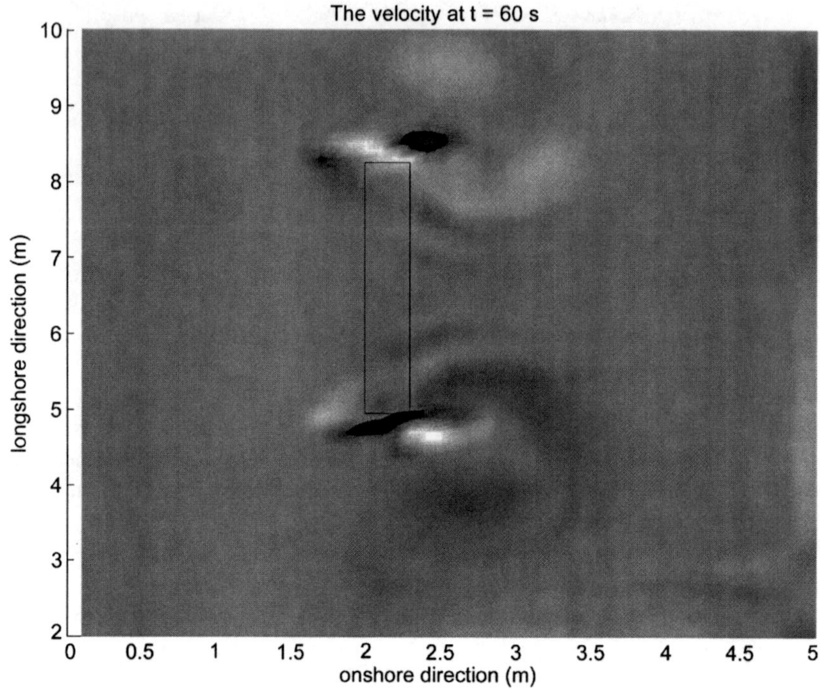


Figure 14: Vorticity pattern from numerical simulation. Contour values increase from negative (black) to positive (white) in the range $-2.5\mathrm{s}^{-1} < \omega < 2.5\mathrm{s}^{-1}$.

of these flows were also analysed in terms of trajectories and statistics of passive tracers like "water particles".

Typical regimes of 2D turbulence, like enstrophy cascading and turbulence shearing, have been found to characterize the flow induced by submerged breakwaters. Enstrophy cascading seems to dominated the flow induced by one single structure while rip currents shearing dominates the flow due to arrays of breakwaters.

Ongoing research is developing along two main lines:

1. from the theoretical/numerical point of view a study has been undertaken aimed at determining a suitable framework for HLES-type computations of coastal flows performed by means of depth-averaged NSWE. The problem, as described by Lesieur [27], is that of modeling a flow which is quasi-2D in the large scales and 3D in the small scales. It is then sensible to assume that the effects of small-scale or sub-grid scale (SGS) motions on larger scale motions can be accounted for in terms of mass/momentum diffusion more or less heuristically defined and depending on eddy mixing coefficients whose size is many orders of magnitude larger than the molecular values. After the pioneering work of Basdevant & Sadourny [53] much research is

being devoted to defining the most suitable form of the diffusive term to be included, for example, in the momentum equation. In general such term is written as:

$$\nu_T(\nabla^2)^\alpha \mathbf{v} \tag{17}$$

in which $\nu_T$ is an effective eddy viscosity and $\alpha$ an integer coefficient of order 1 ($\alpha = 1$ for the standard Laplacian operator). It is thus clear that two orders of problems arise:

i. to define the most suitable approach to compute $\nu_T$ (*i.e.* to find a suitable closure);
ii. to determine the best value for $\alpha$.

In recent applications of NSWE/HLES models to coastal flows the second problem has been somehow sidestepped, much of the work being devoted to find the most suitable closure for the flows under investigation. This is not an easy task since it is not easy to decide whether the SGS model has to consider the small-scale 3D point of view and use an eddy viscosity chosen in accordance with the small-scale dynamics (*i.e.* obeying Richardson's law) or the 2D point of view. However, recent closures have been based on a Smagorinsky approach [54] and $\nu_T$ evaluated as a function of the tensor of velocity deformation of the depth-averaged 2DH flow [55, 56]. In summary Laplacian-type operators are most often used in conjunction with algebraic closures for the eddy viscosity. In order to evaluate suitability of such an approach and, eventually, to devise alternative strategies, we are analysing the fundamental requirements the diffusive term should obey to. In fact, depending on the chosen cutoff wavenumber $k_c$, the parameterization of the small scale has to obey different requirements related with energy and enstrophy cascading. In particular if $k_c$ falls in the enstrophy cascade range the parameterization has to allow for a constant flux of enstrophy and a zero kinetic energy flux through $k_c$. On the contrary, if $k_c$ falls in the the energy cascade range a constant flux of kinetic energy, to be dissipated by friction at the largest scales (bed friction in the NSWE framework), must be allowed to cross $k_c$. This is the reason for which some parameterizations prescribe a friction-dependent eddy viscosity [55]. For the sake of simplicity we started with the simplest parameterization, *i.e.* Laplacian viscous diffusion with constant $\nu_T$, and we are currently evaluating the properties of the numerical solution as a function of the position of $k_c$ in the range of the spectral wavenumbers. Theoretical studies are also devoted to analyse features of enstrophy dissipation [57]. In dependence of the results obtained closures of increasing complexity will be investigated.

2. From the experimental point of view the data coming from the experiments performed at the Bari basin and aimed at the analysis of coastal mixing through the evolution of passive tracers are being analysed. The large amount of data are believed to represent a good starting point both for a theoretical description of the mixing features of the flows due to wave overpassing

submerged structures and for the quantitative evaluation of fundamental parameters like the absolute flow diffusivity. Obviously, the model under construction and above described will be both tested against this valuable data set and also used to reproduce the experimental flows.

## 6 Acknowledgements

We wish to thank J.H. LaCasce and A. Provenzale for the many useful discussions. This research was partially supported within the MIUR PRIN 2002 Project "Influenza di vorticità e turbolenza nelle interazioni dei corpi idrici con gli elementi al contorno e ripercussioni sulle progettazioni idrauliche".

## References

[1] Oltman-Shay, J., Howd, P.A. & Berkemeier, W.A., Shear instabilities of the mean longshore current: field observations, *J. Geophys. Res.-Oceans*, **94**, pp. 18031–18042, 1989.

[2] Peregrine, D.H., Surf zone currents, *Theor. Comp. Fluid Dyn.*, **10**, pp. 295–309, 1998.

[3] Chen, Q., Dalrymple, R.A., Kirby, J.T., Kennedy, A.B. & Haller, M.C., Boussinesq modelling of a rip current system, *J. Geophys. Res.-Oceans*, **104**, pp. 20617–20637, 1999.

[4] Liek, G.A., Roelvink J.A. & Uittenbogaard, R.E., The influence of large horizontal eddies on the depth, shape and extent of scour holes, *SASME Book of Abstracts*, Topic 2.2a, 2000.

[5] Brocchini, M., Mancinelli, A., Soldini, L. & Bernetti, R., Structure-generated macrovortices and their evolution in very shallow depths, *Proc. 28th Int. Conf. Coast. Engng.-ASCE*, **1** pp. 772–783, 2002.

[6] Soldini, L., Lorenzoni, C., Piattella, A., Mancinelli, A. & Brocchini, M., Nearshore macrovortices generated at a submerged breakwater: experimental investigation and statistical modeling, *Proc. 29th Int. Conf. Coast. Engng.-ASCE*, **2**, pp. 1380–1392, 2004

[7] Brocchini, M., Kennedy, A.B., Soldini, L. & Mancinelli, A., Topographically-controlled, breaking wave-induced macrovortices. Part 1. Widely separated breakwaters, *J. Fluid Mech.*, **507**, pp. 289–307, 2004.

[8] Kennedy, A., Brocchini, M., Soldini, L. & Gutierrez, E., Topographically-controlled, breaking wave-induced macrovortices. Part 2. Changing geometries, *J. Fluid. Mech.*, (in print), 2004.

[9] Steijn, R., Roelvink, D., Rakhorst, D., Ribberink J. & van Overeem, J., North Coast of Texel: a comparison between reality and prediction, *Proc. 26th Int. Conf. Coast. Engng.-ASCE*, **2**, pp. 2281–2293, 1998.

[10] Smith, J.A. & Largier, J.L., Observations of nearshore circulation: rip currents, *J. Geophys. Res.-Oceans*, **100(6)**, pp. 10967–10975, 1995.

[11] Lloyd, P.M. & Stansby, P.K., Shallow-water flow around model conical island of small side slope. Part I: surface piercing, *J. Hydraul. Engng.-ASCE*, **123**, pp. 1057–1067, 1997.

[12] Schär, C. & Smith, R.B., Shallow-water flow past isolated topography. Part II: transition to vortex shedding, *J. Atmos. Sci.*, **50**, pp. 1401–1412, 1993.

[13] Schär, C. & Durran, D.R., Vortex formation and vortex shedding in continuously stratified flows past isolated topography, *J. Atmos. Sci.*, **54**, pp. 534–554, 1997.

[14] Lloyd, P.M., Stansby, P.K. & Chen, D., Wake formation around islands in oscillatory laminar shallow-water flows. Part 1. Experimental investigation, *J. Fluid Mech.*, **429**, pp. 217–238, 2001.

[15] Taylor, G.I., Diffusion by continuous movements, *Proc. Lond. Math. Soc.*, **20**, pp. 196–212, 1921.

[16] Fischer, H.B., List, E.J., Koh, R.C.Y., Imberger, J. & Brooks, N.H., *Mixing in inland and coastal waters,* Academic Press, New York, 1979.

[17] Larson, M. & Kraus, N.C., Numerical-model of longshore-current for bar and trough beaches, *J.W.P.C.O.E.-ASCE*, **117**, pp. 326–347, 1991.

[18] Takewaka, S., Misaki, S. & Nakamura, T., Dye diffusion experiment in a longshore current field, *Coast. Engng. J.*, **45**, pp. 471–487, 2003.

[19] Jirka, G.H., Large scale flow structures and mixing processes in shallow flows, *J. Hydr. Res.-IAHR*, **39**, pp. 567–573, 2001.

[20] Soldini, L., Piattella, A., Brocchini, M., Mancinelli, A. & Bernetti, R., Macrovortices-induced horizontal mixing in compound channels, *Ocean Dyn.*, **54**, pp. 333–339, 2004.

[21] Bowen, A.J. & Holman, R.A., Shear instabilities of the mean longshore current, 1. Theory, *J. Geophys. Res.-Oceans*, **94**, pp. 18023–18030, 1989.

[22] Allen, J.S., Newberger, P.A. & Holman, R.A., Nonlinear shear instabilities of alongshore current on plane beaches, *J. Fluid Mech.*, **310**, pp. 181–213, 1996.

[23] Chen, D. & Jirka, G.H., Experimental study of plane turbulent wake in a shallow water layer, *Fluid Dyn. Res.*, **16**, pp. 11–41, 1995.

[24] Dracos, T., Giger, M. & Jirka, G.H., Plane turbulent jets in a bounded fluid layer, *J. Fluid Mech.*, **241**, pp. 587–614, 1992.

[25] Uijttewaal, W.S.J. & Booij, R., Effects of shallowness on the development of free-surface mixing layers, *Phys. Fluids*, **12**, pp. 392–402, 2000.

[26] Kraichnan, R., Inertial ranges in two-dimensional turbulence, *Phys. Fluids*, **10**, pp. 1417–1428, 1967.

[27] Lesieur, M., *Turbulence in Fluids,* Kluwer, Dordrecht, 1987.

[28] Provenzale, A., Transport by coherent barotropic vortices, *Ann. Rev. Fluid Mech.*, **31**, pp. 55–93, 1999.

[29] Richardson, L.F., Atmospheric diffusion shown on a distance neighbour graph, *Proc. Roy. Soc. London A*, **110**, pp. 709–737, 1926.

[30] Er-El, J. & Peskin, R., Relative diffusion of constant-level balloons in the Southern hemisphere, *J. Atmos. Sci.*, **38**, pp. 2264–2274, 1981.

[31] Davis, R.E., Drifter observations of coastal surface currents during CODE: the statistical and dynamical view, *J. Geophys. Res.-Oceans*, **90**, pp. 4756–4772, 1985.

[32] LaCasce, J.H. & Bower, A., Relative dispersion in the subsurface North Atlantic, *J. Mar. Res.*, **58**, pp. 863–894, 2000.

[33] Fong, D.A. & Stacey, M.T., Horizontal dispersion of a near-bed coastal plume, *J. Fluid Mech.*, **489**, pp. 239–267, 2003.

[34] Johnson, D. & Pattiaratchi, C., Transient rip currents and nearshore circulation on a swell dominated beach, *J. Geophys. Res.-Oceans*, **109**, C02026, doi: 10.1029/2003JC001798, 2004.

[35] Lippman, T.C. & Holman, R.A., Quantification of sand bar morphology: a video technique based on wave dissipation, *J. Geophys. Res.-Oceans*, **94**, pp. 995–1101, 1989.

[36] Piattella, A., *On mixing in natural shallow flows,* PhD Dissertation, Polytechnic University of Marche, Italy, 2004.

[37] Piattella, A., Mancinelli, A. & Brocchini, M., Mescolamento indotto da macrovortici in ambiente costiero: analisi numerica e fisica, *Proc. 29th Convegno di Idraulica e Costruzioni Idrauliche*, **3**, pp. 809–814, (in Italian), 2004.

[38] Brocchini, M., Drago, M. & Iovenitti, L., The modelling of short waves in shallow waters. Comparison of numerical models based on Boussinesq and Serre equations, *Proc. 23th Int. Conf. Coast. Engng.-ASCE*, **1**, pp. 76–88, 1992.

[39] Schäffer, H.A., Madsen, P.A. & Deigaard, R., A Boussinesq model for waves breaking in shallow water, *Coast. Engng.*, **20**, pp. 185–202, 1993.

[40] Kennedy, A.B., Chen, Q., Kirby, J.T. & Dalrymple, R.A., Boussinesq modeling of wave transformation, breaking and runup. I: 1D, *J. Waterway, Port Coast. and Ocean Engng.-ASCE*, **126**, pp. 39–47, 2000.

[41] Veeramony, J. & Svendsen, I.A., The flow in surf-zone waves, *Coast. Engng.*, **39**, pp. 93–122, 2000.

[42] Bühler, O. & Jacobson, T.E., Wave-driven currents and vortex dynamics on barred beaches, *J. Fluid Mech.*, **449**, pp. 313–339, 2001.

[43] Pratt, L.J., On inertial flow over topography. Part 1. Semigeostrophic adjustment to an obstacle, *J. Fluid Mech.*, **131**, pp. 195–218, 1983.

[44] Brocchini, M., Bernetti, R., Mancinelli, A. & Albertini, G., An efficient solver for nearshore flows based on the WAF method, *Coast. Engng.*, **43**, pp. 105–129, 2001.

[45] Centurioni, L.R., Dynamics of vortices on a uniformly shelving beach, *J. Fluid Mech.*, **472**, pp. 211–228, 2002.

[46] Lorenzoni, C., Soldini, L., Mancinelli, A., Piattella, A. & Brocchini, M., Macrovortici generati da barriere sommerse: analisi fisica e numerica, *Proc. "La difesa idraulica del territorio-2003"*, **1**, pp. 703–714, (in Italian), 2005.

[47] Lorenzoni, C., Soldini, L., Mancinelli, A., Piattella, A. & Brocchini, M., La circolazione idrodinamica in presenza di barriere sommerse: un'analisi sperimentale, *Proc. 29th Convegno di Idraulica e Costruzio-ni Idrauliche*, **3**, pp. 573–580, (in Italian), 2004.

[48] Haller, M.C., Dalrymple, R.A. & Svendsen, I.A., Rip channel and nearshore circulation, *Coast. Dynamics*, pp. 594–603, 1997.

[49] Haller, M.C., Dalrymple, R.A. & Svendsen, I.A., Experimental study of nearshore dynamics on barred beach with rip channels, *J. Geophys. Res.-Oceans*, **107(C6)**, pp. 14,1–14,21, 2002.

[50] Dronen, N., Karunarathna, A., Fredsøe, J., Sumer, M.B. & Deigaard, R., An experimental study of rip channel flow, *Coast. Engng.*, **45**, pp. 223–238, 2002.

[51] Mancinelli, A., Soldini, L., Brocchini, M., Bernetti, R. & Scalas, P., Modelling the effects of structures on nearshore flows, *Proc. 4th Int. Symp. Waves*, pp. 1715–1724, 2001.

[52] Bennett, A.F., A Lagrangian analysis of the turbulent diffusion, *Rev. of Geophys.*, **25(4)**, pp. 799–822, 2000.

[53] Basdevant, C. & Sadourny, R., Parameterization of virtual scale in numerical simulation of two-dimensional turbulent flows. In "Two-dimensional turbulence", *J. Mec. Theor. Appl.*, suppl., edited by R. Moreau, pp. 243–270, 1983.

[54] Smagorinsky, J., Some historical remarks on the use of nonlinear viscosities, In: *Large eddy simulation of complex engineering and geophysical flows,* ed. B. Galperin and S.A. Orszag, Cambridge University Press, 1993.

[55] Uittenbogaard, R.E., *Model for eddy diffusivity and viscosity related to subgrid velocity and bed topography,* Delft Hydraulics Report, 1999.

[56] Stansby, P.K., A mixing-length model for shallow turbulent wakes, *J. Fluid Mech.*, **495**, pp. 369–384, 2003.

[57] Brocchini, M. & Colombini, M., A note on the decay of vorticity in shallow flows, *Phys. of Fluids*, **16(7)**, pp. 2469–2475, 2004.

# CHAPTER 4

# Large scale circulations in shallow lakes

G. Curto[1], J. Józsa[2], E. Napoli[1], G. Lipari[1] & T. Kramer[2]
[1]*Department of Hydraulic Engineering and Environmental Applications, University of Palermo, Italy.*
[2]*Department of Hydraulic and Water Resources Engineering, Budapest University of Technology and Economics, Hungary.*

## Abstract

In this paper wind-driven horizontal and vertical large scale circulations in shallow lakes are analysed.

As an improved approximation of the external forcing field, the wind speed acceleration due to the abrupt reduction in the surface roughness between the land and the water is quantified along the fetch using a semi-empirical approach which allows the identification of the aerodynamic features and hydrodynamic effects of an *Internal Boundary Layer* (IBL) growing within the bottom of the atmospheric boundary layer.

The consequent fetch-dependence of the wind speed and corresponding wind shear stresses on the lake surface causes the appearance of a wind stress curl, which is responsible, together with changes in bathymetry, for causing strong horizontal circulations. The effects of wind speed changes on the wind-driven flow patterns are analysed both analytically and numerically, showing the need to take these changes into account in order to correctly predict wind-induced water currents in shallow basins.

## 1 Introduction

Shallow lakes have recently been receiving greater attention all over the world. Their unique value and multi-purpose utility have increasingly been recognised which has led to the misuses of a number of them, thus worsening their ecological state even to an alarming extent at places. Furthermore, recent changes in the global climate or, at least, the fact that extreme conditions seem to be more frequent, has also changed the boundary conditions for these vulnerable water bodies. In spite of this, lake studies are still quite moderately financed compared to maritime research, and often only focus on deep lakes. When trying to adapt the results obtained in deep water lakes or shallow coastal seas, one has to cope, nevertheless, with a number of problems due to differences in the prevailing time and space scales found in shallow lakes. In fact, shallow lakes have their own features and need specialist research and management methodologies.

Large scale circulations in shallow lakes are primarily driven by wind acting on the water surface. The air–water interface, in free surface flows subject to wind action, is the chief location of energy and gas (oxygen and carbon dioxide) exchange between the atmosphere and the fluid mass. A full understanding of these exchange processes and of the hydrodynamic features of the mixing-layer between air and water currents [1, 2] is therefore important in order to evaluate and predict lake water quality.

Part of the momentum of the wind over the lake is transferred to the water at the lake surface generating waves, turbulence, drift currents and Langmuir circulations, as well as large scale circulations and seiche. This momentum flux indirectly drives the exchange processes at the lake bottom, mixing within the water body, and interactions between the littoral and the pelagic zones. In fact, the more shallow the lake, the more efficient the influence of the external surface forces on the bottom in general [3].

Both horizontal and vertical circulations can be observed in shallow lakes under wind action. In particular horizontal circulations can be highlighted through the analysis of the vorticity equation of the depth averaged horizontal velocities [4]. In this formulation three different sources of vorticity are recognised: the Coriolis effect due to the Earth's rotation, changes in bathymetry and wind stress curl.

The latter effect has so far been related mainly to large scale changes in weather systems, which are responsible for changes in the wind speed. These effects are thus commonly accounted for only in the analysis of very large lakes [5]. As another source of irregularity, large scale topographic features upstream of a lake can also result in spatially varying wind field over the lake [6, 7], playing a role also in medium or small lakes. A reasonable estimation of this effect needs dense enough wind measurements network, preferably coupled with some sort of mezo-scale atmospheric boundary layer model of appropriate vertical and horizontal resolutions. However, measurements are seldom dense enough in space to form a firm basis for wind field reconstruction in themselves, so that accounting for the effect of topographic features upstream the lake is very difficult. In this paper this effect thus will not be considered.

Another source of wind field irregularity is the acceleration of air flow crossing the shoreline towards the lake due to the abrupt change in surface roughness between the land and water, resulting in a fetch dependent wind speed and an appreciable wind stress curl. The horizontal space scale of this change may be negligible in large lakes, however, it is often comparable to the horizontal dimensions of small lakes. The effect of the change in roughness on the wind speed is restricted to a relatively shallow region of the atmospheric boundary layer, resulting in the formation of an Internal Boundary Layer. Although this effect is well known in meteorological literature [8–11], until recently [3], it has been neglected in numerical simulations of large scale lake circulations (except for some heuristic attempt *e.g* by [12, 13]), where a constant wind speed is often assumed. A comparative analysis of the magnitude of wind stress curl due to topographical features and roughness change cannot be easily performed in general terms since both effects depend on a number of features (orography of the region near the lake shore, land roughness, *etc*). The paper thus will focus only on the effect of the roughness change between land and water which can be parameterized depending only on the land roughness and fetch.

Wind speed changes are also related to the temperature difference between land and water (*Thermal Boundary Layer*) [10, 14], but this effect is not addressed in the present study.

In this paper the sources of horizontal and vertical circulations are analysed. Our attention will be restricted to homogeneous water (barotropic conditions), so that the effects of stratification are not taken into account. The Coriolis effect is also neglected in the paper, in order to focus only on the comparison of the effects of changes in bathymetry and wind stress curl due to roughness changes between land and water.

The general description of large scale circulation in shallow lakes is confirmed by two- and three-dimensional numerical simulations in schematic lakes with different shape and bathymetry, subject to the wind action of different speed and direction. 3D simulations are performed using an in-house finite-volume code second-order accurate both in time and space [15]. Although quasi-3D equations employing the hydrostatic pressure assumption can be suitably used for shallow flows, fully 3D simulations are performed in order to better describe the vertical circulation patterns near the lake shoreline. In the code an implicit discretization of vertical turbulent terms is employed [16], while the other terms are treated explicitly. A fractional-step method is used for the time advancement, and the free surface elevation is calculated at each time step using the kinematic boundary condition. For a detailed description and validation of the numerical code, which is not provided in this paper, the reader is referred to [15].

In the following section the equations describing the motion in free-surface water bodies are reviewed and analysed in order to describe the vertical and horizontal (Section 3) circulations processes in shallow lakes. A semi-empirical treatment of the Internal Boundary Layer is then introduced in Section 4 following the general description of Taylor and Lee [17]. Finally, in Section 5 the results of numerical simulations showing the relative importance of the different vorticity sources in lakes are reported and conclusions are drawn.

## 2 Equations and physical processes

The motion of water in unstratified, non-rotating lakes is governed by the momentum and mass conservation laws (Navier-Stokes and continuity equations), which can be written in the conventional summation approach as

$$\frac{\partial u_i}{\partial t}+\frac{\partial u_i u_j}{\partial x_j}-\nu\frac{\partial^2 u_i}{\partial x_j \partial x_j}+\frac{1}{\rho}\frac{\partial p}{\partial x_i}+g\delta_{i3}=0 \quad (1)$$

where $t$ is time, $u_i$ is the $i$-th component of the velocity, $x_i$ the $i$-th axis (with the axis $x_3$ vertical and oriented upward), $\nu$ is the kinematic viscosity, $\rho$ is the water density, $p$ is the pressure, $g$ is acceleration due to the gravity and $\delta_{ij}$ is the Kronecker function.

Since the density is not usually constant in lakes, depending on the temperature and the concentration of suspended or dissolved substances, a "barotropic" and a "baroclinic" pressure are frequently distinguished. The former is the hydrostatic pressure obtained assuming the water density to be constant, while the latter accounts for the density variations in the vertical water column. The analysis in the paper will be restricted to barotropic conditions, which hold for shallow lakes or for well mixed deep lakes, neglecting the problems related to the thermal stratification.

The total pressure is thus given by the sum of the barotropic pressure, the atmospheric pressure at the air–water interface (that will be assumed to be zero for the whole air–water interface) and the "modified pressure" $q$ which accounts for the non-hydrostaticity of the pressure.

Introducing the above assumptions the Navier-Stokes equations can be re-written as

$$\frac{\partial u_i}{\partial t} + \frac{\partial u_i u_j}{\partial x_j} - \nu \frac{\partial^2 u_i}{\partial x_j \partial x_j} + \frac{1}{\rho}\frac{\partial q}{\partial x_i} + g\frac{\partial \eta}{\partial x_i} = 0 \tag{2}$$

which hold for "single valued" water surface, in which only one value of $\eta$ exists for each water column (breaking waves for instance cannot be represented).

The system of Navier-Stokes and continuity equations can only be solved numerically, but even the most powerful parallel computers have not so far been able to manage the huge number of unknowns resulting from the discretization of the equations on grids as fine as the smallest scales of motion. Since non linear instabilities occur due to the excess of inertia forces with respect to the stabilizing viscous stresses, the kinetic energy of the large scale motion is in fact transferred to the largest turbulent eddies, whose dimensions are comparable with the characteristic domain length (in a non stratified lake, typically the water depth H). The turbulent cascade of energy then drives energy to smaller eddies, until the dimension of the "Kolmogorov scale" is reached, where all the energy is dissipated by viscous stresses [18]. Since the Kolmogorov scale is several orders of magnitude smaller than the characteristic domain length (typically with a ratio smaller than $10^{-4}$), a three-dimensional computation of the whole spectrum of the scales of motion in a natural basin would require grids with more than $10^{12}$ cells or calculus points, which is currently unfeasible.

Engineers and geophysicists, however, are frequently not interested in small scale water motion. The Navier-Stokes equations are thus usually averaged in time in order to separate turbulent fluctuations from large scale organized motion. This separation can only be obtained if the largest turbulent time scale is reasonably smaller than the time scale of the large scale "mean" motion, which allows for the selection of a time interval in which the large scale quantities can be assumed to be constant.

The Reynolds Averaged Navier-Stokes are thus obtained

$$\frac{\partial \overline{u}_i}{\partial t}+\frac{\partial \overline{u}_i\overline{u}_j}{\partial x_j}-\nu\frac{\partial^2 \overline{u}_i}{\partial x_j \partial x_j}+\frac{1}{\rho}\frac{\partial \overline{q}}{\partial x_i}+g\frac{\partial \overline{\eta}}{\partial x_i}+\frac{\partial \overline{u_i' u_j'}}{\partial x_j}=0 \tag{3}$$

where the bar denotes averaged quantities and the prime denotes turbulent fluctuation. The closure of the problem now requires the introduction of "turbulence models" which express the "correlations" $\overline{u_i' u_j'}$ between fluctuating velocities in terms of averaged quantities.

A review of turbulence models employed in the analysis of lakes and reservoirs is beyond the aims of this paper. Here only the conventional approach based on the introduction of the "eddy viscosity" concept is discussed, in which the correlations $\overline{u_i' u_j'}$ are expressed as

$$\overline{u_i' u_j'}=-\nu_t\left(\frac{\partial \overline{u}_i}{\partial x_j}+\frac{\partial \overline{u}_j}{\partial x_i}\right)+\frac{2}{3}\delta_{ij}k \tag{4}$$

where $\nu_t$ is the eddy viscosity, $\delta_{ij}$ is the Kronecker delta and $k$ is the turbulent kinetic energy ($k=\overline{u_1' u_1'}+\overline{u_2' u_2'}+\overline{u_3' u_3'}$). Assumption eqn. (4) is based on the hypothesis that an analogy exists between the molecular motion responsible for the development of viscous stresses and the turbulent motion which causes momentum exchanges within the water mass. The eddy viscosity $\nu_t$ has the dimension of a length multiplied by a velocity and, in contrast to the kinematic viscosity, is a property of the flow and changes in space and time. Several models have been proposed in the literature for calculating the eddy viscosity, from simple algebraic parameterizations to the introduction of additional transport equations (for a review, see Rodi [19]).

Horizontal topographic scales are much larger than the vertical ones in lakes, coastal waters and in most natural water basins, so that computational grids are usually quite anisotropic. Horizontal small scale circulations thus cannot be numerically resolved owing to the coarseness of the grid, resulting in the need for modelling subgrid scale motions in the horizontal directions. This task is usually approached by introducing horizontal eddy viscosity much larger than the vertical one, which is commonly assumed to be constant. In this paper this approach will be used, defining a constant horizontal eddy viscosity $\nu_{t,h}$ and a space dependent vertical eddy viscosity $\nu_{t,v}$.

This approach results in the equations

$$\frac{\partial \overline{u}_i}{\partial t}+\frac{\partial \overline{u}_i\overline{u}_j}{\partial x_j}-\nu_{t,h}\left(\frac{\partial^2 \overline{u}_i}{\partial x_1^2}+\frac{\partial^2 \overline{u}_i}{\partial x_2^2}\right)-\frac{\partial}{\partial x_3}\left(\nu_{t,v}\frac{\partial \overline{u}_i}{\partial x_3}\right)$$
$$+\frac{1}{\rho}\frac{\partial \overline{q}}{\partial x_i}+g\frac{\partial \overline{\eta}}{\partial x_i}=0 \tag{5}$$

where the kinematic viscosity is assumed to be negligible compared to the horizontal and vertical eddy viscosities and the turbulent kinetic energy $k$ is absorbed in the modified pressure $q$.

As previously mentioned, horizontal length scales are much larger than the vertical ones in lakes and correspondingly vertical velocities are often negligible with respect to the horizontal ones. This observation results in a simplification of the eqn. (5), which is obtained by depth-averaging the horizontal equations and neglecting vertical accelerations (shallow water approximation). Depth-averaged equations read [19, 20]

$$\begin{aligned}
&\frac{\partial U_1}{\partial t} + \frac{\partial U_1 U_1}{\partial x_1} + \frac{\partial U_1 U_2}{\partial x_2} + g\frac{\partial}{\partial x_1}(H + z_B) - \frac{1}{\rho H}\left(\frac{\partial H\tau_{11}}{\partial x_1} + \frac{\partial H\tau_{12}}{\partial x_2}\right) \\
&\quad + \frac{\tau_{s1} - \tau_{b1}}{\rho H} + \frac{1}{\rho H}\frac{\partial}{\partial x_1}\int_{z_b}^{z_b+H} \rho(u_1 - U_1)^2 dx_3 \\
&\quad + \frac{1}{\rho H}\frac{\partial}{\partial x_2}\int_{z_b}^{z_b+H} \rho(u_1 - U_1)(u_2 - U_2) dx_3 = 0 \\
&\frac{\partial U_2}{\partial t} + \frac{\partial U_1 U_2}{\partial x_1} + \frac{\partial U_2 U_2}{\partial x_2} + g\frac{\partial}{\partial x_2}(H + z_B) - \frac{1}{\rho H}\left(\frac{\partial H\tau_{12}}{\partial x_1} + \frac{\partial H\tau_{22}}{\partial x_2}\right) \\
&\quad + \frac{\tau_{s2} - \tau_{b2}}{\rho H} + \frac{1}{\rho H}\frac{\partial}{\partial x_1}\int_{z_b}^{z_b+H} \rho(u_1 - U_1)(u_2 - U_2) dx_3 \\
&\quad + \frac{1}{\rho H}\frac{\partial}{\partial x_2}\int_{z_b}^{z_b+H} \rho(u_2 - U_2)^2 dx_3 = 0
\end{aligned} \tag{6}$$

where $U$ and $V$ are the depth-averaged horizontal velocity components, $H$ is the water depth, $z_b$ is the depth of the bottom, $\tau_{ij} = -\rho\overline{u_i u_j}(i = 1, 2)$ are the horizontal turbulent stresses and $\tau_s$ and $\tau_b$ are the shear stresses at the water surface and at the bed. The former depends on the wind action at the air–water interface, while the second can be expressed as a function of the depth-integrated velocity and the bottom roughness. In eqn. (6) the last two terms account for the vertical changes in horizontal velocities. They are named dispersion terms and are usually modelled in analogy with the turbulent terms, by introducing a *dispersion coefficient* D. It is important, however, to note that the dispersion terms account for transport processes related to velocity changes along the depth and are not related in any way to the turbulence [19].

## 3 Vertical and horizontal circulations in lakes

Water motions in lakes are primarily induced by wind. Wind stress generates a drift current near the air–water interface, resulting in the upwelling of the free surface downwind and in its downwelling upwind (wind setup). A pressure gradient acting against the wind is thus established. Although near the surface the wind stress always prevails over the pressure gradient, at increasing depths the shear stresses become weaker and weaker, resulting in the pressure effects overpowering them. In flat bottom basins a zero net flux is obtained at the steady state in each water column, with a drift current driven by the wind near the free surface and a counter-current near the bottom, which is required to respect the condition of continuity.

In natural water basins, characterised by more complex bathymetries, the relative effect of the wind shear stress and the pressure gradient is more difficult to identify. The force per unit volume exerted by the wind on a water column in the $i$-th direction can be expressed as the ratio $\tau_{s,i}/H$, where $H$ is the water depth, while the pressure gradient action on a unit volume is $\gamma \partial \eta / \partial x_i$, where $\gamma$ is the unit weight of water. While the latter term is independent of water depth, the former reduces inversely with increasing depth. Thus, where the water depth is lower than the basin mean value, the wind action prevails over the pressure gradient, resulting in a current aligned with the wind along the whole water column. On the contrary, where the depth is greater, the force per unit volume exerted by wind is overpowered by the pressure counteraction and a net flux is obtained in the direction against the wind. At the free surface the current is still directed with the wind but at increasing depth the current direction is reversed, resulting in a depth averaged velocity in the direction opposite to the wind.

The steady state is obtained after an oscillatory motion is established (*seiching*), due to the alternating prevalence of drift currents and countercurrents, until viscous stresses dissipate the excess energy [21]. In large lakes the inertia of the water masses is able to maintain oscillations distinguishable for long periods (even several days). Since winds change in speed and direction with a frequency higher than the inverse of the time required to achieve steady conditions, unsteady states are the more frequent in large lakes.

In Section 1 the depth averaged Reynolds equations were shown, which are particularly useful when dealing with the analysis of horizontal circulations. A clear depiction of the horizontal circulation processes and their sources in lakes is in fact obtained by the vertical vorticity equation, resulting from the application of the *curl* operator to eqn. (6). Defining the vorticity of the horizontal depth-averaged velocity field as

$$\zeta = \frac{\partial V}{\partial x_1} - \frac{\partial U}{\partial x_2} \tag{7}$$

after a little algebra and some simplifications the following equation is obtained

$$H\frac{\partial \zeta}{\partial t} + HU \cdot \nabla\zeta + \zeta U \cdot \nabla H + \frac{1}{\rho}\left(\nabla \times \tau_s + \frac{1}{H}\tau_s \times \nabla H\right) - \frac{1}{\rho}\left(\nabla \times \tau_b + \frac{1}{H}\tau_b \times \nabla H\right) + HD\nabla^2\zeta = 0 \tag{8}$$

Equation (8) allows an analysis of the terms affecting the time rate of change of the vorticity: in particular the second and the last terms express the transport processes (due to convection and dispersion, respectively), the third expresses the changes in vorticity due to depth variation (accounting for strengthening or weakening vorticity when transported toward deeper or shallower zones), the fourth expresses the increase of the vorticity due to the wind action, the fifth is the dissipative effect of the bottom stress. It can be noted that only the *wind stress curl* and the vector product of the wind stress by the depth gradient have an active role in introducing vorticity in the depth averaged velocity field, while the other terms only transport or dissipate vorticity [5, 6].

The relative effect of the various terms in eqn. (8) cannot be easily identified using length scale analysis, since their effect strongly depends on the direction of the wind and of the water currents with respect to the bottom gradients. The effect on the vorticity of synoptic spatial changes in the wind action (resulting in a curl of the shear stresses different from zero) is usually accounted for in oceanographic literature, but is usually assumed to be negligible in lake circulation analysis since the weather systems responsible for changes in the wind speed are usually far larger than the lake extensions, with the exception of the largest lakes in the world [4].

What is important to consider especially for small lakes, nevertheless, is the fact that the air current crossing the shoreline towards the lake is accelerated due to the abrupt change in surface roughness between land and water, resulting in a fetch-dependent wind speed and an appreciable wind stress curl.

In the next section a semi-empirical theory is reported [17], which received a number of experimental confirmations [9–11], allowing the estimation of wind speed acceleration due to the different roughness of the land and the water surface.

## 4 The atmospheric boundary layer and variations due to roughness and temperature changes passing from land to water

The wind drag coefficient $C_{10}$, relating the wind speed at the anemometric height of 10m $W_{10}$ to the wind shear velocity is defined as

$$C_{10} = \frac{{u^*}^2}{W_{10}^2} \tag{9}$$

and can be calculated by Wu's formula [22]

$$C_{10} = (0.8 + 0.065 W_{10}) \cdot 10^{-3} \tag{10}$$

($W_{10}$ in m/s) holding for a large range of wind velocities. As reviewed by Wuest and Lorke [23], different estimates of $C_{10}$ can be used for weak (<5m/s) and strong winds [24–27], with a minimum for a wind speed of approximately 5m/s.

The wind speed 10m above the water level can thus be used to calculate the wind shear stress by using the wind drag coefficient.

However, the air current over the water surface accelerates due to the lower roughness of the water relative to the land. This effect increases with fetch, causing the formation of an Internal Boundary Layer (IBL) in the lower part of the Atmospheric Boundary Layer, characterised by velocities higher than those encountered at the same level on the land.

An estimate of the IBL height $\delta_b$ has been proposed by Taylor & Lee [17], where $\delta_b$ depends on the *fetch F* and on the water surface roughness $z_{0,w}$ as

$$\delta_b(F) = 0.75 z_{0,w} \left( \frac{F}{z_{0,w}} \right)^{0.8} \tag{11}$$

The wind action over the water surface produces waves which increase the air friction and act as additional roughness. The roughness $z_{0,w}$ of the water surface thus depends on the height of the waves (the significant wave height is usually used, which is defined as the average height of the highest third of the waves). A measure of the roughness of the water surface has been given by Charnock [28], who found the relationship for the vertical wind speed profile

$$W_z \approx u^* \left[ \frac{1}{k} \ln \left( \frac{gz}{u^{*2}} \right) + K \right] \tag{12}$$

where $u^*$ is the wind shear velocity, $g$ the acceleration due to gravity, $k$ the von Kármán's constant, $z$ the height above the water surface and $K$ a constant to be tuned. The term $u^{*2}/g$ is known as the *waveheight scale*, which is a measure of the roughness of the surface waves. Using the eqn. (12) the relationship can be thus written

$$z_{0,w} = \alpha \frac{u^{*2}}{g} \tag{13}$$

in which the value $\alpha = 0.0185$ is proposed by Wu [22] (that corresponds to $K = 9.97$ in eqn. (12)). Equation (12) is a very simple model which disregards the age of the waves and relates their height only to the local shear velocity of the wind. Although more complex relations can be considered (an extension of the Charnock relation by a parameterisation of the Charnock parameter $\alpha$ with wave age as additional parameter has been proposed for istance by Johnson *et al* [29]), eqn. (12) is very popular in the literature [30] and will be used in this paper, while the effect of the wave age will be addressed in a future research.

It can be then assumed that both below and above the IBL height a logarithmic profile holds for the wind velocity,

$$W_z(F) = \frac{u^*(F)}{\kappa} \ln \left( \frac{z}{z_{0,w}} \right) \quad \text{for } z \leq \delta_b(F) \tag{14}$$

and

$$W_z(0) = \frac{u^*_{land}}{\kappa} \ln \left( \frac{z}{z_{0,land}} \right) \quad \text{for } z \geq \delta_b(F) \tag{15}$$

where $k$ is the Kármán constant, $W_z(0)$ and $W_z(F)$ are the wind speed on the land and on the water for a given fetch, respectively, and finally $z_{0,land}$ and $u^*_{land}$ are the land roughness and the wind shear stress on the land.

Since at the height $z = \delta_b(F)$, the condition holds

$$W_{\delta_b}(0) = W_{\delta_b}(F) \tag{16}$$

the relationship is obtained

$$u^* = u^*_{land} \frac{\ln\left[\frac{\delta_b(F)}{z_{0,land}}\right]}{\ln\left[\frac{\delta_b(F)}{z_{0,w}}\right]} \tag{17}$$

Introducing eqns. (11) and (13) into eqn. (17) results in the implicit relationship

$$u^* \ln\left[0.75\left(\frac{Fg}{0.0185u^{*2}}\right)^{0.8}\right] = u^*_{land} \ln\left[0.75\left(\frac{0.0185u^{*2}}{g}\right)^{0.2} \frac{F^{0.8}}{z_{0,land}}\right] \tag{18}$$

which allows the wind shear velocity to be obtained as a function of the fetch and of the land roughness.

Finally, wind shear stresses can be easily obtained by

$$\tau_s = \rho_{air} u^{*2} \tag{19}$$

# 5 Numerical analysis of wind induced hydrodynamic fields in shallow basins

## 5.1 Model investigations in a fictitious sample lake

The understanding of the phenomena based on the model presented so far can be further improved by considering a simple, fictitious lake geometry and carrying out modelling experiments systematically in specific conditions.

A 2D numerical model to discretize eqn. (6) by standard finite difference method on equidistant Cartesian grid is used, in which the conventional Manning-type quadratic bottom shear stress law is adopted and a constant dispersion coefficient $D$ is introduced. The assumption of a quadratic dependence of the bed shear stress from the local velocity via a Chezy friction coefficient is not correct when the wind stress exceeds the bed shear stress, as it can locally happen in wind-induced flows. We nevertheless reckon, on the basis also of previous computations not reported here, that the general patterns described below are not strongly influenced by the assumed quadratic relationship.

The basic bottom topography of the basin can be seen in fig. 1. It is 1.5m deep in the middle, then becomes gradually shallow approximately 1m toward the shore. 10m/s NW steady state winds are applied both with spatially constant and IBL-based distribution.

Figure 2 shows the IBL-based distribution of the wind shear stress for the NW wind. The distribution of the two main vorticity sources in this situation is presented

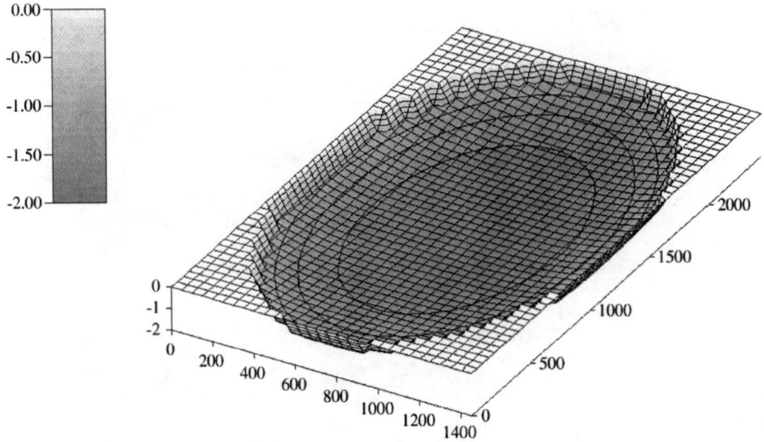


Figure 1: Bottom topography of the sample lake displayed on the 50m cell size finite difference grid. Depths and lengths in meters.

in fig. 3. It can be seen that they compete with each other in the upwind near-shore zone. Under the given conditions the term related to the wind stress curl proves stronger there, resulting in one single, basin-wide clock-wise circulation (fig. 4), in sharp contrast with the typical two-cell circulation traditionally obtained under uniform external forcing.

Figure 5 tackles another important issue, namely the changes of the flow pattern due to significant rising or lowering of the lake water level. It can be seen that dropping the level by 0.7m strengthens the vorticity source due to the near-shore relative bottom gradient to an extent compensating and even exceeding its counterpart, which then results in the appearance of a twin gyre. On the contrary, rising the water level by the same amount makes the one-gyre circulation even more pronounced. Note that three-dimensional effects were disregarded here.

## 5.2 3D simulations in a rectangular basin with a flat bottom and an inclined bottom

Having shown how the wind induced currents in water basins are influenced by the dependence on the fetch of the wind shear stresses, using 2D simulations, in this section the results of 3D numerical simulations of wind induced flows in shallow example lakes will be shown and analysed in order to highlight the existence of both vertical and horizontal wind-induced circulations in shallow lakes and to compare their relative magnitude.

A finite volume numerical code is used for the simulations, which is second order accurate both in time and space and is able to solve the momentum and mass conservation equations for incompressible fluids on structured grids. A fractional step method is used to overcome the incompressible pressure-velocity decoupling. A detailed description of the numerical code is given in [15]. A $k$-$\epsilon$ turbulence model

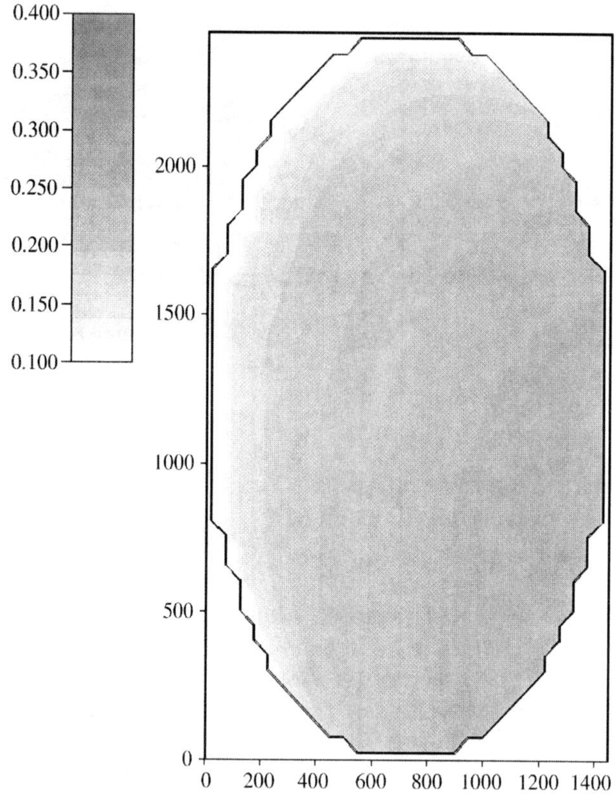


Figure 2: Wind shear stress field in Pascal over the lake according to the IBL development corresponding to $W_{10} = 10$m/s NW wind and 0.15m roughness height at the upwind shore. Lengths in meters.

in the standard formulation ($c_\mu = 0.09, \sigma_k = 1.0, \sigma_\epsilon = 1.3, c_{\epsilon 1} = 1.92, c_{\epsilon 2} = 1.44$) [19] is used to represent the vertical turbulent viscosity while the horizontal turbulent viscosity is assumed to be constant ($\nu_{t,h} = 0.5\text{m}^2$/s). The use of this kind of turbulence model implies the assumption of the presence of an equilibrium layer close to the upper and the lower boundaries, where a logarithmic law of the wall is assumed to hold. This hypothesis is no longer valid when the shear stress at the bottom tends to vanish, which in the analysed case occurs only occasionally. We thus think that the use of more refined turbulence modelling, although it would be more correct on the theoretical point of view, could moderately change the results of the simulations and the related conclusion drawn in the paper.

Results from numerical simulations obtained in a rectangular basin 4000m long and 2000m wide, with a bottom roughness of 0.004m are shown. A flat bottomed shape basin has been considered as well as an inclined bottom with slopes in the direction $x_2$ spanning between $10^{-4}$ and $10^{-3}$. The mean depth of the basin has been fixed at 2m in all cases.

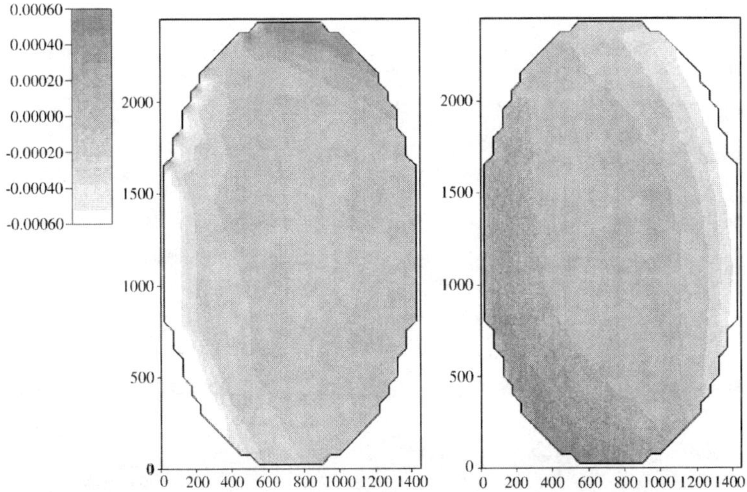

Figure 3: $\nabla \times \tau_s$ (left) and $(\tau_s \times \nabla H)/H$ (right) fields in Pa/m over the lake according to the IBL development corresponding to $W_{10}$=10m/s NW wind and 0.15m roughness height at the upwind shore. Lengths in meters.

The numerical domain has been discretised using $40 \times 20 \times 10$ cells in the directions $x_1, x_2$ and $x_3$, respectively, with a refinement of the grid near the lateral walls, the bottom and the free surface. Since the time to achieve the steady-state was relatively variable in the cases considered, the simulations were performed for 12 hours, when stationarity was almost attained. The free slip condition was used for the lateral boundaries while the logarithmic wall-law was used at the bottom. Wind-induced shear stresses at the free-surface were obtained using the water shear velocity, which is related to the air shear velocity by equation:

$$u_w^* = u_a^* \sqrt{\frac{\rho_{air}}{\rho_{water}}} \tag{20}$$

The *kinematic* boundary condition

$$\frac{\partial \eta}{\partial t} + u_1 \frac{\partial \eta}{\partial x_1} + u_2 \frac{\partial \eta}{\partial x_2} = u_3 \tag{21}$$

was then used to relate free surface movements to the velocities.

The simulations used both constant wind shear stresses and wind shear stresses obtained from the semi-empirical theory.

The simulations performed in the flat bottomed basin with $\alpha = 0°$ (the wind direction will be indicated hereinafter by the angle $\alpha$ with the direction $x_1$ parallel to the longest side of the basin) resulted in negligible depth-averaged velocities in the whole basin (less than 0.5mm/s) after 12 hours. This result was expected, since the vorticity equation of the depth averaged velocities shows that in these conditions (flat bottom, $\alpha = 0$) no horizontal circulation can develop. Despite the

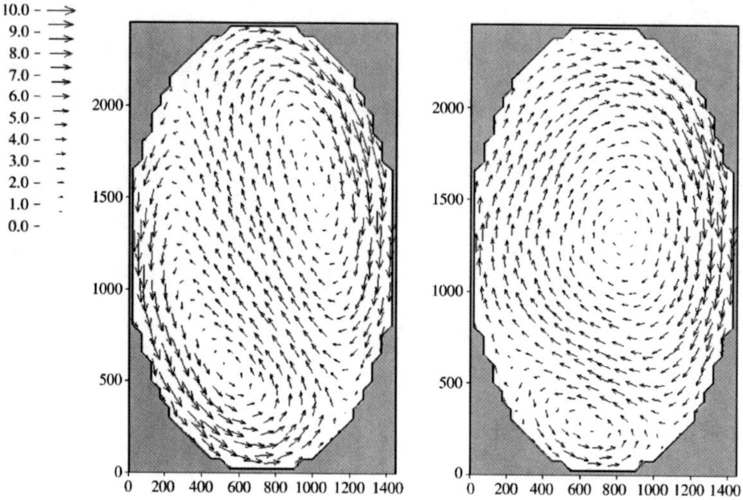


Figure 4: Modelled steady-state flow pattern, induced by wind shear stress field uniform over the lake corresponding to $W_{10}$=11m/s NW wind (left); and by a wind shear stress field according to the IBL development corresponding to $W_{10}$=10m/s NW wind and 0.15m roughness height at the upwind shore (right). Flow velocity in cm/s, lengths in meters.

three-dimensionality of the simulations performed, depth-averaged velocity fields are analysed here, since they allow a clear identification of horizontal circulations, which are mainly influenced by the dependence of the wind stresses on the fetch.

The results did not change for increasing values of the angle $\alpha$ when neglecting the dependence of the wind stress on the fetch. Accounting for the fetch dependence of the shear stresses, however, resulted in quite different velocity patterns, due to the growth of the curl of the wind field. In fig. 6 the depth-averaged velocity fields calculated with the wind shear stresses obtained from eqns. (18) and (19) are shown for $\alpha = 45°$. Two counter-rotating vortices with vertical axis are clear, driven by the wind stress curl.

The results of 3D simulations allow the identification and the comparison of horizontal and vertical circulations since they provide a complete description of the velocity field, which was not the case for the 2D analysis. To this aim two indices are defined here to estimate the horizontal and vertical circulations in the vertical transverse mid-plane of the basin starting from the differences between local velocities and depth-averaged and width-averaged velocities, respectively:

$$C_V = \frac{\sum_{p,q} \left[u_1(p,q) - \mu(q)\right]^2 \cdot A_{pq}}{A_{sez}} \tag{22}$$

$$C_H = \frac{\sum_{p,q} \left[u_1(p,q) - \mu(p)\right]^2 \cdot A_{pq}}{A_{sez}} \tag{23}$$

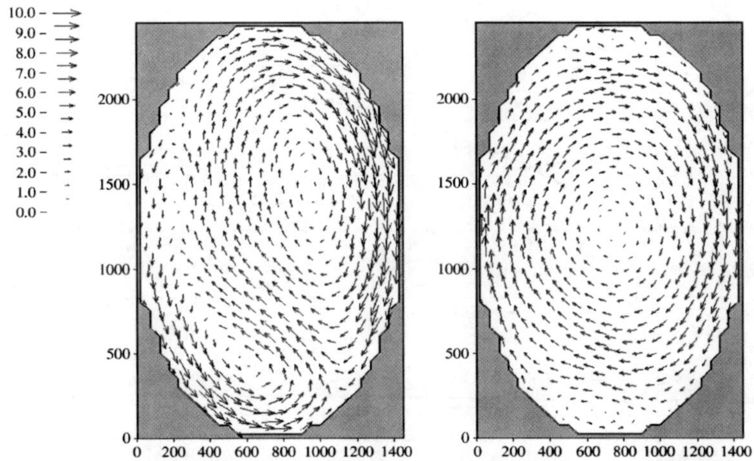


Figure 5: Effect of 0.7m drop (left) and 0.7m rising (right) of the overall lake water level. Modelled steady-state flow patterns induced by wind shear stress field according to the IBL development corresponding to $W_{10}$= 10m/s NW wind and 0.15m roughness height at the upwind shore. Flow velocity in cm/s, lengths in meters.

where the summation spans all the cells in the section $x_1 = 2000$m (which are identified by the indices $p$ and $q$ in the lateral and vertical directions, respectively), $u_1(p,q)$ is the local streamwise velocity in the same cells, $A_{pq}$ is the area of their faces with normal $x_1$, $\mu(q)$ and $\mu(p)$ are the averages of the streamwise velocities

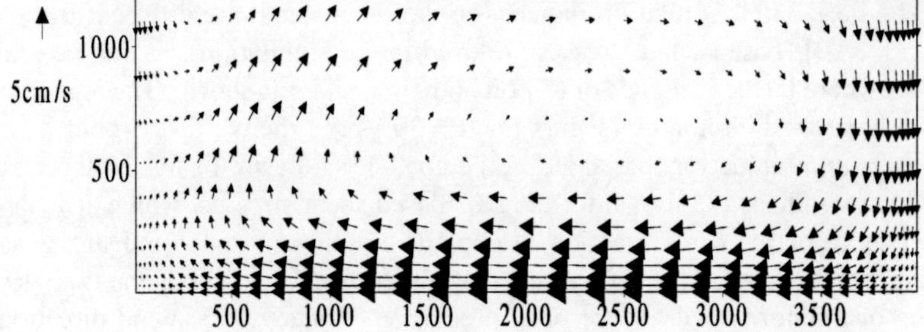


Figure 6: Depth averaged velocities in the flat bottom basin for $\alpha = 45°$. Shear stresses from the semi-empirical theory. Lengths in meters.

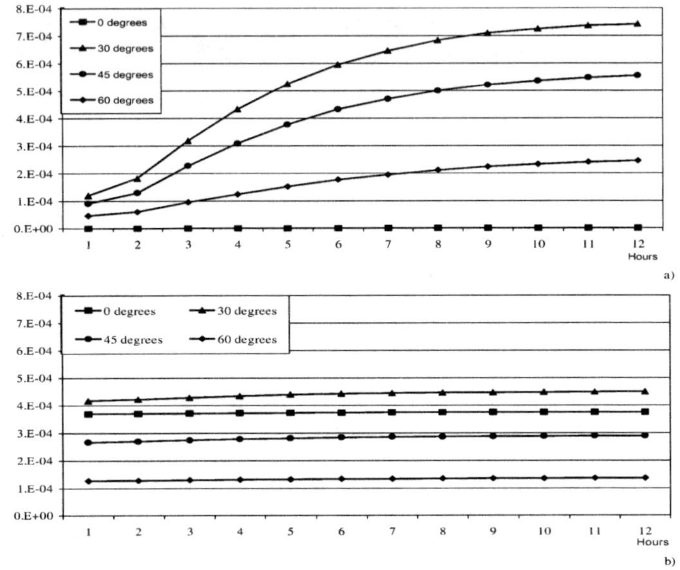


Figure 7: Indices of the horizontal and vertical circulations as a function of both the angle $\alpha$ and the time: (a) $C_O$ with $\tau_s$ from the semi-empirical theory; (b) $C_V$ with $\tau_s$ from the semi-empirical theory.

$u_1$ in the vertical and horizontal directions, respectively, and finally $A_{sez}$ is the area of the whole section.

The values of the index $C_H$ in the flat bottom basin under the influence of a constant wind stress resulted to be near zero whatever the wind direction, since the velocities only depend on the depth and are independent of the $x_2$ coordinate, which made each term of the summation in eqn. (23) equal to zero. The vertical circulations as parameterized by the index $C_V$ were, however, different from zero, with decreasing values for angles spanning from $\alpha = 0°$ to $\alpha = 90°$ since the index is calculated using the $u_1$ component of the velocity, which becomes lower when the angle between the wind direction and the axis $x_1$ is increased to 90°.

The horizontal circulation indices, on the other hand, are different from zero when the IBL-based wind stresses are used in the simulations, as can be seen in fig. 7a, where the dependence of $C_H$ on both time and $\alpha$ is shown. The higher value of the horizontal circulation occurs for $\alpha = 30°$ since the wind stress curl reaches the maximum value for that angle in the considered basin. In fig. 7b the vertical circulation indices are shown for the IBL-based shear stresses, which also show a dependence on the wind direction, as explained in the constant wind stress case.

An interesting feature which can be observed from fig. 7a, b is the time dependence of the horizontal and vertical circulation. Whatever the wind direction, in fact, the vertical circulation indices change only slightly (less than 10%) during the 12 simulated hours, while the horizontal circulations on the contrary show a clear increase during the simulated time. It can thus be concluded that the achievement

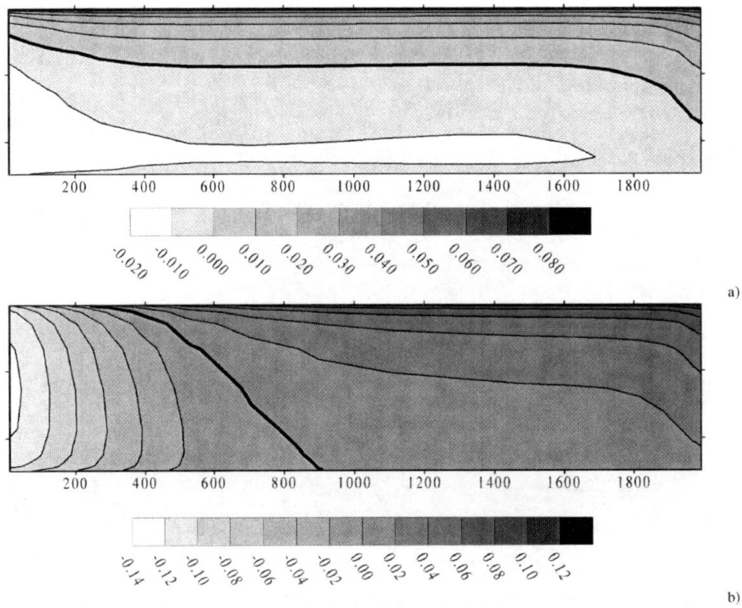


Figure 8: Isolines of the $u_1$ component of the velocity in the vertical transverse midplane ($x_1 = 2000$m) for the constant wind stress (a) and for the wind stresses obtained from eqns. (18) and (19) (b) with $\alpha = 45°$. Velocities in m/s, lengths in meters. Vertical scales distorted.

of stable horizontal circulations is much slower than for the vertical ones so that in the first hours after the wind starts vertical circulations prevail over the horizontal ones, while at the steady state (nearly achieved after 12 hours) in the simulated basin the situation is inverted.

This observation and a general comparison between horizontal and vertical circulations also extending to strongly time-dependent conditions can be very useful, since two-dimensional models are only able to describe horizontal circulations, while 3D models allow the representation of vertical circulation.

In order to complete the description of the velocity fields in the flat bottom basin the isolines of the $u_1$ component of the velocity in the vertical transverse mid-plane ($x_1 = 2000$m) are shown in fig. 8a,b for the constant wind stress (case a) and the IBL-based wind stresses obtained from eqns. (18) and (19) (case b) with $\alpha = 45°$. The isolines (and particularly the one corresponding to null velocities, plotted as a bold line in the figures) are nearly horizontal in case (a), showing the virtual absence of horizontal circulations. In case (b), on the contrary, the isolines are much more inclined, since both horizontal and vertical circulations occur in the basin (a vertical zero-point line would represent the absence of vertical circulations and the clear prevalence of horizontal ones).

In order to identify the effect of a bottom slope on the three-dimensional velocity field, simulations have also been performed in a rectangular basin with a bottom

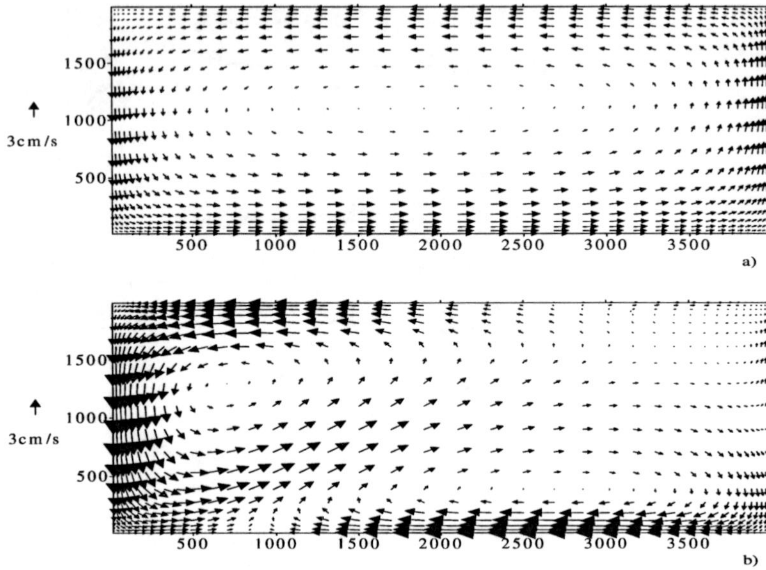


Figure 9: Depth averaged velocities in the bottom slope basin ($i = 0.0004$) for the constant wind stress (a) and for the wind stresses obtained from eqns. (18) and (19) (b) with $\alpha = 30°$. Lengths in meters.

slope of 0.0004 in the direction $x_2$ (depth ranging between $H = 1.6$m at $x_2 = 0$ and $H = 2.4$m at $x_2 = 2000$). The depth averaged velocities under the influence of a fetch dependent wind with $\alpha = 45°$ are shown in fig. 9. Whilst a constant wind stress resulted in a counter-clockwise gyre, a double gyre is observed in fig. 9 due to the combined effects of the wind stress curl and of the bottom slope. The former, in fact, strengthens the counter-clockwise gyre in the upper left corner of the basin (which is driven by the bathymetry) while it creates a clockwise gyre in the lower right corner which prevails over the counter rotating gyre induced by the bottom slope.

The indices $C_H$ and $C_V$ are plotted in figs. 10a and b, showing a prevalence of the vertical circulations over the horizontal ones and, again, a slower growth of the horizontal circulations with respect to the vertical ones.

Finally, in fig. 11a and b the isolines of the $u_1$ component of the velocity in the vertical transverse mid-plane ($x_1 = 2000$m) are shown for the constant wind stress (case a) and for the IBL-based forcing (case b). In (case a) the zero-point line is inclined upward from left to right showing that the current is nearly aligned with the wind in the shallower part of the basin and is directed in the opposite direction, except in a very thin surface layer, in the right deeper part. When the fetch-dependence of the wind is accounted for, however, the current is aligned with the wind in the central part of the section considered (almost down to the bottom), while it flows in the opposite direction near the lateral walls, where positive velocities (*i.e.* driven by wind) occur only very close to the free-surface.

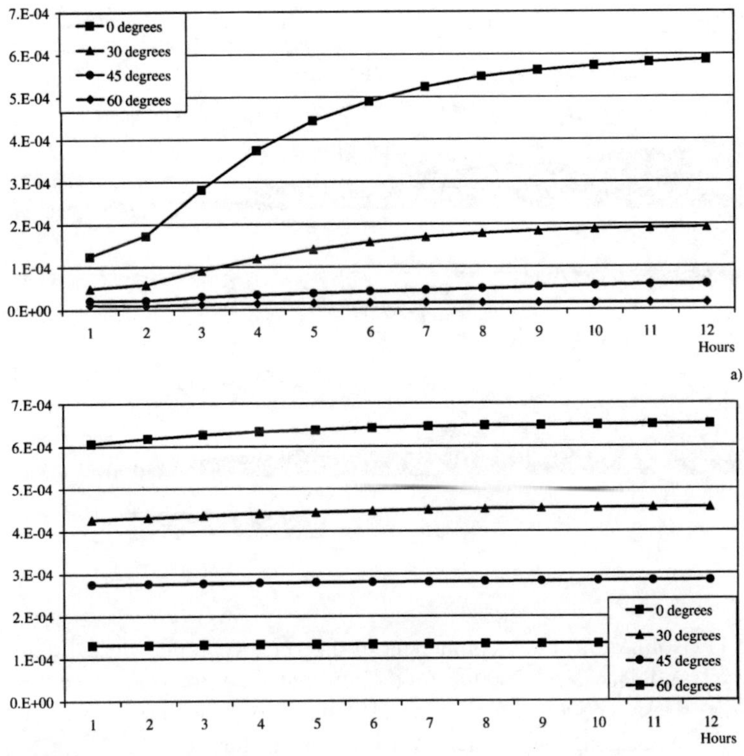


Figure 10: Indices $C_v$ and $C_H$ in the bottom slope basin ($i = 0.0004$) as a function of the time and the wind direction.

## 6 Conclusions

The vorticity equation of the depth-averaged velocities has been derived in order to identify the terms contributing to horizontal circulatory motions in shallow lakes. In particular the effect of the wind stress curl due to the abrupt change in surface roughness experienced by the air current when passing the lake shoreline is analysed.

A semi-empirical theory allowing the calculation of the wind speed on the water surface as a function of the fetch using the wind speed measured on the land is reported and discussed. The theory also allows the wind shear stress acting on the water surface, the water surface roughness and the height of the Internal Boundary Layer which develops in the lower part of the atmospheric boundary layer to be obtained. Numerical simulations of the wind-induced water currents in shallow lakes were then performed using both 2D and 3D models. The former allows the clear identification of the bulk horizontal circulatory motions at low computational costs, while the latter, though requiring more computational time, also results in

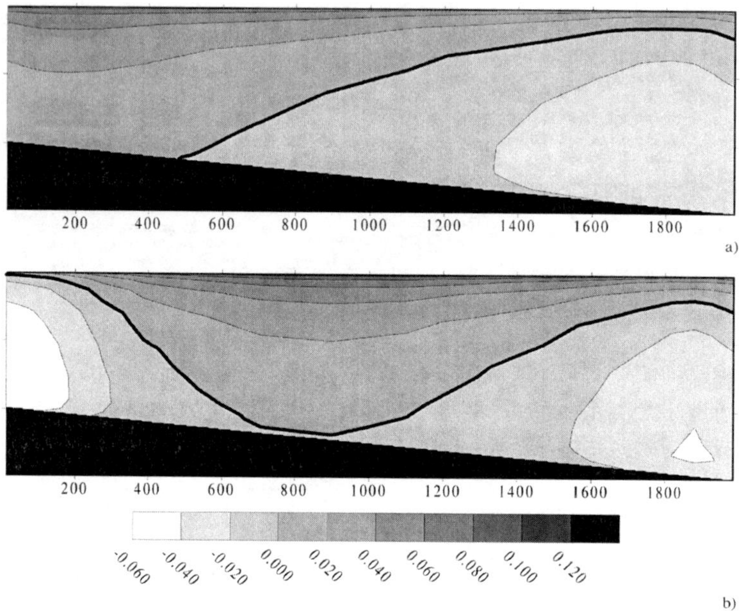


Figure 11: Isolines of the $u_1$ component of the velocity in the vertical transverse midplane ($x_1 = 2000m$) for the constant wind stress (a) and for the wind stresses obtained from eqns. (18) and (19) (b) with $\alpha = 45°$. Velocities in m/s, lengths in meters. Vertical scale distorted.

a good representation of the vertical circulation, which allowed us to compare the magnitude of horizontal and vertical circulations as well as their relative role in lake-wide water exchange mechanisms.

The results obtained show the need to properly account for the wind speed changes as predicted by the IBL theory (or by numerical simulations) in the simulation of wind driven currents in shallow lakes.

## 7 Acknowledgments

The authors thank the anonymous referees for their valuable comments on the first version of the manuscript.

This study has received financial support by the Italian Ministry of Scientific and Technology Research, project PRIN 2002 "Influence of vorticity and turbulence in interactions of water bodies with their boundary elements and effect on hydraulic design."

## References

[1] Andreasson, P., Energy conversion in turbulent wind-induced countercurrent flow. *J. Hydraul. Res.*, **30(6)**, pp. 783–800, 1992.

[2] Wu, J., Wind-induced drift currents. *J. Fluid Mech.*, **68**, pp. 49–70, 1974.

[3] Józsa, J., *Shallow Lake Hydrodynamics. Theory, measurement and numerical model applications.* Lecture Notes of the IAHR Short Course on Environmental Fluid Mechanics, Budapest, 7–16 June, 2004.

[4] Simons, T., *Circulation models of lakes and inland seas.* Canadian Bulletin of Fisheries and Aquatic Sciences, n. 203, Ottawa, Canada, 1980.

[5] Schwab, D. & Beletsky, D., Relative effects of wind stress curl, topography, and stratification on large-scale circulation in Lake Michigan. *J. Geophys. Res.*, **108**, 2003.

[6] Strub, P. & Powell, T., Wind-driven surface transport in stratified closed basins: direct versus residual circulations. *J. Geophys. Res.*, **91**, pp. 8497–8508, 1986.

[7] Lemmin, U. & D'Adamo, N., Summertime winds and direct cyclonic circulation: observations from Lake Geneva. *Ann. Geophysicae*, **14**, pp. 1207–1220, 1996.

[8] Stull, R., *An introduction to Boundary Layer Meteorology.* Kluwer Acad. Pub., 1991.

[9] Doran, J. & Gryning, S., Wind and temperature structure over a land-water-land area. *J. Appl. Meteorol.*, **26**, pp. 973–979, 1987.

[10] Garratt, J., The internal boundary layer – a review. *Bound-Lay. Meteorol.*, **50**, pp. 171–203, 1990.

[11] Klipp, C. & Mahrt, L., Conditional analysis of an internal boundary layer. *Bound-Lay. Metereol.*, **108**, 2003.

[12] Podsetchine, V. & Schernewski, G., The influence of spatial wind inhomogeneity on flow patterns in a small lake. *Water Res.*, **33(15)**, pp. 3348–3356, 1999.

[13] Józsa, J., Sarkkula, J. & Tamsalu, R., Calibration of modelled shallow lake flow using wind field modification. *Proc. VIII. Int. Conf. on Comput. Methods in Water Res.* pp. 165–170, 1990.

[14] Venkatram, A., Internal boundary-layer development and fumigation. *Atmos. Environ.*, **11**, pp. 479–482, 1977.

[15] Cioffi, F., Gallerano, F. & Napoli, E., Three-dimensional numerical simulation of wind-driven flows in closed channels and basins. *J. Hydraul. Res.*, **43(3)**, pp. 290–301, 2005.

[16] Wang, Y. & Hutter, K., A semi-implicit semispectral primitive equation model for lake circulation dynamics and its stability performance. *J. Comput. Phys.*, **139**, pp. 209–241, 1998.

[17] Taylor, P. & Lee, R., Simples guidelines for estimating wind speed variations due to small scale topographic features. *Climat. Bull., Canadian Meteorol. and Ocean. Soc.*, **18(2)**, pp. 3–32, 1984.

[18] Monin, A.S. & Yaglom, A.M., *Statistical Fluid Mechanics: Mechanics of Turbulence.* MIT Press, Cambridge, England, 1972.

[19] Rodi, W., *Turbulence models and their application in hydraulics – a state of the art review.* Int. Assoc. for Hydraulic Research. Delft, 1984.

[20] Kuipers, J. & Vreugdenhil, C., *Calculation of two-dimensional horizontal flow.* Delft Hydraulic Laboratory Report S 163, part I, 1973.

[21] Wu, J. & Tsanis, I., Numerical study of wind-induced water currents. *J. Hydraul. Eng. ASCE*, **121**, pp. 388–395, 1995.

[22] Wu, J., Wind-stress coefficients over sea surface from breeze to hurricane. *J. Geophys. Res.*, **87**, pp. 9704–9706, 1982.

[23] Wuest, A. & Lorke, A., Small scale hydrodynamics in lakes. *Annu. Rev. Fluid Mech.*, **35**, pp. 373–412, 2003.

[24] Wu, J., The sea surface is aerodynamically rough even under light winds. *Bound.-Lay. Meteorol.*, **69**, pp. 149–158, 1994.

[25] Jones, I. & Toba, Y., *Wind stress over the Ocean*. Cambridge Univ. Press UK, 2001.

[26] Yelland, M. & Taylor, P., Wind stress measurements from the open ocean. *J. Phys. Oceanogr.*, **26**, pp. 541–558, 1996.

[27] Bradley, E., Coppin, P. & Godfrey, J., Measurements of sensible and latent heat flux in the western equatorial Pacific ocean. *J. Geophys. Res.*, **96**, pp. 3375–89, 1991.

[28] Charnock, H., Wind stress on a water surface. *Q. J. Roy. Meteor. Soc.*, **81**, pp. 639–640, 1955.

[29] Johnson, H., Hjstrup, J., Vested, H. & Larsen, S., On the dependence of sea surface roughness on wind waves. *J. Phys. Oceanogr.*, **28**, pp. 1702–1716, 1998.

[30] Tombrou, M. & Founda, D., Wind profile diagnosis from surface routine metereological data from a coastal area. *Bound. Lay. Meteorol.*, **87**, pp. 217–231, 1998.

# CHAPTER 5

# Multiple states in open channel flow

A. Defina & F.M. Susin
*Department IMAGE, Padua University, Italy.*

## Abstract

Steady flow regimes in a free surface flow approaching an obstacle are described and extensively discussed. Attention is focused on the phenomenon of hydraulic hysteresis, and a simple one-dimensional theory to predict its occurrence in a supercritical channel flow is proposed. It is shown that in many cases knowledge of the Froude number of the undisturbed approaching flow and of a geometric characteristic of the obstacle allows for a reliable prediction of the flow state. In the region of multiple regimes, however, the previous history of the flow must also be known.

Three different obstacles in a rectangular channel are considered, namely a sill, a vertical sluice gate, and a circular cylinder, and the theoretical boundaries of the hysteresis region are specified for each obstacle. The experimental results show that the theoretical predictions are consistent with experiments in the case of obstacles that do not affect channel width *(i.e.* sills and gates). On the contrary, in the case of channel contraction, a further parameter, which the presented theory does not account for, was found to affect the behavior of the flow, namely the ratio of undisturbed flow depth to contraction width.

Finally, in the case of a vertical sluice gate it was found that hysteresis develops in a subcritical undisturbed approaching flow as well.

## 1 Introduction

The occurrence of steady rapidly varied flow in the vicinity of short obstacles is not unusual in open channels. Prediction of flow characteristics as a function of undisturbed approaching flow conditions and obstacle geometry is a primary objective, mainly related to design purposes. In fact, this information must usually be known in order to establish whether or not a constriction is severe enough to influence the upstream flow and to accurately estimate any possible increase in flow depth upstream of the obstacle.

Different kinds of obstacles exist. Any sudden and local change in flow boundaries that obstructs regular flow can be considered an obstacle. Obstruction can be due to variations in bed topography (*e.g.* sills and sharp- or broad-crested weirs), changes in channel width (*e.g.* bridge piers and contractions), or restrictions in flow depth due to the presence of underflow gates. Obstacles that combine two or more of these obstructions may also exist.

Interaction between the flow and the obstacle produces effects that are independent from both the type of obstacle and the undisturbed flow regime, at least from a qualitative point of view. In fact, for both supercritical and subcritical undisturbed approaching flow conditions two general classes of rapidly varied flow can be identified. In the first one, referred to as "weak interaction" (WI), the flow continues undisturbed when passing the obstacle, *i.e.* the flow regime does not change at the obstacle even though flow depth varies locally. Moreover, in this case the specific energy of the flow anywhere along the channel is greater than the minimum specific energy required to pass the obstruction. In the second class, referred to as "strong interaction" (SI), not only do flow depth variations occur in the channel but transition does as well, *i.e.* flow regime changes from supercritical to subcritical and/or *vice versa.* In this case, the specific energy just upstream of the obstacle is equal to the minimum specific energy required to pass the obstacle.

On the contrary, the actual flow configuration in the vicinity of the obstacle depends on the undisturbed flow regime, *i.e.* whether the undisturbed flow approaching the obstacle is supercritical or subcritical. In particular, if a supercritical undisturbed flow approaches the obstacle, the following two different configurations may be established. In a WI class rapidly varied flow, supercritical conditions persist both upstream and downstream of the obstruction and at the obstruction as well. On the contrary, SI causes the upstream flow to undergo a transition from supercritical to subcritical regime and undisturbed conditions are only restored downstream of the obstacle.

The occurrence of both these different steady-state configurations can be predicted as a function of fundamental flow and obstacle parameters, at least within a simple one-dimensional theoretical approach. Nevertheless, a unified theory of rapidly varied flow has not been developed yet. This may be the reason why even specialists only marginally know a somewhat surprising feature that may arise when an undisturbed supercritical flow approaches an obstacle. The theoretical boundaries of steady flow regimes in the vicinity of an obstruction clearly show that a region in the space of fundamental parameters exists where both the above flow states, *i.e.* WI and SI, may occur for the same external steady conditions. The state that is actually established depends on the history of the flow, *i.e.* on the way in which flow conditions have evolved up to the current ones, thus implying the hysteretic character of the flow.

The occurrence of hysteresis in a supercritical flow approaching a sill has been recognized and widely investigated both experimentally and theoretically over the past few decades [1–8]. On the contrary, much less effort has been devoted to the study of hydraulic hysteresis for other types of obstacles, such as channel constriction. Nevertheless, some experimental studies reported in the literature qualitatively have confirmed the hysteretic behavior of a supercritical flow through channel constriction [9, 10]. More recently, the existence of hysteresis in flow under a sluice gate has been demonstrated both theoretically and experimentally [11].

Here, our primary objectives are to examine hydraulic hysteresis in general terms, *i.e.* independently of the type of obstacle, and review significant results concerning specific obstructions. A theoretical approach is presented which uses simple relationships expressing energy balance and momentum conservation to infer a criterion able to predict the conditions in which hysteresis occurs in a supercritical flow. This general criterion is then applied to some specific obstacles, namely a sill, a vertical sluice gate, and a circular pier. The theoretical predictions are compared with available experimental data. Finally, some conclusions concerning the main features of the investigated phenomenon are reported, and situations where the proposed theoretical approach fails are highlighted.

## 2 One-dimensional approach

A simple criterion to identify hysteresis occurrence in a rapidly varied flow has been recently proposed by Defina and Susin [11]. Here, we shortly recall the fundamental aspects of this theoretical approach and show that the existence of hysteresis emerges when sufficient and necessary conditions for WI and SI interaction occurrence are both considered.

We consider a one-dimensional supercritical undisturbed channel flow of velocity $v_0$ and depth $y_0$ approaching a generic obstacle. The frictional effects of the boundaries are considered and the pressure is assumed to be hydrostatic except in the proximity of the obstacle. A non-horizontal channel is considered. The bottom slope is assumed to be small enough that $cos(\vartheta) \simeq 1$, $\sin(\vartheta) \simeq tan(\vartheta)$, $\vartheta$ being the angle of the channel bed to the horizontal.

To describe the phenomenon of hysteresis, it is convenient to use the concept of specific energy $E$, *i.e.* mechanical energy per unit weight of flow relative to the bottom of the channel. Thus, for the *undisturbed approaching flow*

$$E_0 = y_0 + \frac{v_0^2}{2g} \tag{1}$$

where $g$ is gravity. Moreover, the minimum specific energy required to pass the obstacle is denoted with $E_{min}$.

It is well known that if $E_0$ is lower than $E_{min}$, then the approaching flow does not have enough energy to pass the obstacle and, as a consequence, it necessarily undergoes a transition from supercritical to subcritical regime. A stationary hydraulic jump occurs far from the obstacle and a subcritical backwater profile is established after the jump. The backwater profile, along which $E$ increases, allows for the minimum specific energy to be achieved at the section just before the obstacle. In this situation, a "strong interaction" between the flow and the obstacle occurs. Hence, the sufficient condition for SI (*i.e.* supercritical-subcritical transition in the upstream reach of the channel, fig. 1b) to occur is

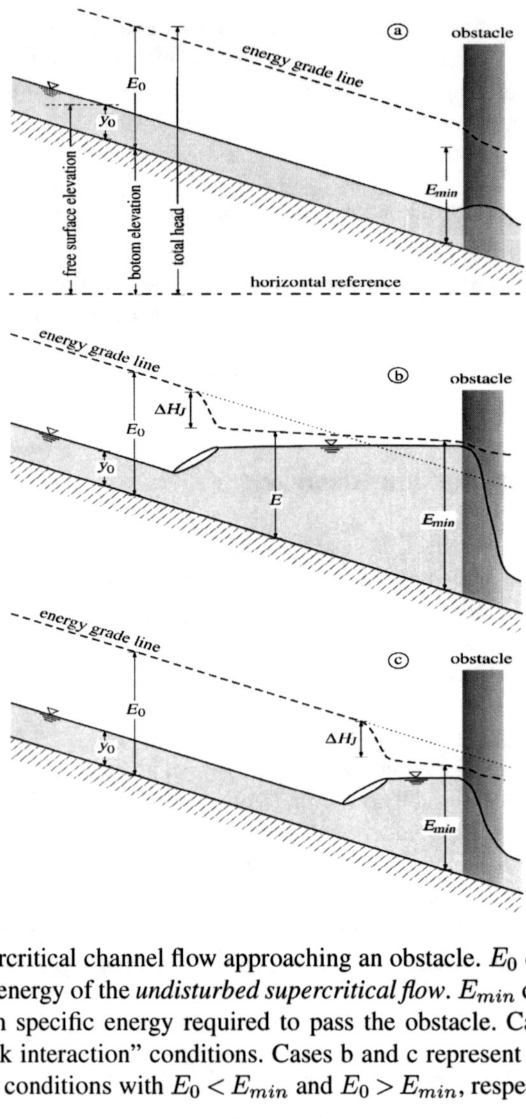


Figure 1: Supercritical channel flow approaching an obstacle. $E_0$ denotes the specific energy of the *undisturbed supercritical flow*. $E_{min}$ denotes the minimum specific energy required to pass the obstacle. Case a represents "weak interaction" conditions. Cases b and c represent "strong interaction" conditions with $E_0 < E_{min}$ and $E_0 > E_{min}$, respectively.

$$E_0 < E_{min} \tag{2}$$

The necessary condition for SI to occur can be inferred by the use of the total energy conservation equation between the flow just upstream of the obstacle and the flow at the obstacle [11]. As long as a hydraulic jump is established in the upstream reach of the channel (fig. 1b and c), we have

$$E_0 - \Delta H_J \leq E_{min} \tag{3}$$

where $\triangle H_J$ denotes the energy loss at the hydraulic jump.

It is now worth noting that conditions complementary to eqns. (2) and (3) give the necessary and sufficient conditions for WI to occur, respectively (Figure 1a). In fact, if supercritical conditions are maintained at the obstacle, then the specific energy of the *undisturbed approaching flow* must be greater than $E_{min}$, *i.e.*

$$E_0 \geq E_{min} \tag{4}$$

Moreover, if the following condition holds

$$E_0 - \Delta H_J > E_{min} \tag{5}$$

then the specific energy of the flow just upstream of the obstacle is certainly greater than the minimum specific energy required to pass the obstacle, *i.e.* the flow is supercritical everywhere.

On the basis of the above considerations, we can now state that if the specific energy of the *undisturbed supercritical approaching flow*, $E_0$, satisfies the conditions expressed by eqns. (3) and (4), then both WI and SI flow regimes can exist. Moreover, for a given undisturbed flow condition and given geometric characteristics of the obstacle, $E_{min}$ may depend on the flow regime being established just upstream of the obstacle. Hence, if the minimum specific energy for the supercritical (WI) and subcritical (SI) regimes just upstream of the obstacle are denoted with $E_l$ and $E_L$, respectively, both steady flow configurations, *i.e.* with or without upstream transition, are possible and stable as long as the following constraint is met

$$E_l \leq E_0 \leq E_L + \Delta H_J \tag{6}$$

In this situation, the "history" of the flow plays a crucial role, as it determines the state that is actually established in the vicinity of the obstruction [5, 8]. For this reason, the behavior of the flow is said to be "hysteretic".

In the following sections, three different types of obstacles are examined. Condition eqn. (6) is specified for each obstacle, and the amplitude of the hysteresis domain is expressed in terms of the fundamental flow parameters and geometrical characteristics of the obstruction.

## 3 Flow over a sill

We will now develop condition eqn. (6) for the case of flow over a sill in a rectangular channel in terms of the Froude number of the undisturbed approaching flow, $F_0$, and the non-dimensional step height $\Delta z/y_c$, where $y_c$ is the critical depth for a given flow rate and channel width. In this case (fig. 2), the threshold values for specific energy are

$$E_l = \Delta z + E_c + \Delta H_l \qquad E_L = \Delta z + E_c + \Delta H_L \tag{7}$$

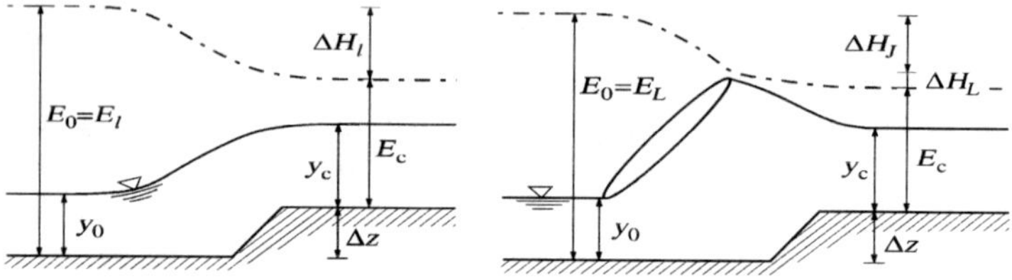


Figure 2: Steady flow regimes for a supercritical undisturbed flow approaching a sill, with notation. Limit conditions for the weak interaction regime (left) and strong interaction regime (right) are shown.

where $\triangle z$ is the height of the sill, $E_c$ is the critical specific energy, and $\triangle H_l$ and $\triangle H_L$ are the energy losses just before the section at which the critical condition for WI and SI regimes, respectively, is attained.

Combining eqns. (6) and (7), and recalling that

$$E_c/y_c = 3/2, \quad E_0/y_c = F_0^{-2/3} + F_0^{4/3}/2, \quad \text{and}$$
$$\triangle H_j/y_c = \frac{F_0^{-2/3}}{16}\frac{(\sqrt{1+8F_0^2}-3)^3}{(\sqrt{1+8F_0^2}-1)} \tag{8}$$

the following expression can be easily found for the lower and upper boundaries of the hysteresis region

$$\frac{\triangle z_l}{y_c} = F_0^{-2/3} + F_0^{4/3}/2 - 3/2 - \triangle H_l/y_c \tag{9}$$

$$\frac{\triangle z_L}{y_c} = F_0^{-2/3} + \frac{F_0^{4/3}}{2} - \frac{3}{2} - \frac{F_0^{-2/3}}{16}\frac{(\sqrt{1+8F_0^2}-3)^3}{(\sqrt{1+8F_0^2}-1)} - \triangle H_L/y_c \tag{10}$$

The above result has already been proposed in previous studies [3–6] which neglected energy losses $\triangle H_l$ and $\triangle H_L$. In eqns. (9) and (10), the nondimensional energy losses $\triangle H_l/y_c$ and $\triangle H_L/y_c$ both depend on $F_0$ and the geometry of the step. Although these losses can be correctly evaluated only through experiments, a rough estimation based on mass and momentum conservation equations is possible, and given here below. Let us first consider the case of the WI regime. Using the notation adopted in fig. 3a and the concept of momentum function $M$ (*i.e.* the sum of pressure force and momentum flux, per unit width and unit weight of fluid, at a given section of the flow [12]), the equilibrium condition in the horizontal direction is

$$M_u - M_d - m_S = 0 \tag{11}$$

where $M_u$ and $M_d$ denote the upstream and downstream values of $M$, respectively, and $m_S$ is the force, in the flow direction, exerted by the upward face of the step (per unit width and unit weight of fluid). Assuming that at the section immediately

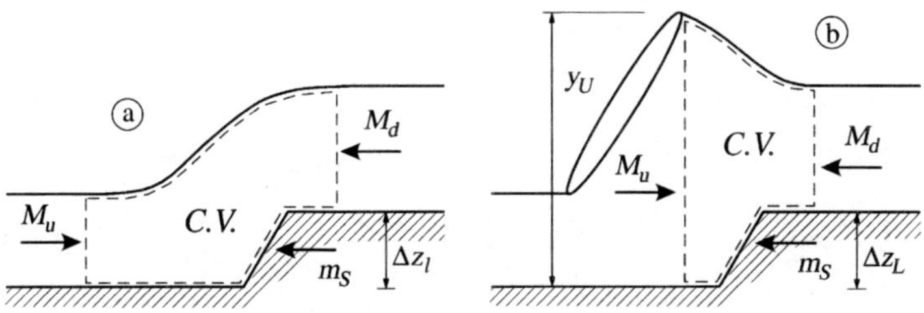


Figure 3: Control volume (CV) selected for the application of the momentum balance equation in the case of weak interaction regime (a) and strong interaction regime (b).

upstream of the step the flow depth is $y_c + \triangle z_l$ and the pressure is hydrostatic [3], then $m_s$ is given as

$$m_S \cong (y_c + \triangle z_l/2)\triangle z_l \tag{12}$$

Noting that when the WI $\rightarrow$ SI limit condition occurs $M_u = M_0$ and $M_d = M_c$, and recalling that

$$\frac{M_0}{y_c^2} = \frac{F_0^{-4/3}}{2} + F_0^{2/3} \qquad \frac{M_c}{y_c^2} = \frac{3}{2} \tag{13}$$

eqn. (11) can be rearranged to read

$$\left(\frac{\triangle z_l}{y_c} + 1\right)^2 = 2F_0^{2/3} + F_0^{-4/3} - 2 \tag{14}$$

Finally, combining eqns. (9) and (14) gives

$$\frac{\triangle H_l}{y_c} = F_0^{-2/3} + \frac{F_0^{4/3}}{2} - \frac{1}{2} - \sqrt{F_0^{-4/3} - F_0^{2/3} - 2} \tag{15}$$

The above procedure is also used to compute energy loss for the SI regime (fig. 3b). In this case, the horizontal force exerted by the upward face of the step, $m_S$ is

$$m_S \cong (y_U - \triangle z_L/2)\triangle z_L \tag{16}$$

where the water depth just upstream of the obstacle, $y_U$, is the conjugate of the undisturbed flow depth, *i.e.*

$$\frac{y_U}{y_c} = \frac{F_0^{-2/3}}{2}\left(-1 + \sqrt{1 + 8F_0^2}\right) \tag{17}$$

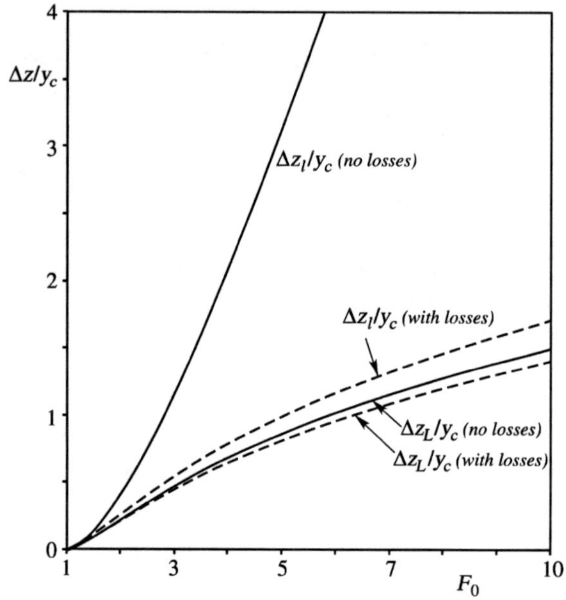


Figure 4: Plot of limit conditions for both weak and strong interaction in the plane ($\triangle z/y_c$, $F_0$), when energy losses are neglected (solid lines) and included (dashed lines).

The equilibrium condition expressed by eqn. (11) gives

$$\frac{\triangle z_L}{y_c} = \frac{F_0^{-2/3}}{2}\left(\sqrt{1+8F_0^2}-1\right) - \sqrt{3 - F_0^{4/3}\left(\sqrt{1+8F_0^2}-1\right)/2} \quad (18)$$

and combining eqns. (18) and (10) then gives

$$\frac{\triangle H_L}{y_c} = \frac{F_0^{-2/3}-6}{4} + \frac{F_0^{-2/3}}{\sqrt{1+8F_0^2}-1} + \sqrt{3 - \frac{F_0^{4/3}}{2}\left(\sqrt{1+8F_0^2}-1\right)} \quad (19)$$

The plot of eqns. (9) and (10) is shown in fig. 4. Solid lines are used when energy losses are neglected, dashed lines when energy losses are expressed by eqns. (15) and (19). In both cases, three different regions can be distinguished in the plane ($\triangle z/y_c$, $F_0$). The first region, extending above the curve $\triangle z_L/y_c$, corresponds to WI conditions with a supercritical flow extending along the whole channel. In the second region, below the curve $\triangle z_l/y_c$, the flow necessarily undergoes a supercritical to subcritical transition upstream of the step (*i.e.* SI). Finally, the hysteresis region, in which both stable states are possible, lies between these two curves.

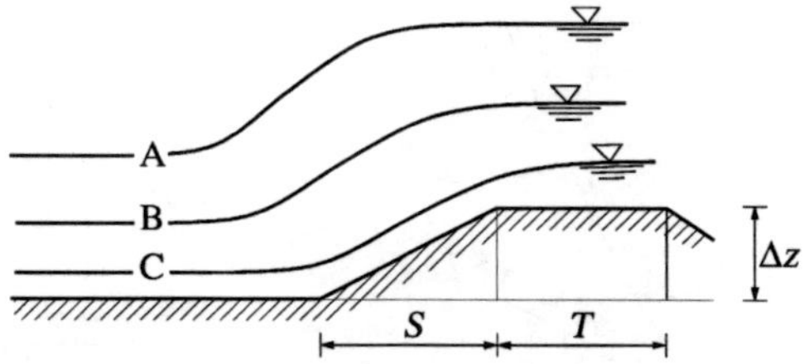


Figure 5: Steady flow profiles for a supercritical flow over a step, with notation.

It is worth noting that when energy losses are neglected, the hysteresis region is considerably wide (fig. 4). On the contrary, when energy losses are considered, the WI→SI limit condition moves towards the SI→WI limit condition and the hysteresis region is much smaller. Surprisingly, most fluid mechanics textbooks propose eqn. (9) with $\triangle H_l = 0$ as the limit condition for WI↔SI.

Before we go any further, it is important to emphasize that the assumptions introduced in order to obtain eqns. (15) and (19) allow for a reliable prediction of the qualitative behavior of the hysteresis boundaries, but do not always apply for a quantitative analysis. In particular, $m_S$ is reasonably approximated as previously reported only when the flow just upstream of the obstacle is subcritical (*i.e.* for the SI configuration). Indeed, in this case losses are rather small and do not significantly affect the inviscid solution. On the contrary, when the flow just upstream of the obstacle is supercritical, energy losses strongly depend on the shape of the step and on the ratio $y_0/S$, where $S$ is the length of the upstream face of the step (fig. 5). Equation (12) for $m_s$ allows for a reliable evaluation of energy losses provided that the ratio $y_0/S$ is not much smaller than one (fig. 5, profile A). Otherwise, *i.e.* when $y_0 \ll S$ (fig. 5, profile C), eqn. (12) strongly overestimates $m_s$ and thus the amount of energy lost by the current in passing the step as well.

It is also worth highlighting that the present theory assumes hydrostatic pressure distribution. On the contrary, at both SI→WI and WI→SI limit conditions free surface slope and curvature across the step are not negligibly small. However, their effects can be accounted for by introducing a suitable *equivalent energy loss or gain* as suggested by Marchi [13], to be included in the terms $\triangle H_l$ and $\triangle H_L$.

By analyzing the experimental data available in the literature and comparing them with the present theoretical results we can provide an adequate explanation of the above statements.

Figure 6 compares our theoretical results with the experimental data of Muskatirovic and Batinic [3]. It can be observed that when the step is not severe, *i.e.* $S = 2\triangle z$ (fig. 6a), experimental points collapse onto the theoretical curves which include energy losses, thus confirming the reliability of the assumptions made to compute $\triangle H_l$ and $\triangle H_L$. When $S = 0$ (fig. 6b), Muskatirovic and Batinic [3] do not observe any hysteresis. The experimental points at the WI→SI and SI→WI limit conditions collapse onto a single curve which approximately corresponds to the theoretical upper limit, $\triangle z_L/y_c$, implying that eqn. (19) (weakly) underestimates losses.

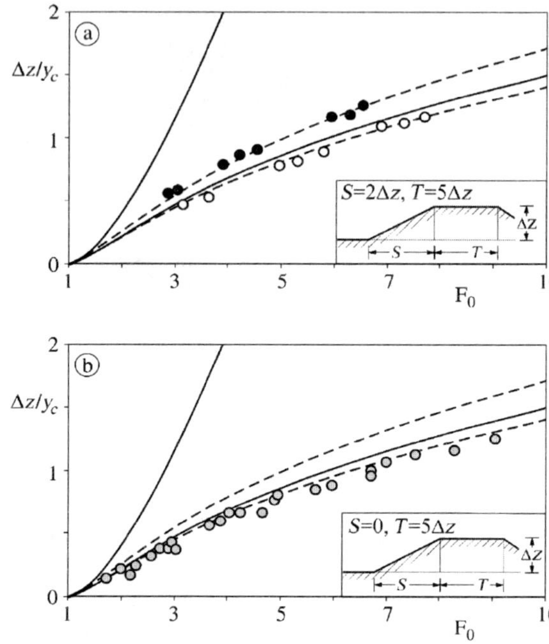


Figure 6: Experimental data of Muskatirovic and Batinic [3]. (a) Full symbols denote the WI→SI limit condition, open symbols denote the SI→WI limit condition; (b) symbols denote the WI↔SI limit condition. The curves are the same as those in fig. 4.

An interesting set of experimental data is provided by Baines and Whitehead [8]. They measured the limits for SI→WI and WI→SI in a short upward sloping channel. This device somewhat resembles a step with a gently sloping upstream face (*i.e.* with large $S$) and $T = 0$ (fig. 7). When the flow is supercritical along the channel (fig. 7a), then the free surface profile qualitatively corresponds to profile C of fig. 5 and no energy losses should be included. Nevertheless, in this case, the upstream face of the step is rather long. Consequently, bottom friction is not entirely negligible and a small energy loss has to be incorporated in $\triangle H_l/y_c$.

Figure 8 compares experimental findings by Baines and Whitehead [8] with present theoretical curves. The experimental points describing the WI→SI limit are arranged just below the theoretical curve with $\triangle H_l/y_c = 0$, in agreement with the above discussion about energy losses affecting the flow in this case. On the contrary, the experimental behavior of the SI→WI limit is somewhat surprising. The measured points are arranged above the theoretical curve, implying that the flow energy increases along the channel. A close inspection of the free surface profile when the flow is subcritical along the whole channel and the hydraulic jump is attached to the downstream face of the rounded sluice gate (fig. 7b), shows that near the downstream end of the channel both the free surface slope and curvature

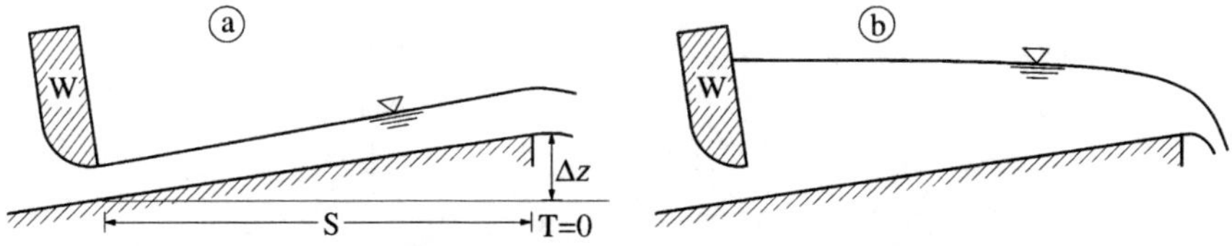


Figure 7: Experimental device adopted by Baines and Whitehead [8] to study hydraulic hysteresis. Flow along the short upward sloping channel is supercritical (a) and subcritical (b).

are negative and large. As previously discussed, their effects can be accounted for by introducing a suitable equivalent energy gain to be included in $\triangle H_L/y_c$.

The following conclusions can be drawn from the above discussion. First of all, we can see that the theory presented here gives reliable predictions of hysteresis limits, provided that energy losses $\triangle H_l$ and $\triangle H_L$ are properly evaluated. Moreover, it is worth pointing out that most fluid mechanics textbooks propose eqn. (9) with $\triangle H_l = 0$ as the WI↔SI limit condition . This disagrees with the present results and experimental evidence, which suggest that, in most cases, the inviscid solution given by eqn. (10) is a more suitable approximation for the WI→SI limit.

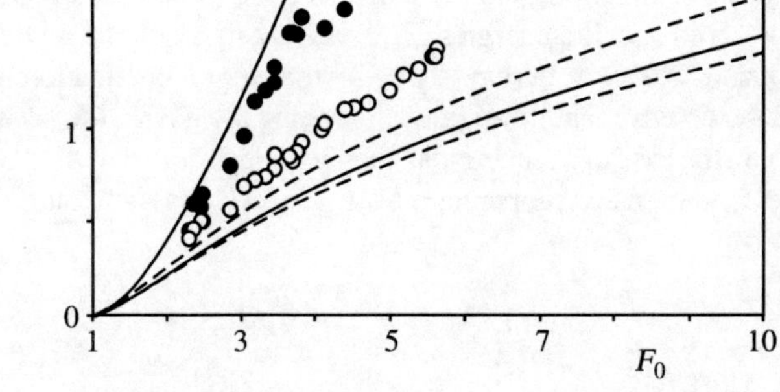


Figure 8: Present theoretical limits of the hysteresis region compared to the experimental data of Baines and Whitehead [8]. Full symbols denote the WI→SI limit, open symbols denote the SI→WI limit. The curves are the same as those in fig. 4.

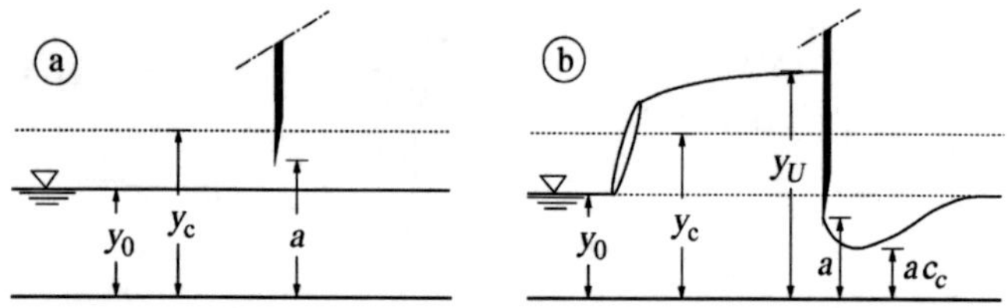


Figure 9: Sketch of possible steady flow configurations when an undisturbed supercritical flow approaches a vertical sluice gate. (a) Undisturbed flow conditions along the channel (WI); (b) free outflow conditions at the gate with an upstream transition from supercritical to subcritical flow (SI).

## 4 Flow under a sluice gate

The hysteretic behavior of a flow under a vertical sluice gate was thoroughly described both theoretically and experimentally by Defina and Susin [11] and, to our knowledge, this is the only investigation on this topic. Here, we will briefly recall the theoretical aspects, experimental apparatus and procedure. We will focus particular attention on comparing theoretical predictions and experimental results.

Figure 9 shows possible steady flow configurations when an undisturbed supercritical flow approaches a vertical sluice gate. In the WI case (fig. 9a), the height of the gate opening is sufficiently larger than the undisturbed water depth so that smooth undisturbed flow conditions exist along the channel. In the SI case (fig. 9b), transition from supercritical to subcritical regime takes place upstream of the gate and, at the obstacle, the free outflow condition is established (*i.e.* the fluid issues from under the gate as a jet of supercritical flow with a free surface open to the atmosphere). The hysteretic behavior of the flow concerns the possibility that, for a given $F_0$, both SI and WI configurations may exist for the same opening of the gate.

In both the above situations, energy losses at the gate can usually be neglected. Hence, the threshold specific energies $E_L$ and $E_l$, as well as the related openings of the gate $a_L$ and $a_l$, can be derived by using the Bernoulli equation between the upstream and the downstream flows (the latter being the flow at the *vena contracta*, in the SI case). In the case of a rectangular channel, the flow is shown to be hysteretic as long as the nondimensional opening of the gate, $a/y_c$ is such that

$$a_l/y_c \leq a/y_c \leq a_L/y_c \tag{20}$$

where $a_l/y_c$ and $a_L/y_c$ are related to the Froude number of the undisturbed approaching flow $F_0$ as expressed by the following equations

$$a_l/y_c = F_0^{-2/3} \tag{21}$$

$$\left(\frac{a_L}{y_c}c_c\right) + \frac{1}{2}\left(\frac{a_L}{y_c}c_c\right)^{-2} = \frac{4F_0^{4/3}}{1+\sqrt{1+8F_0^2}} + \frac{2F_0^{4/3}}{(-1+\sqrt{1+8F_0^2})^2} \tag{22}$$

In eqn. (22) $c_c$ is the contraction coefficient (*i.e.* the ratio of the flow depth at the *vena contracta* to the height of the gate opening). A reliable evaluation of $a_L/y_c$ requires a reliable evaluation of $c_c$ as well. The following approximate equation will from here on be adopted

$$c_c = 1 - r(\zeta)\sin(\zeta) \tag{23}$$

$$a_L/y_U = 1 - r(\zeta)[1-\cos(\zeta)] \tag{24}$$

$$r(\zeta) = 0.153\zeta^2 - 0.451\zeta + 0.727 \tag{25}$$

where $\zeta$ is a dummy parameter and $y_U$ is the flow depth immediately upstream of the gate. A detailed discussion about the evaluation of $c_c$ and eqns. (21) to (25) is reported in ref. [11].

Figure 10 shows the behavior of $a_l/y_c$ and $a_L/y_c$ in the plane ($a/y_c$, $F_0$). Possible flow regimes in regions bounded by these two curves are also specified. It is worth pointing out that the ascending branch $AM$ of the curve $a_L/y_c$ is replaced with the horizontal line $A'M$: $a/y_c = (a/y_c)_{max} \cong 1.15$ (see Appendix A of Defina and Susin [11] for details). Hence, the boundaries of the hysteresis region do not merge at $F_0 = 1$. This suggests that the dual solution domain in the ($a/y_c$, $F_0$) plane may close in the region of subcritical undisturbed approaching flows (*i.e.* $F_0 < 1$). Indeed that is exactly what is found to happen once the theoretical analysis of the flow under a sluice gate is extended to subcritical undisturbed flow condi-

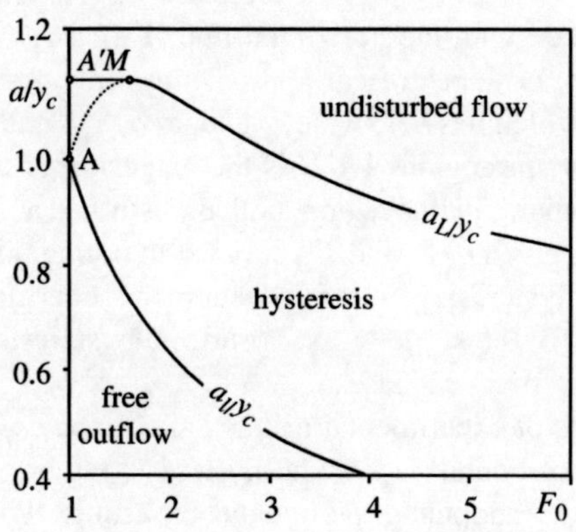


Figure 10: Steady flow regimes in the vicinity of a vertical sluice gate for a supercritical undisturbed approaching flow on the $F_0 - a/y_c$ diagram.

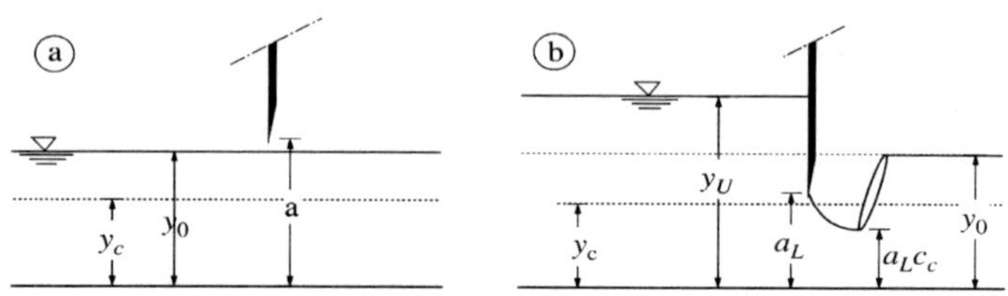


Figure 11: Sketch of possible steady flow configurations when an undisturbed subcritical flow approaches a vertical sluice gate. (a) undisturbed flow conditions along the channel (WI); (b) free outflow conditions at the gate with downstream transition from supercritical to subcritical flow (SI).

tions. In this case, three different steady states may be established in the vicinity of the gate: undisturbed flow conditions when WI occurs, and either free or submerged outflow in the case of SI.

The flow continues undisturbed as long as the gate does not touch the free surface (fig. 11a). Therefore, when WI conditions are established in the channel, the non-dimensional gate opening is certainly greater than $a_l/y_c$, where $a_l/y_c$ depends on $F_0$ as given by eqn. (21). Once the gate has touched the free surface, either submerged or free outflow is established, depending on the opening of the gate. The limit value of the gate opening (denoted with $a_L/y_c$) between these two configurations is such that the related flow depth at the *vena contracta* equals the conjugate depth of the downstream undisturbed subcritical flow (fig. 11b). Hence, the momentum balance between the *vena contracta* and the downstream flow gives

$$\left(\frac{a_L}{y_c} C_c\right) = \frac{F_0^{-2/3}}{2}\left(-1 + \sqrt{1 + 8F_0^2}\right) \tag{26}$$

Curves $a_l/y_c$ and $a_L/y_c$ for $F_0 = 1$ are plotted in fig. 12. The branch $AM'$ of the curve $a_L/y_c$ is replaced with the horizontal line $A'M'$: $a_L/y_c = (a_L/y_c)_{max} \cong 1.15$ as was the case for supercritical approaching flow. As a consequence, the region $AA'B$ exists, which lies below the boundary $a_L/y_c$ and above the boundary $a_l/y_c$. In other words, the region $AA'B$ is the extension of the hysteresis region into the subcritical domain. In this region, both undisturbed and free outflow steady states are possible for given $a/y_c$ and $F_0$. It is worth noting that the ranges of both $a/y_c$ and $F_0$ in which hysteresis is found to occur in the subcritical domain are rather wide ($1 \leq a/y_c < 1.15$, $0.8 < F_0 \leq 1$). Therefore, hysteresis can be expected to manifest itself in many practical cases.

An extensive series of experimental results [11] can be used to compare experimental and theoretical boundaries of the hysteresis region. We will first give a brief description of the adopted experimental procedure. Possible flow regimes were investigated for different values of the gate opening while maintaining a fixed flow rate (*i.e.* a fixed undisturbed Froude number $F_0$). Quasi-uniform flow conditions were initially established in the flume with the gate opening larger than the

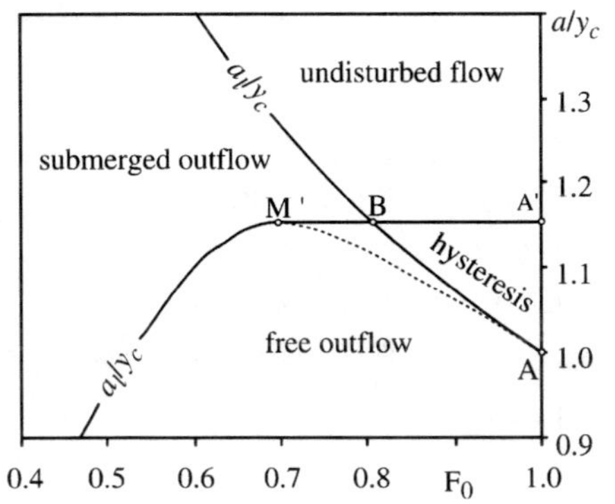


Figure 12: Steady flow regimes in the vicinity of a vertical sluice gate for a subcritical undisturbed approaching flow on the $F_0 - a/y_c$ diagram.

undisturbed flow depth. In order to examine the hysteresis phenomenon, the gate was first lowered until it touched the free-surface, so that SI conditions suddenly occurred, and then gradually raised at small steps. The gradual lifting of the gate was protracted until the WI flow configuration was suddenly restored in the channel.

The experimental undisturbed Froude number ranged from about 0.72 to about 4.57. In all the experiments, the WI→SI limit condition occurred as soon as the gate touched or slightly passed the free-surface. In particular, for $F_0 > 0.8$ free outflow conditions were established, while for $F_0 < 0.8$ the outflow was submerged. This occurrence not only confirms the behavior of the limit $a_l/y_c$ theoretically predicted (actually this result was somewhat obvious) but it also substantiates the physical meaning of the branch $A'M'$ in fig. 12. $F_0 \cong 0.8$ was also found to be the smallest undisturbed Froude number for which hysteresis occurred. In fact, for $F_0 < 0.8$, undisturbed conditions were found to be restored as soon as the gate was raised just above the undisturbed flow depth. On the contrary, for $F_0 > 0.8$, the SI→WI limit condition only occurred when the opening of the gate was much larger than the undisturbed flow depth.

The hysteretic character of the flow is clearly shown in fig. 13, where the behavior of the non-dimensional flow depth just upstream of the gate, $y_U/y_c$ is described as a function of the gate opening $a/y_c$. Experimental conditions for $F_0 = 0.98$ are plotted. The existence of two different steady configurations in the vicinity of the gate for the same $a/y_c$ but different previous states of the flow is evident.

Figure 14 gives a comprehensive plot of all the experimental conditions at which flow reversion from SI to WI suddenly occurred. The theoretical boundaries of the hysteresis region are plotted as well. The experimental points $(F_0, a_L/y_c)$ agree well with the upper theoretical boundary $a_L/y_c$ both qualitatively and quantitatively. In particular, they clearly exhibit the horizontal trend in the range $0.81 < F_0 < 1.65$.

It is worth pointing out that, as previously observed, the reliability of the theoretical limit $a_L/y_c$ is strictly related to the equation adopted to evaluate the contraction coefficient $c_c$. In other words, the more physically based $c_c$ is, the more reliable

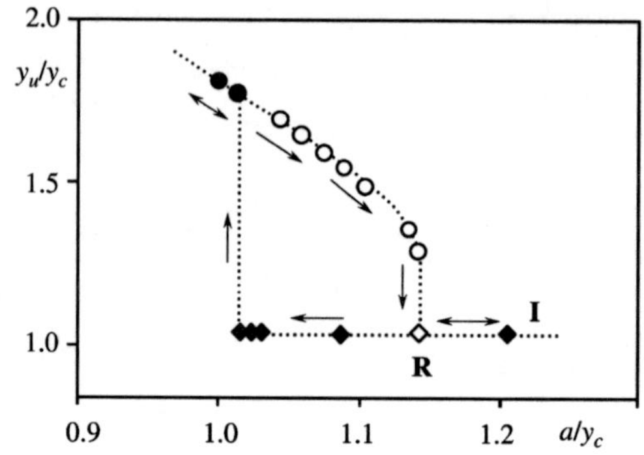


Figure 13: Experimental hysteresis loop in the $a/y_c - y_U/y_c$ diagram. The Froude number of the undisturbed approaching flow is $F_0 = 0.98$. The sequence of experimental points goes clockwise from I (initial undisturbed flow) to R (restored undisturbed flow). The circles and diamonds denote free outflow and undisturbed flow, respectively; the full and open symbols denote the lowering stage and raising stage, respectively.

$a_L/y_c$ is. For example, if the theoretical expression given by Cisotti and Von Mises (as reported by Gentilini [14]) is adopted for $c_c$, then $a_L/y_c$ behaves as shown in fig. 15 for $F_0 \geq 1$. In this case, gravity effects on the issuing flow are neglected, and both $c_c$, and $a_L/y_c$ are thus rather poorly estimated. As a consequence, only a qualitative agreement between theoretical predictions and experimental results is found, as shown in fig. 15.

Finally, we can state that the simple theoretical approach outlined in Section 1 also applies fairly well to the case of flow under a vertical sluice gate. In this case, an

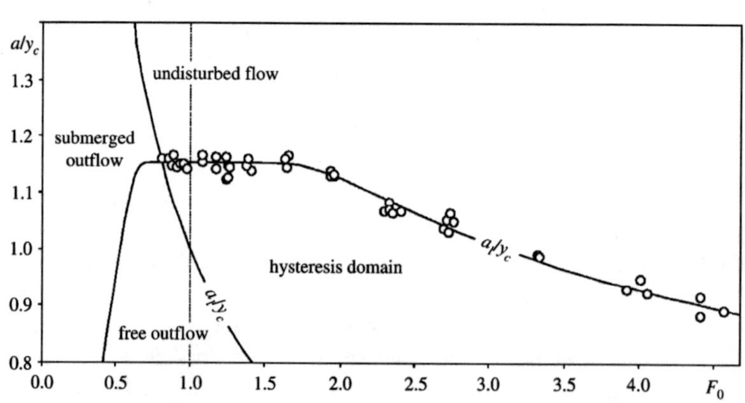


Figure 14: Comparison between the theoretical (–) and experimental (○) upper boundary of the hysteresis region. The lower boundary is also plotted for completeness.

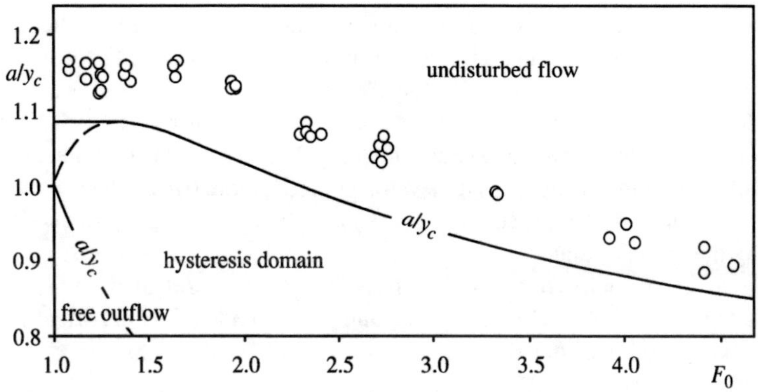


Figure 15: Theoretical upper boundary of the hysteresis region ($a_L/y_c$) when $c_c$ is computed according to Cisotti and Von Mises formula. Experimental data are also plotted for comparison. Symbols are the same as in fig. 14.

accurate evaluation of the contraction experienced by the flow issuing from under the gate is required in order to obtain reliable predictions of the upper hysteresis limit.

## 5 Flow past a vertical circular cylinder

The theoretical evaluation of minimum specific energies $E_l$ and $E_L$ for the case of channel constriction is far from being an easy task. In this case, in fact, the flow in the vicinity of the obstacle assumes a two-dimensional character (in plan). In addition, free surface slope and curvature, *i.e.* non-hydrostatic pressure distribution along the vertical, strongly affect the flow.

Therefore, we can expect minimum specific energy as determined according to the one-dimensional approach to poorly match experimental conditions. The one-dimensional approach uses only one geometric parameter to describe the obstruction, namely the ratio of obstructed channel width $b$ to the full channel width $B$ (fig. 16c). The shape, length, and position of the obstacle with respect to channel axis and flow depth are neglected. However, these geometric characteristics are very important in determining the limit conditions for both WI and SI configurations, as will be shown in the present section.

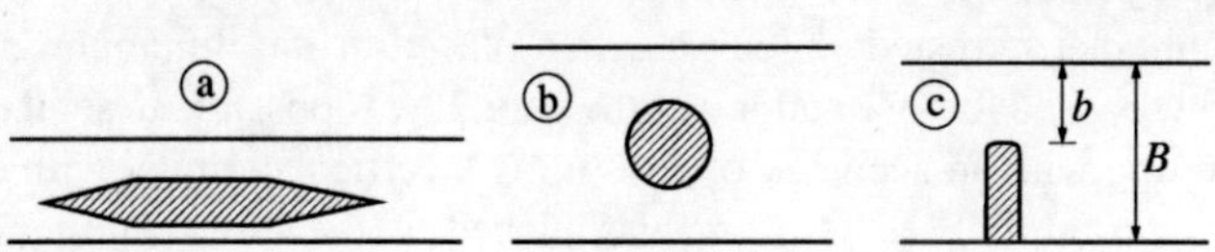


Figure 16: Plan view showing some examples of channel constrictions characterized by the same value of the ratio *b/B*.

In spite of its limitations, the one-dimensional approach can nonetheless provide an initial, even if rough, approximation of limit conditions that can then be compared with true, experimental values. We will now examine the case of a flow around a vertical circular cylinder placed along the axis of a rectangular channel (fig. 16b), both theoretically and experimentally. In this case, the fundamental non-dimensional parameters adopted to develop condition eqn. (6) are the Froude number of the undisturbed approaching flow, $F_0$ and the ratio $b/B$, where $b = B - D$, $D$ being the cylinder diameter.

Due to the smooth shape of the obstacle, energy losses at the obstacle are assumed to be negligible. Hence, both $E_l$ and $E_L$ coincide with the critical specific energy at the section of maximum contraction, $E_{cr}$. Recalling that

$$\frac{E_{cr}}{y_c} = \frac{3}{2}(b/B)^{-2/3} \tag{27}$$

and combining eqns. (6) and (27), the theoretical boundaries of the hysteresis domain can be easily found

$$(b/B)_l = \sqrt{\frac{27}{8}} F_0 \left(1 + \frac{F_0^2}{2}\right)^{-3/2} \tag{28}$$

$$(b/B)_L = \sqrt{\frac{27}{8}} F_0 \left(1 + \frac{F_0^2}{2} - \frac{\left(\sqrt{1+8F_0^2} - 3\right)^3}{16\left(\sqrt{1+8F_0^2} - 1\right)}\right)^{-3/2} \tag{29}$$

Equations (28) and (29) are plotted in fig. 17. As was the case for the obstacles examined in the previous sections, three different regions can be distinguished in the plane of fundamental parameters. The first region, extending above the curve $(b/B)_L$, corresponds to the WI configuration, with a supercritical flow extending along the whole channel. In the second region, below the curve $(b/B)_l$, the flow necessarily undergoes a supercritical to subcritical transition upstream of the obstacle (*i.e.* SI). Finally, the hysteresis region, in which both stable states are possible, lies between these two curves.

The effects of the physical processes not included in the above theory were examined experimentally. The apparatus is sketched in fig. 18. The experiments were performed in a 0.38m wide, 0.5m high, and 20m long tilting flume with Plexiglas walls, whose bottom slope could be adjusted to a maximum of 5%. Water was supplied by a constant head tank which maintained very steady flow conditions.

A vertical, sharp crested sluice gate was placed at the flume entrance, and a gear system was used to raise and lower the gate. It was possible to set the height of the gate opening with an accuracy of ±0.2mm. A vertical cylinder with a diameter in the range $0.060\text{m} < D < 0.205\text{m}$ was placed at the test section, which was located approximately 3m downstream of the channel inlet. One ultra-sonic transducer, movable along the channel axis upstream of the cylinder, measured flow depths with an accuracy of ±0.5mm. Water depth was accurately measured at three

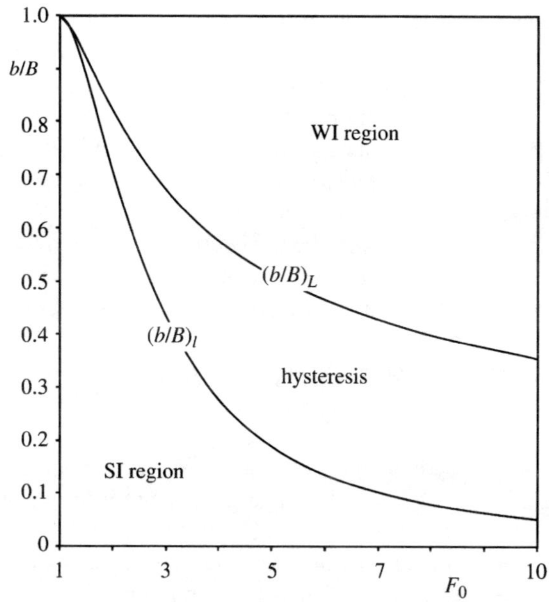


Figure 17: Plot of the limit conditions for both weak and strong interaction in the plane $(b/B, F_0)$.

positions, namely at 0.3m, 1.0m, and 2.0m upstream of the cylinder. This made it possible to extrapolate the undisturbed water depth at the test section, $y_0$, with an accuracy of $\pm 1.0$mm.

Each run was conducted according to the following procedure, while maintaining a fixed flow rate in the range $0.01\text{m}^3/s < Q < 0.06\text{m}^3$/s. Initially, the height of the sluice gate opening was small enough to ensure supercritical flow conditions from the gate to the end of the flume (*i.e.* WI). Then, the gate opening was increased by small increments so that, at each step, a new steady flow configuration with slightly increased $y_0$ (*i.e.* slightly decreased $F_0$) was established in the channel.

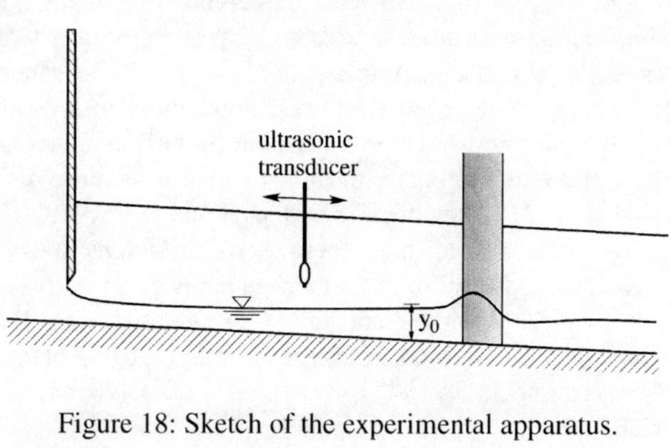


Figure 18: Sketch of the experimental apparatus.

The gate raising was protracted until a stationary hydraulic jump was established upstream of the cylinder, *i.e.* until SI conditions occurred. The experimental value of the undisturbed Froude number at the WI→SI limit condition was estimated to be the average of the two undisturbed Froude numbers measured just before and after the jump formed. This limit corresponds to the curve $(b/B)_l$ in fig. 17.

Once the SI configuration was established, the gate opening height was decreased by small steps until the jump was swept away and vanished, *i.e.* WI conditions were restored. The experimental value of the undisturbed Froude number at the SI→WI limit condition was estimated to be the average of the two undisturbed Froude numbers measured just before and after the jump vanished. This limit corresponds to the curve $(b/B)_L$ in fig. 17.

Before discussing the present experimental results, we will give a qualitative description of the flow pattern observed in the vicinity of the obstacle during the experiments.

At the beginning of each run, the flow was supercritical along the whole channel apart from a small area behind the bow shock wave produced by the cylinder (fig. 19a). In these conditions, the detached shock front experienced a regular reflection at the wall. Moreover, the fluid behind the detached shock was driven toward the channel axis by the expansion fan emanating from the cylinder. This produced a nearly straight secondary shock front behind the cylinder which intersected the primary reflected shock.

At Froude numbers lower than the initial $F_0$, the reflection of the detached shock wave at the wall became irregular. A Mach stem was established, which was approximately normal to the wall (fig. 19b) although somewhat distorted due to wall friction. Nevertheless, it was clearly recognizable. On the contrary, the slipstream, if there was any, could not be identified. This was possibly related to free surface curvature effects, which diffused the shock wave fronts thus obscuring the pattern of the shock waves.

Further decreases in the Froude number resulted in a progressive increase in the Mach stem height until a hydraulic jump was established upstream of the cylinder (fig. 19c).

The rather complex flow pattern observed in the vicinity of the obstacle at the WI→SI limit condition suggests that the one-dimensional approach might be inadequate to infer minimum specific energy $E_l$ in terms of only two fundamental parameters (*i.e.* $b/B$ and $F_0$). Actually, the behavior of the boundaries of the hysteresis region is controlled by at least two competing factors: (i) effects of free surface slope and curvature, which are measured by the parameter $y_0/b$, and (ii) energy losses and two-dimensional effects, which are difficult to discern and are controlled by the ratio $b/B$. The effects of both $y_0/b$ and $b/B$ will now be analyzed.

A comprehensive plot of all the experimental conditions at which the flow configuration changed from WI to SI and *vice versa* is given in fig. 20. Each plot in fig. 20 refers to a given cylinder diameter (*i.e.* to a given ratio $b/B$) and contains experimental boundaries of the hysteresis region in the $(y_0/b, F_0)$ plane. Theoretical undisturbed Froude numbers $F_{0l}$ and $F_{0L}$ provided by eqns. (28) and (29) are shown in fig. 20 as well.

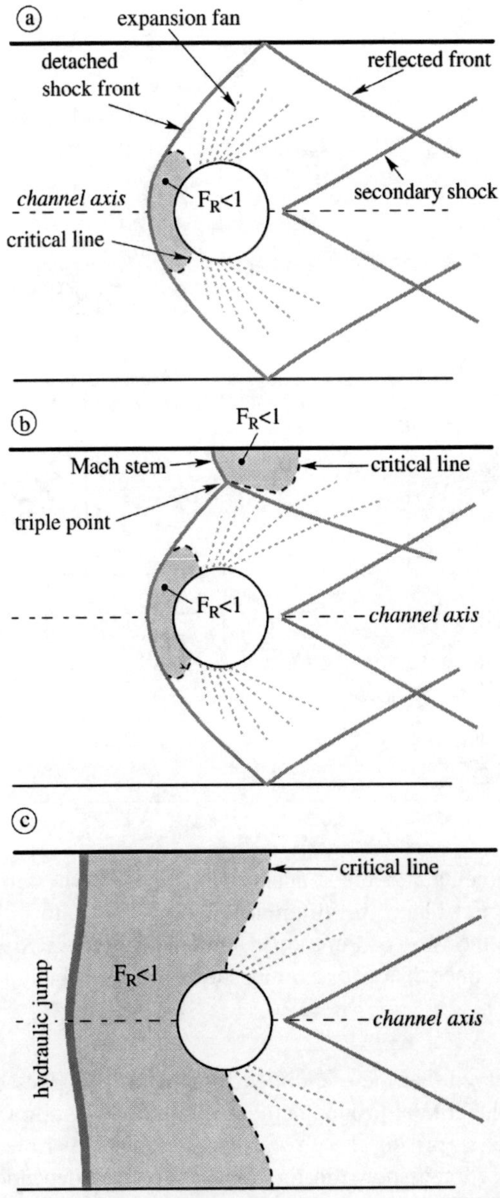


Figure 19: Sketch of the front pattern developing in the vicinity of the cylinder.

For the case $b/B = 0.74$, the cylinder diameter was $D = 0.1$m. However, in order to obtain experimental values of $y_0/b$ up to 0.6, a pair of cylinders of diameter $D = 0.05$m aligned normal to the flow direction was also used. Points with $y_0/b > 0.27$ refer to this experimental configuration.

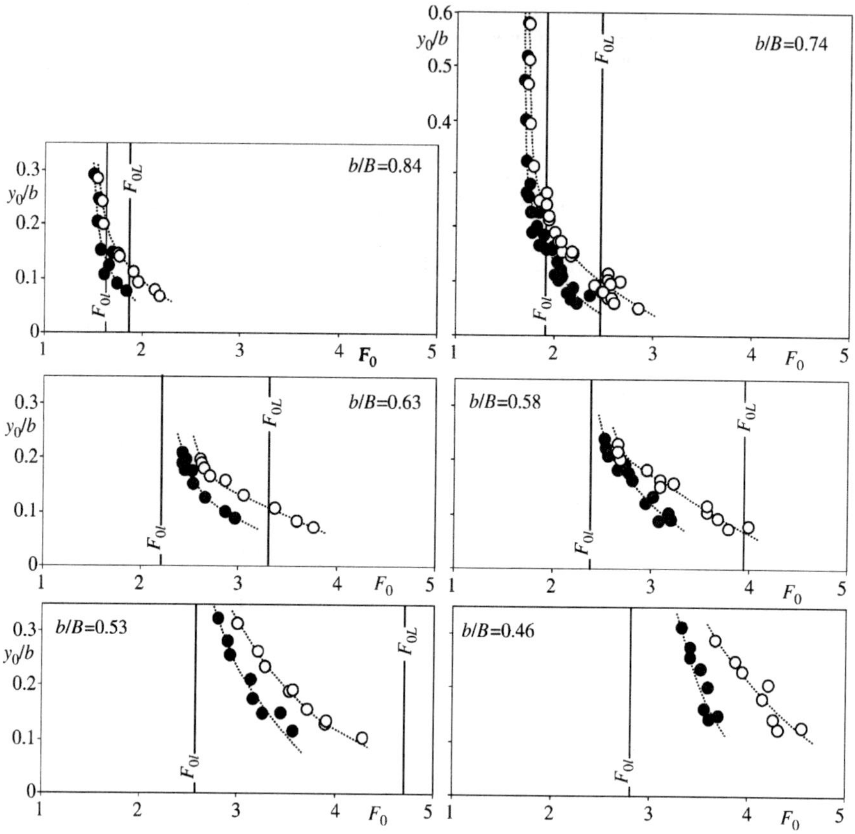


Figure 20: Experimental and theoretical boundaries of the hysteresis region in the $(y_0/b,\ F_0)$ plane for different values of the ratio $b/B$. Full symbols denote the WI→SI limit, open symbols denote the SI→WI limit. Solid vertical lines denote one-dimensional theoretical boundaries.

It can be observed that the one-dimensional approach gives the correct order of magnitude of the critical Froude numbers at the SI→WI limit conditions. This result is somewhat surprising if we recall the complex experimental flow pattern previously discussed and shown in fig. 19. Anyway, the dependence of both the upper and lower boundaries of the hysteresis region on the free surface slope and curvature is rather evident. At small values of $y_0/b$, *i.e.* when energy losses prevail over free surface slope and curvature effects, experimental values of the undisturbed Froude number at the SI→WI limit conditions are greater than those predicted by the inviscid one-dimensional model, as expected. At greater values of $y_0/b$, effects due to free surface slope and curvature prevail. As observed in Section 3, these effects act as an equivalent gain of energy. Actually, both experimental boundary curves shift towards the left side of the $(y_0/b,\ F_0)$ plane as $y_0/b$ increases.

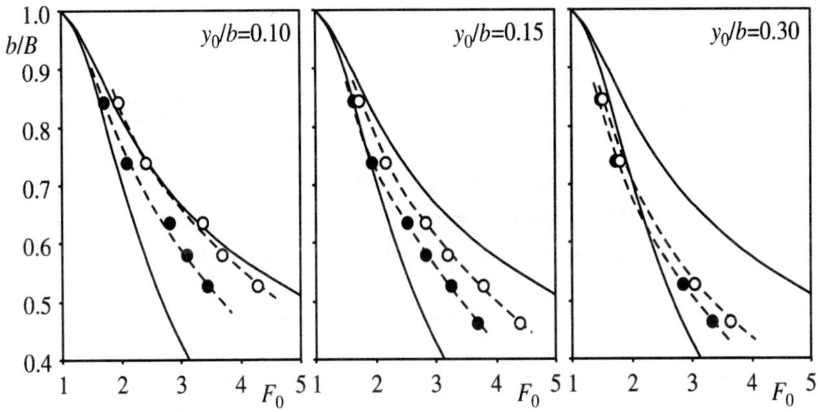


Figure 21: Experimental (symbols) and theoretical (solid lines) boundaries of the hysteresis region in the $(b/B,\, F_0)$ plane. $y_0/b = 0.1$ (left), $y_0/b = 0.15$ (center), and $y_0/b = 0.3$ (right).

The above discussion is confirmed in fig. 21, which shows the behavior of the boundaries of the hysteresis region in the standard $(b/B,\, F_0)$ plane for three different values of $y_0/b$.

Figure 22 shows the behavior of experimental minimum specific energies $E_l$ and $E_L$, normalized with the one-dimensional theoretical values $E_{l_{1D}}$ and $E_{L_{1D}}$, respectively, as a function of the ratio $y_0/b$.

All the points show a similar trend driven by free surface slope and curvature effects: energy ratios $E_l/E_{l_{1D}}$ and $E_L/E_{L_{1D}}$ decrease with $y_0/b$ increasing, at least in the range $0 < y_0/b < 0.3$. However, experimental data for $b/B = 0.74$ suggest that the above ratios may approach a constant value as $y_0/b$ is further increased.

Finally, it is worth recalling that when $y_0/b \to 0$, the effects related to the two-dimensional character of the flow are very important in establishing the lower limit $E_l$. Indeed, fig. 22 shows that the differences between the experimental results and one-dimensional predictions increase with $b/B$ decreasing. Moreover, these differences only slightly decrease with $y_0/b$ increasing, thus suggesting that the two-dimensional character of the flow affects the one-dimensional solution even at moderately high water depths.

On the contrary, when considering the upper boundary of the hysteresis region, experimental $E_L$ slightly differs from $E_{L_{1D}}$, and only a weak dependence of $E_L/E_{L_{1D}}$ on $b/B$ can be recognized.

In this section the occurrence of hysteresis for the case of channel constriction was discussed. In particular, the results of an in-depth theoretical and experimental study of the case of a flow around a vertical circular cylinder were presented. Although this type of obstacle is very simple, as it is characterized by just one length scale, *i.e.* the diameter $D$, the experimental results showed that the threshold specific energies $E_l$ and $E_L$ (*i.e.* the hysteresis boundaries) are dependent on parameters $F_0$, $b/B$, and $y_0/B$ in a rather complex way. It was also shown that the solution

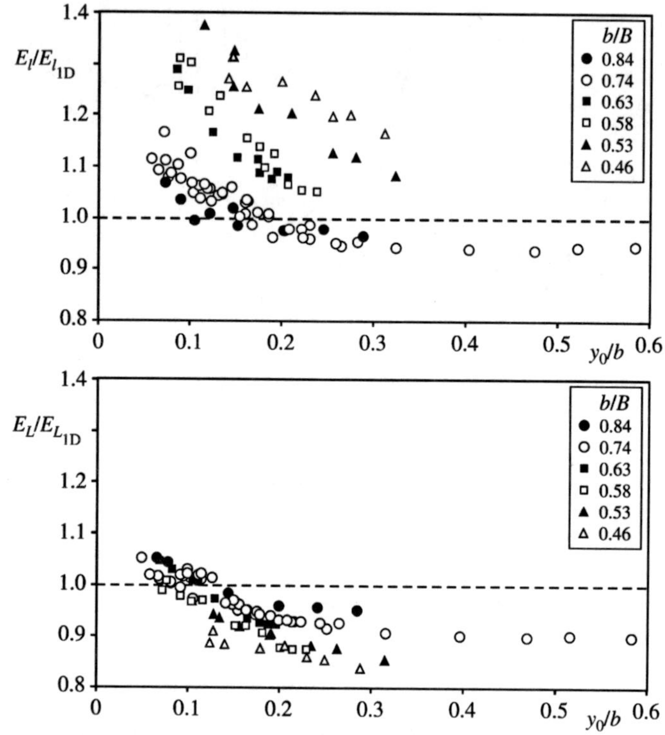


Figure 22: Ratio of experimental to theoretical minimum specific energy as a function of $y_0/b$. The upper and lower plots refer to the lower and upper hysteresis boundaries, respectively.

provided by the one-dimensional approach is not very different from the experimental one, although the extension of the experimental hysteresis region was found to be much smaller than the one predicted by the one-dimensional model.

## Conclusions

The hysteretic behavior of a free surface steady flow approaching an obstacle was examined. A simple one-dimensional theoretical approach to predict conditions for the occurrence of hydraulic hysteresis and to evaluate the boundaries of the hysteresis region for a supercritical undisturbed approaching flow was proposed.

The theoretical approach was described in detail for three different obstacles in a rectangular channel, namely a sill, a vertical sluice gate, and a vertical circular cylinder.

In all cases, the theoretical boundaries of the hysteresis region were found to be a function of the Froude number of the undisturbed approaching flow, $F_0$, and of a geometric parameter characteristic of the type of obstacle.

For the case of flow under a vertical sluice gate, it was also shown that the hysteretic behavior is not characteristic only of supercritical undisturbed approaching flows but pertains to subcritical undisturbed approaching flows as well provided that the undisturbed Froude number is greater than approximately 0.8.

The reliability of theoretical predictions of hysteresis was tested through comparison with experimental data, either measured by the authors or available in the literature.

It was found that for all investigated obstacles the one-dimensional approach correctly describes the hysteretic behavior of the flow, at least qualitatively. On the contrary, in order to make reliable quantitative predictions, the effects of the physical processes developing at the obstacle had to be evaluated as accurately as possible. In particular, energy losses at the obstacle, effects due to free surface slope and curvature, and contraction phenomena were found to play a crucial role in determining hysteresis boundaries.

Although the above effects can sometimes be suitably evaluated by approximated theoretical expressions, it was shown that in most cases proper experimental investigations are required to correctly predict the amplitude of the hysteresis region and the behavior of hysteresis boundaries as well.

## 6 Acknowledgments

The authors wish to thank Micoli, Trevisan, and Panelli for their valuable contribution to the experimental investigations. Helpful reviews by the anonymous referees are kindly acknowledged. This work was supported by MURST and University of Padova under the National Research Program, PRIN 2002, Influenza di vorticità e turbolenza nelle interazioni dei corpi idrici con gli elementi al contorno e ripercussioni sulle progettazioni idrauliche.

## References

[1] Abecasis, F.M. & Quintela, A.C., Hysteresis in steady free-surface flow, *Water Power*, **4**, pp. 147–151, 1964.

[2] Mehrotra, S.C., Hysteresis effect in one- and two-fluid systems, *Proceedings of the V Australian Conference on Hydraulics and Fluid Mechanics,* University of Canterbury, Christchurch, New Zealand, **2**, pp. 452–461, 1974.

[3] Muskatirovic, D. & Batinic, D., The influence of abrupt change of channel geometry on hydraulic regime characteristics, *Proceedings of the 17th IAHR Congress,* Baden Baden, **A**, pp. 397–404, 1977.

[4] Pratt, L.J., A note on nonlinear flow over obstacles, *Geophys. Astrophys. Fluid Dynamics*, **24**, pp. 63–68, 1983.

[5] Baines, P.G., A unified description of two-layer flow over topography, *J. Fluid Mech.*, **146**, pp. 127–167, 1984.

[6] Austria, P.M., Catastrophe model for the forced hydraulic jump, *J. Hydraul. Res.*, **25(3)**, pp. 269–280, 1987.

[7] Lawrence, G.A., Steady flow over an obstacle, *J. Hydraul. Eng. ASCE*, **8**, pp. 981–991, 1987.

[8] Baines, P.G. & Whitehead, J.A., On multiple states in single-layer flows, *Phys. Fluids*, **15(2)**, pp. 298–307, 2003.

[9] Becchi, I., La Barbera, E. & Tetamo, A., Studio sperimentale sul rigurgito provocato da pile di ponte, *Giornale del Genio Civile*, **111(6-7-8)**, pp. 277–290, 1973.

[10] Salandin, P., Indagine sperimentale sulla localizzazione del risalto a mezzo di quinte, *Giornale del Genio Civile*, **131(1-2-1)**, pp. 47–71, 1995.

[11] Defina, A. & Susin, F.M., Hysteretic behavior of the flow under a vertical sluice gate, *Phys. Fluids*, **15(9)**, pp. 2541–2548, 2003.

[12] Henderson, F.M., *Open channel flow,* MacMillan Publishing Co., New York, pp. 67–75, 1966.

[13] Marchi, E., Effetti locali dovuti alla pendenza e alla curvatura del pelo libero in un restringimento, *Atti del XXIV Convegno di Idraulica e Costruzioni Idrauliche,* Napoli, Italy, **T2**, pp. 35–44, 1994.

[14] Gentilini, B., Efflusso dalle luci soggiacenti alle paratoie piane inclinate e a settore, *L'Energia Elettrica*, **6**, pp. 361–380, 1941.

# CHAPTER 6

# Flow induced excitation on basic shape structures

S. Franzetti[1], M. Greco[2], S. Malavasi[1] & D. Mirauda[2]
[1] *Department I.I.A.R., Politecnico di Milano, Italy.*
[2] *Department I.F.A., Basilicata University, Italy.*

## Abstract

The study of flow-induced excitation on structures and obstacles is one of the main topics of fluid dynamics related to the practical interests in a large number of engineering applications *e.g.* aerodynamic, mechanical, civil, naval, *etc.* New design and project techniques have offered hazardous solutions, resulting in structures that are even more slender and flexible. This has led to a number of situations of self-excited vibration due to the interaction between flow fields and structures. Forces coming from this mechanism depend upon both the incoming flow and the structure motion, giving rise to a strong non-conservative force field, which may eventually lead to a growing structure motion. The aim of this chapter is to offer an overture about the phenomenon of the fluid–structure interaction. Because of the importance that the cylindrical and spherical shapes have in the practical applications and the generalizations that these shapes allow, in this chapter the fluid–structure interaction is mainly referred to these basic shapes.

## 1 Introduction

Flow-induced excitations of bodies, obstacles and structures in steady or unsteady flows, are at present both a relevant field of research as well as the subject of important studies of theoretical and experimental nature.

International literature reports several studies and contributions relating to such topics for the quasi two-dimensional systems and are summarized in the works of Sarpkaya [1], Ramberg & Griffin [2], Bearman [3], as well as in the papers of Blevins [4] and Naudascher & Rockwell [5].

From the 1970s up to the 1980s, the research was mainly focused on the study and analysis of flow fields and vortex structures generated downstream of the bodies. The emphasis of the results was aimed at a clear definition of the kinetic characteristics of the currents, related also to the different geometries of the flow field, through a range of values of several dimensionless parameters governing the process, such as Reynold's number, Strouhal's number and Keulegan-Carpenter's number (Bearman [6], Keulegan & Carpenter [7], Sarpkaya [1]).

Subsequently, from the 1980s on, the development of new acquisition and visualization techniques for describing flow field structures, as well as the increase in computational capacities for data processing, allowed the research to study by implementing physical experiments the direct assessment of the effects induced by the flow fields on the bodies (Blackburn & Henderson [8], Lin *et al* [9], Sheridan *et al* [10]) and the dynamic responses of the obstacle. In these studies the body is thought as a boundary condition for the flow field.

Only in the last few years, the description of the "interaction" between body and flow presents a different approach. It is focused on the possibility of explaining the different behaviors of bodies in water, and in fluids in general, by looking at the system as a whole. From this point of view, the body, thought of as a "structural system", does not represent only one of the boundary conditions for the flow field but, due to its geometrical and mechanical characteristics, plays a relevant role in governing the dynamics of the process as well.

In order to point out the main active phenomena in the flow–structure interaction processes, the present chapter deals with the analysis of the sources of excitation acting on the structure, both external and self-excitation, and the dynamic response of the obstacle.

## 2 The source of excitation – kinematic implications of flow structure on induced excitation

### 2.1 Fluid–structure interaction

The immersion of a solid body in a turbulent flow induces distortions that are connected to strong kinematic and dynamic instability. The fluid dynamic forces, due to the fluid–structure interaction, can be analyzed in terms of mean and instantaneous components; the latter is responsible for the excitation of vibrations.

According to the dominant excitation mechanism involved (Naudascher [11]) the sources of such vibrations can be classified into four groups (fig. 1): (EIE) Extraneously Induced Excitation caused by fluctuating velocities or pressures which are independent of any flow instabilities originating from the structure and from structural motion, with the exception of added-mass effect; (IIE) Instability-Induced Excitation caused by an instability of the flow due to the presence of the structure; (MIE) Movement-Induced Excitation due to fluctuating forces arising from movements of the vibrating structure; (EFO) Excitation due to Fluid Oscillation caused by a fluid oscillator becoming excited in one of its natural modes. In any of the first three cases (EIE), (IIE) (MIE), the exciting forces may or may not be affected by the simultaneous excitation due to fluid oscillation (EFO). However, even if the sources of excitation are usually studied according to the above classifications, the

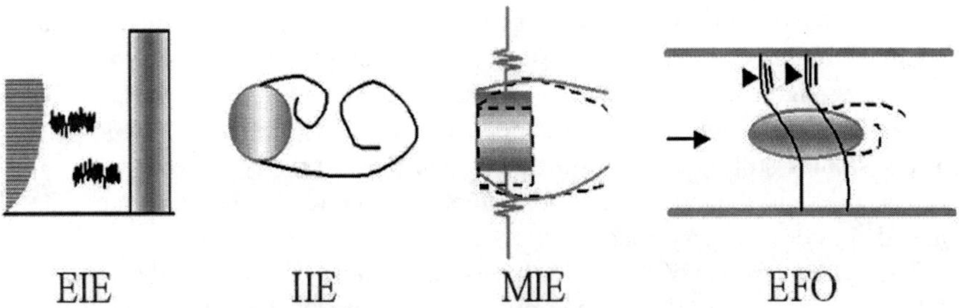


Figure 1: Examples of sources of excitation mechanisms (Naudascher [11]).

excitation of flow-induced vibrations in a real system is very often complex, since EIE, IIE, MIE and EOF may occur simultaneously.

In fig. 1, the Extraneously Induced Excitation (EIE) is represented by the instabilities of the incoming flow due to its turbulent level, which is not affected by the characteristic of the structure. Other sources of EIE are: earthquakes, machines and machine parts, two-phase flow and oscillating flow. The Instability-Induced Excitation (IIE) is depicted by the vortex shedding downstream to a stationary circular cylinder; the shape of the obstacle mainly affects the kinematic characteristics of the wake. In the case of the Movement-Induced Excitation (MIE), the obstacle is not stationary; its movements interact with the vortex shedding evolution or induce vortex shedding. In this situation the structure behaves like a body oscillator (see Section 3): the transverse movements of the structure induces distortions on the flow field which in turn induces the self-excitation of the structure. Finally, the Excitation due to Fluid Oscillation (EFO) mechanism is represented by standing gravity waves generated between a long pier and the walls of a flume. Flow-induced excitation can be enhanced by the EFO mechanism especially if one of EFO frequencies assumes the natural body-oscillator frequency, the dominant frequency of flow instability, or both (refer to the relevant literature for an extensive discussion on the effects of fluid oscillations: Guilmineau & Queutey [12], Lam & Dai [13], Yan [14]).

The structure of an external flow around an immersed body and the way in which the flow can be described and analyzed often depends on the geometry of the body. Three main categories of bodies are usually considered: (a) two-dimensional objects (infinitely long and of constant cross-sectional size and shape); (b) axisymmetrical bodies (formed by rotating their cross-sectional shape around the axis of symmetry), and (c) three-dimensional bodies that may or may not be symmetrical. In practice there can be no truly two-dimensional bodies, however, many objects are sufficiently long so that the end effects of considering the body as two-dimensional are negligible.

Another classification of body shapes can be made depending on whether the body is streamlined or blunt. Flow characteristics depend strongly on the amount of streamlining present. In general, streamlined bodies (*e.g.* airfoils, racing cars, *etc.*) have a little effect on the surrounding fluid in comparison with the effect of blunt bodies (*e.g.* buildings, parachutes, *etc.*).

The geometrical shape of structures or vehicles is very complex and the fluid-dynamic efficiency of the shape is usually studied employing a physical model.

Even though the fluid–structure interaction mechanism has been extensively studied on basic shapes, it still presents open questions; therefore a large part of current studies still concern basic shapes (cylinders, prisms, spheres, *etc.*), also because the fluid-dynamic studies on complex structures are hardly extendible to other shapes or boundary conditions.

The cylinder is one of the basic shapes most studied because of the simplicity of its form and because this form mimics a large number of practical applications. Long-span bridges, tall buildings, tall towers, cables, and so on, are examples of flexible cylindrical structures that are very sensitive to vortex-induced vibrations. The characteristic elongation in the transverse direction makes the cylindrical shapes very sensitive to the induced excitation of the flow. To be able to find the appropriate countermeasures required to control the fluid-dynamic response, the generation mechanism of this response should be clarified first. This mechanism is the flow pattern around the obstacle. Because of the complexity of the flow pattern, in order to understand the fluid–structure interaction mechanism, usually basic cross-sections such as 2-D rectangular cylinders, H-shaped cylinders and circular cylinders have been investigated and mainly in smooth flows.

In smooth steady flow conditions the cross-sectional dimensions of a stationary cylinder are generally the main characteristics responsible for the flow pattern deformations and consequently for the induced excitation on the cylinder. The sensitivity of the flow pattern on the main parameters that affect the phenomenon (such as Reynolds number, turbulent intensity, aspect ratio *etc.*) depends on the cross-sectional form of cylinder.

While vortex shedding principally depends on the Reynolds number (fig. 2) for a circular cross-sectional form, the vortex shedding mechanism is more complex for a sharp edge cross-section. For a rectangular sharp edged cross-section, the aspect ratio (L/D) generally represents the main parameter to be taken into account.

The forces that the flow induces on a circular cylinder affected by the flow field structures shown in fig. 2, both on the mean and fluctuating components are highlighted in figs. 3 and 4. The figures detail the drag coefficient and the Strouhal number of a circular cylinder versus Reynolds number. The different values of Strouhal number are justified by the characteristics of the boundary layer on the cylinder and of the near wake. In a subcritical range of Re ($150-300 < \mathrm{Re} < 1 \times 10^5$–$1.3 \times 10^5$), the near wake passing a stationary smooth cylinder is laminar, the vortices shed periodically and the force fluctuations correspond to a spectrum of extremely narrow bandwidth. In the post critical range of Reynolds number ($1 \times 10^5$–$1.3 \times 10^5 < \mathrm{Re} < 3.5 \times 10^6$), the boundary layer on the cylinder becomes turbulent downstream from a laminar separation bubble, and the near-wake becomes less regular; the force spectra in this post critical range are rather broadband. In the transcritical range $\mathrm{Re} > 3.5 \times 10^6$, finally, the boundary layer becomes turbulent upstream of separation and there is an apparent return of well-defined periodicity of both vortex shedding and force fluctuation with $\mathrm{Sh} \approx 0.3$ (fig. 4c).

Figure 2: Flow patterns for flow past a smooth circular cylinder at various Reynolds numbers: (a) Re $\approx 1.5 \times 10^{-1}$, no separation; (b) Re $\approx 1.5 \times 10^{1}$, steady separation bubble; (c) Re $\approx 1.5 \times 10^{2}$, oscillating Karman vortex street wake; (d) Re $\approx 2.5 \times 10^{4}$, laminar boundary layer with wide turbulent wake; (e) Re $\approx 3.2 \times 10^{5}$, turbulent boundary layer with narrow turbulent wake (Munson *et al* [15]).

## 2.2 Sharp-edged rectangular cylinder

In the case of a sharp-edged elongated rectangular cylinder, the flow detaches on the upstream (primary separation) and downstream corners (secondary separation), and the flow distortion is affected by several parameters. The vortices generated close to the body develop and shed from it, creating an unsteady wake. The characteristics of the wake are dependent on the Reynolds number of the obstacle Re $= U_0 D/\nu$, but the aspect ratio (L/D) and the incidence angle between cylinder and flow have the main influence on the vortex shedding and therefore on the structural excitation. Figure 5 shows the mean flow field around a rectangular cylinder (L/D = 3) in unbounded flow, numerically obtained by Yu & Kareem [16] at Re = $1 \times 10^{5}$.

The primary separation of the shear layer occurs on the leading edge. The separated flow initially diverges from the body, with an angle dependent on the separation pressure, and then curves toward the cylinder surface. When the reattachment occurs, the flow makes a region of recirculation known as "separation

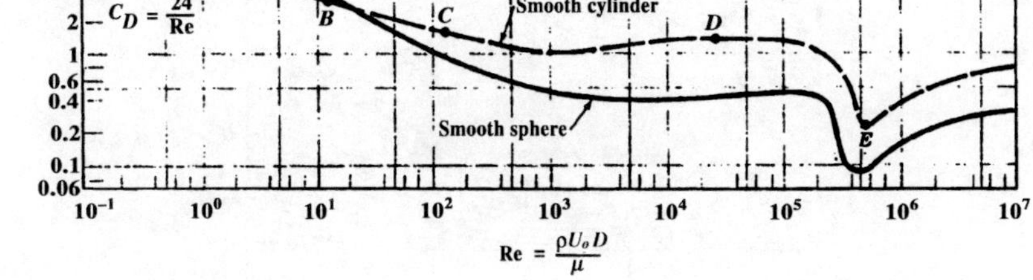


Figure 3: Drag coefficient as a function of Reynolds number for smooth circular cylinder and smooth sphere (Munson *et al* [15]).

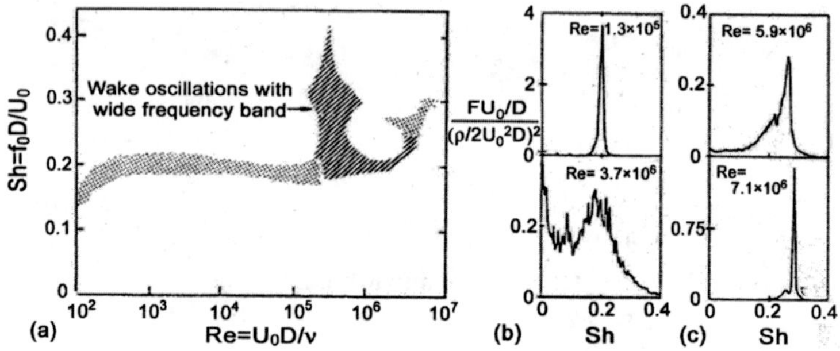


Figure 4: Strouhal number, Sh, of vortex shedding (a) and spectra of the lift force component (b, c) from a stationary, smooth circular cylinder in low-turbulence cross flow (Naudascher [11]).

bubble". In such conditions the secondary separation at the downstream edge of the cylinder causes the roll up of the flow in the rear face generating a secondary vortex that periodically sheds from the body surface. In some cases, when the reattachment of the primary separation is unsteady, the two turbulent structures interact and the vortex shedding becomes more complex. To simplify the description of the phenomenon, the main vortex shedding regimes have been defined and classified on the basis of the characteristics of the main vortices involved. In the case of steady flow conditions and rigid obstacle, Naudascher & Wang [17] give the following classification:

LEVS (Leading-Edge Vortex Shedding): the flow separation occurs at the leading-edge with formation of vortices dominating the near wake of the body (fig. 6a);

TEVS (Trailing-Edge Vortex Shedding): a decisive flow separation at the trailing-edge occurs and vortex-shedding is analogous to the von Kármán street behind circular cylinders (fig. 6c);

ILEV (Impinging Leading-Edge Vortices): a flow separation at the leading-edge and impingement of the leading-edge vortices at the side surfaces and/or edges of the body are present (fig. 6b);

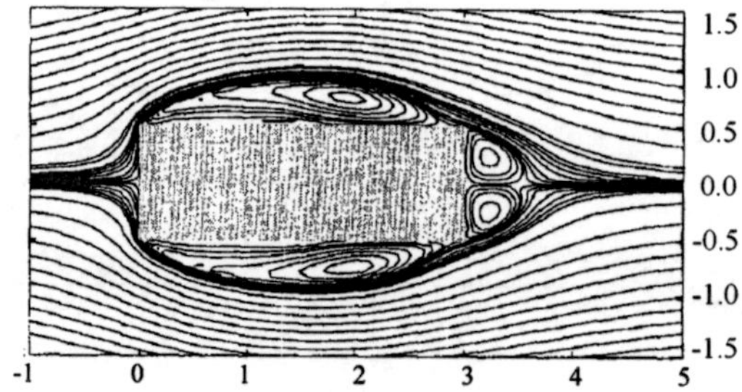

Figure 5: Mean flow field numerically obtained in unbounded flow around a rectangular cylinder, L/D = 3, Re = 1 × $10^5$ (Yu & Kareem [16]).

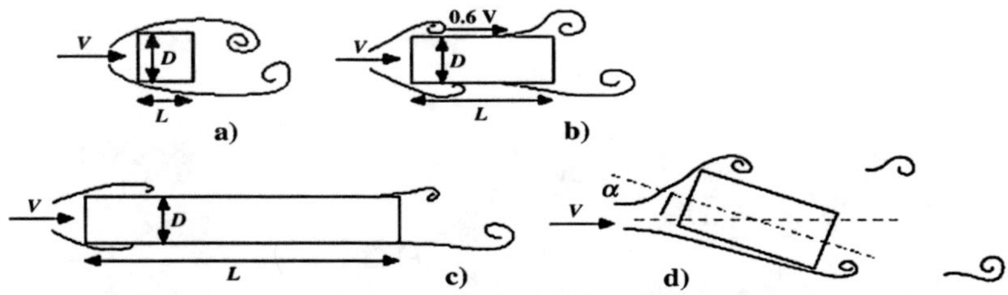


Figure 6: Mechanism of vortex-shedding for IIE source condition on prismatic 2D obstacle (Deniz & Staubli [18]).

AEVS (Alternate-Edge Vortex Shedding): both the leading-edge and the trailing-edge mechanisms are present (fig. 6d).

Each vortex type allows a specific dynamical state. Under the flow conditions above mentioned and for a wide range of Reynolds numbers, the aspect ratio (L/D) affects the vortex shedding and the loading on the structure significantly. When L/D < 2 (fig. 6a), only the primary separation occurs because the shear layer separates at the leading edge and involves the whole side of the cylinder (LEVS); in this range of L/D Bearman & Trueman [19] observed that the formation of the vortex close to the cylinder enlarges the drag coefficient of the obstacle ($C_D$). The minimum distance between rear cylinder face and vortex formation occurs for L/D = 0.64, which corresponds to the maximum value of $C_D$.

When L/D > 6 (fig. 6c), the flow separated at the leading edge reattaches permanently; consequently the trailing edge separation (TEVS) dominates the vortex shedding. For L/D ≈ 2.8 (fig. 6b), the literature indicates a complex situation of possible unstable reattachment (Shimada & Ishiara [20], Yu & Kareem [16]). When $\alpha \neq 0$ (fig. 6d), the symmetry of the flow structure is compromised. Consequently, on one the upper side of the cylinder LEVS prevails and on the lower, TEVS. When this condition occurs, the vortex shedding is characterized by the AEVS regime.

The vortex shedding behavior is well described by the Strouhal number (Sh = $f_0 D/U_0$, where $f_0$ is the dominant frequency of the vortex shedding).

The chart in fig. 7 reports the Sh obtained by several Authors with rectangular cylinders of various L/D immersed in unbounded flows. In fig. 7, two discontinuities of Sh values are evident. The first occurs when L/D is approximately equal to 2.8, the second occurs when L/D = 6. At L/D≈2.8, the flow pattern is bounded between the flow separation type LEVS and the flow reattachment type ILEV. The data dispersion in fig. 7 and the presence of more than one dominant frequency for a specific aspect ratio mainly depend on the upstream flow characteristics. At the latter critical aspect ratio L/D = 6, the flow pattern is bounded between the unsteady flow reattachment type ILEV and the completely steady flow reattachment type TEVS.

For a wide range of conditions, several studies show the negligible influence of Reynolds number when it exceeds the value Re = $1 \times 10^4$. On the contrary, the influence of Re is not negligible for Re < $1 \times 10^4$ (Okajima [21]). In fig. 8, the Sh versus Re is reported with an aspect ratio L/D = 3.

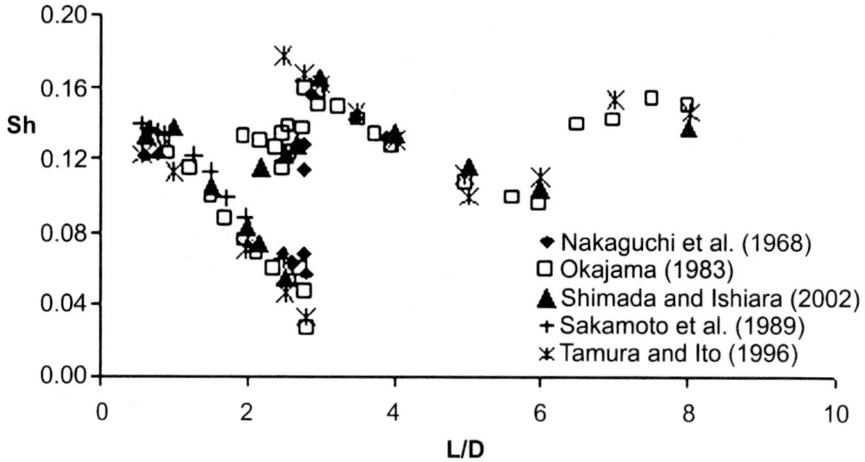


Figure 7: Variations of Strouhal number according to L/D ratio for rectangular cylinders in unbounded flow (Shimada & Ishiara [20]).

The free stream turbulent level of the flow passing a circular cylinder, $\mathrm{Tu} = u'_{rms}/U_0$ (where $u'_{rms}$ is the standard deviation of the inflow velocity on x direction), significantly affects the flow pattern and the excitation induced on the cylinder. As shown in fig. 9a, turbulence decreases $C_D$ at subcritical Re and

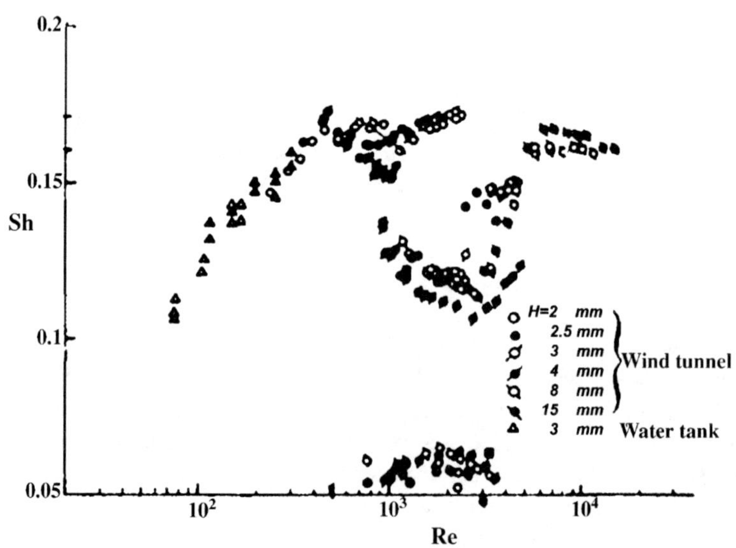


Figure 8: Variation of Strouhal number with Reynolds number for rectangular cylinders with L/D = 3 (Okajima [21]).

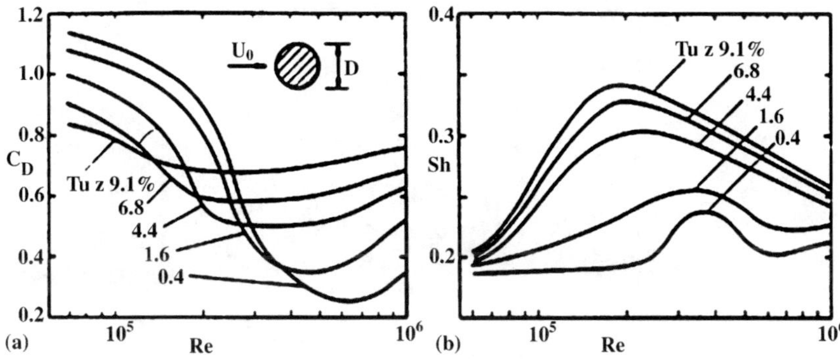


Figure 9: Effect of the free-stream turbulence, Tu, on (a) mean drag coefficient, $C_D$, (b) and on Strouhal number, Sh, for a smooth circular cylinder (Naudascher [11]).

increases it in the supercritical range. The rise in value of Strouhal number in the transition range (fig. 4a) occurs at smaller Re as Tu increases (fig. 9b).

In the case of a sharp-edged rectangular cylinder, free stream turbulence level (Tu) has received a great attention in literature because it significantly influences the structure and the development of the shear layer separated off the upstream corners (Haan *et al* [22], Lin & Melbourne [23], Noda & Nakayama [24], Saathoff & Melbourne [25]). The main effect of Tu is to shift the reattachment point. An increase of Tu leads to a progressive shortening of bubble formation and, thus, to a possible strong modification of vortex shedding (Nakamura *et al* [26]). Noda & Nakayama [24] observed that turbulence shakes the shear layer over a distance comparable with the turbulence scale. The main effects of turbulence occur when L/D is in a range of values near the critical value L/D = 2.8. In this range of L/D, the reattachment of the leading edge separation is not stable. In this situation, the turbulent inflow with the length scale of the same order as D acts by moving the position of the separated shear layer off the downstream corners, promoting the reattachment.

The behavior of vortex shedding is significantly affected also by the presence of boundaries that limit the evolution of the wake. The presence of boundaries are relevant in a large number of civil applications (*e.g.* buildings, bridges, pipelines, *etc.*)

The study of boundary effects has principally been considered in aerodynamic applications especially in terms of blockage ratio ($\gamma_b$), defined as the ratio between the frontal area of the body and the cross-section of the flow without obstacle.

Both for a circular or rectangular cylinder, significant changes in Sh values can occur when the flow confinements are changed. In general, the increasing of the blockage induces an acceleration of the flow near the object, which locally increases the flow velocity (solid blockage) and increases the energy losses in the wake and in the boundary layer (wake blockage). For bluff bodies, the effects of blockage (solid and wake) can be very remarkable and its influence changes the values of

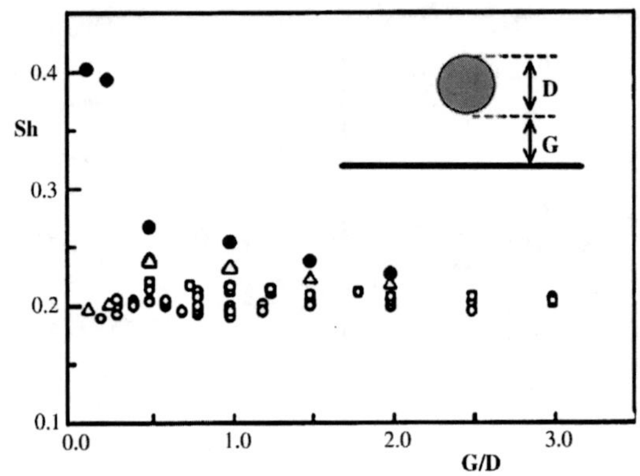


Figure 10: Strouhal number results versus elevation ratio: • Price *et al* [27], $\delta/\mathrm{D}=0.45$, Re = 1900; △ Price *et al* [27], $\delta/\mathrm{D}=0.45$, Re = 4900; □ Angrilli *et al* [28], Re = 2860, 3820, 7640, $\delta/\mathrm{D}=0.2$, 0.4, 0.5; ○ Bearman & Zdravkovich [29], Buresti & Lanciotti [30], Taniguchi & Miyakoshi [31] and Lei *et al* [32], $1.3 \times 10^4 \leq \mathrm{Re} \leq 1.4 \times 10^5$, $0.1 \leq \delta/\mathrm{D} \leq 1.64$. (Price *et al* [27]).

both force coefficients and Strouhal number. These effects are usually taken into account using the following expressions:

$$C_{Fc} = C_F(1-\gamma_b)^{n_{C_F}}; \quad \mathrm{Sh}_c = \mathrm{Sh}(1-\gamma_b)^{n_{sh}} \tag{1}$$

where $C_{F_c}$ and $\mathrm{Sh}_c$ are the corrected force coefficient and Strouhal number, respectively and $n_{C_F}$ and $n_{\mathrm{sh}}$ are experimental coefficients ($0 \leq n_{C_F}, n_{\mathrm{sh}} \leq 1$).

The presence of a significant asymmetry of the boundary conditions has also remarkable effects on the structure excitation. These effects are summarized in fig. 10, where the Strouhal number of a circular cylinder is plotted against the elevation ratio, G/D, of the cylinder above a wall (G is the elevation above the wall and D is the diameter of circular cylinder). The effects on the cylinder excitation are emphasized by low Re values and affected by the boundary layer thickness ($\delta$) above the wall.

The influence of a solid surface on the dynamic effects for a rectangular cylinder has been less considered in literature. A recent study (Cigada *et al* [33]) highlights that for a rectangular cylinder with aspect ratio L/D = 3, the presence of a solid surface significantly affects both the force coefficients and the vortex shedding even if the cylinder is placed at relevant elevation, G, from the surface. The solid boundary affects the lift coefficient, $C_L$, up to G/D ≃ 3.5; in the range $3.5 \geq \mathrm{G/D} \geq 1$, $C_L$ decreases toward the negative value $C_L = -1$ then increases up to $C_L = 1$ in the range ($1 \geq \mathrm{G/D} \geq 0$). The influence on drag coefficient seems limited in the range $1 \geq \mathrm{G/D} \geq 0$, where $C_D$ decreases from its typical unbounded value up to $C_D = 0.7$.

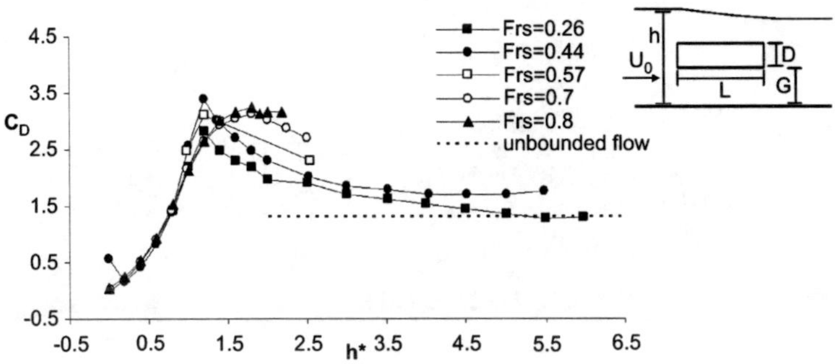


Figure 11: Dependence of the experimental drag coefficient on $h^*$ and $\mathrm{Fr}_s$ (G/D = 2.33), together with the (constant) value of $C_D = 1.3$ for unbounded flow (Malavasi & Guadagnini [34]).

In hydraulic applications the influence of a free surface is usually considered as one of the main parameters in modelling structure excitation, relevant in the study and assessment of the vulnerability of river bridges under partial or total submergence conditions. Experimental studies, carried out by Denson [35] and Malavasi & Guadagnini [34], highlight the significant influence of the free surface in terms of force coefficients. The influence of the free surface on the flow field structure around the obstacle was provided in Malavasi *et al* [36].

Figure 11 shows the behavior of $C_D$ versus $h^*$ for different value of the Froude number, $\mathrm{Fr}_s$, for a rectangular cylinder with L/D = 3 placed at an elevation G/D = 2.33 from the bottom of a hydraulic channel, where $C_D$ is computed as:

$$C_D = \frac{F_D}{0.5\rho U_0^2 D}, \tag{2}$$

$h^*$ is the dimensionless parameter of the cylinder submersion:

$$h^* = \frac{h - G}{D}, \tag{3}$$

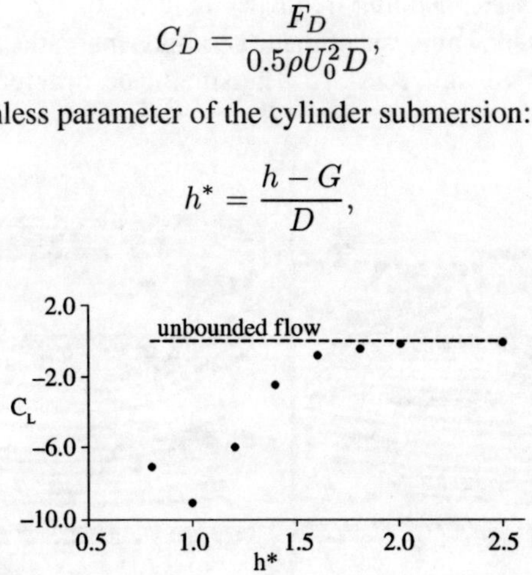


Figure 12: Lift coefficient versus $h^*$ with $\mathrm{Fr}_s = 0.26$, together with the reference value of $C_L = 0$ corresponding to the unbounded flow condition (Malavasi *et al* [36]).

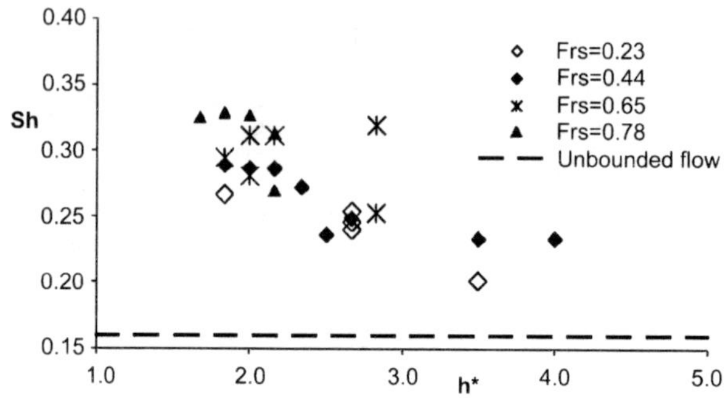


Figure 13: Strouhal number, Sh, versus $h^*$ (G/D = 2.33).

and $\mathrm{Fr}_s$ is the Froude number referred to the cylinder thickness:

$$Fr_s = \frac{U_0}{\sqrt{gD}}. \tag{4}$$

As $h^*$ increases toward 1, $C_D$ increases independently from $\mathrm{Fr}_s$; with further increases of $h^*$, $C_D$ reaches its peak value then decreases and tends to an asymptotic value. The peak of $C_D$ seems to depend on the observed value of Froude number and $h^*$.

The free surface also drastically influences the lift coefficient, as shown in fig. 12 where $C_L$ is plotted versus $h^*$ ($\mathrm{Fr}_s = 0.26$). The lift coefficient presents a negative peak for $h^* = 1$, after which the absolute value of $C_L$ increases tending to zero (unbounded flow condition) as $h^*$ increases.

In fig. 13, the vortex shedding frequency in terms of Strouhal number and calculated by the frequency analysis of the lift component on the rectangular cylinder is plotted versus $h^*$ for different $\mathrm{Fr}_s$. The significant difference between the ex-

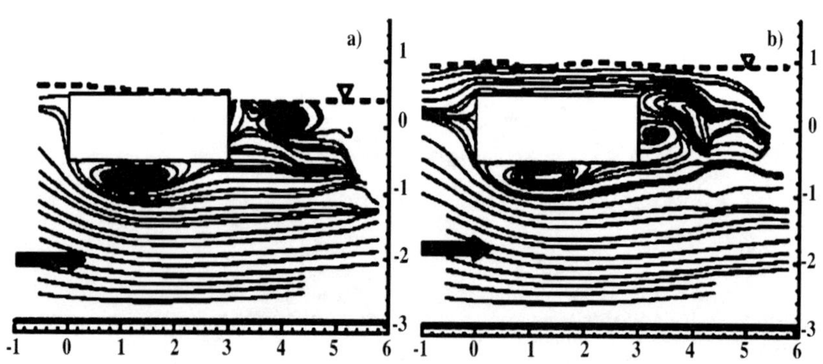


Figure 14: Mean flow field reconstruction by velocity field measured around the cylinder with $\mathrm{Fr}_s = 0.26$, G/D = 2.33 and $h^* = 1.0$ (a) and $h^* = 1.4$ (b) (Malavasi *et al* [36]).

perimental values and the reference value of the unbounded condition may be explained in the features of the confinement of the flow. As shown for example in figs. 14a and 14b, the upper confinement of the flow interferes with the leading-edge separation, changing the structure and the characteristic of the vortex shedding. The asymmetry imposed by the free surface limits the separation on the topside of the cylinder, thus the lack of equilibrium on the vertical loading direction induces significant variation in the $C_L$ value from $C_L = -9$ to $C_L = -2$ as shown in fig. 12.

## 3 Dynamic response of the structure

### 3.1 Basic equations

The analysis of the interaction between flow and structure may also be put forth using the behavior of the body as a reference point for characterizing the processes. This allows us to evaluate the dynamic response of the oscillating obstacle, compared to vortex-induced vibrations phenomenon and the main characteristics of the flow field.

The equation of motion generally used to represent the vortex-induced vibrations of a body oscillator, in steady and unsteady flows, is proposed as follows:

$$m\ddot{x} + B\dot{x} + Cx = F(t), \tag{5}$$

where $x$ is the displacement of the body towards the main flow or in a transversal direction (fig. 15a), $m$ is the total structural oscillating mass, $B$ is the structural damping, $C$ is the spring constant and $F$ is the acting fluid force. In this case, the body oscillator is treated as a discrete-mass system free to vibrate in one/two directions and the fluid force assumes a sinusoidal form:

$$F = F_0 \sin(\omega_s t + \varphi_s), \tag{6}$$

where $\omega_s = 2\pi f_s$ is the circular frequency and $f_s$ is the frequency of fluid force.

The solution of eqn. (5) is composed by the solution of the homogenous equation given by:

$$x = e^{-\zeta\omega_n t} x_0 \cos(\omega_d t - \varphi) \quad \text{with} \quad \omega_d = \omega_n \sqrt{1 - \zeta^2}, \tag{7}$$

and the particular solution:

$$x = x_0 \cos(\omega_s t - \varphi), \tag{8}$$

where $\omega_n = 2\pi f_n$ is the natural circular frequency and $f_n$ is the natural frequency of the system.

The frequency $f_n$, can be generally calculated under the hypothesis of perfect joint taking into consideration the contribution of the added mass according to the following relationship:

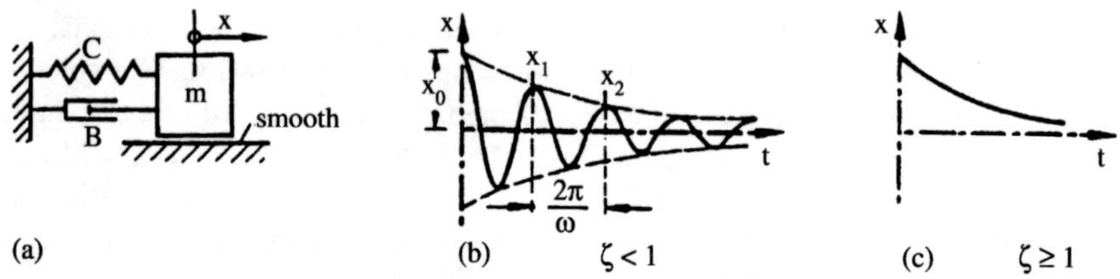


Figure 15: (a) Simple body oscillator with linear damping (*i.e.* resistance proportional to velocity); (b, c) histograms of responses for an underdamped ($\zeta < 1$) and an overdamped ($\zeta \geq 1$) case (Naudascher & Rockwell [5]).

$$f_n = \frac{1}{2\pi}\sqrt{\frac{C + C'}{m + m_a}}, \tag{9}$$

where $C'$ represents the added stiffness, which is usually included in the spring constant of the system and $m_a$ the related added mass.

The damping factor or damping ratio, $\zeta$, is defined as:

$$\zeta = \frac{B}{2m\omega_n} = \frac{B}{2\sqrt{(m + m_a)C}}. \tag{10}$$

For the underdamped case ($\zeta < 1$) the damping factor can be calculated by the exponentially decaying response for the initial condition $t = 0$, $x = x_0$ and $\varphi = 0$ (fig. 15b), as mentioned by Naudascher & Rockwell [5]. In the case $0 < \zeta < 1$ the coefficient has been obtained through the following equation:

$$\phi = \ln\frac{x_n}{x_{n+1}} = \frac{2\pi\zeta}{\sqrt{1 - \zeta^2}}, \tag{11}$$

where $\phi$ is called the logarithmic decrement.

For $\zeta \geq 1$, the displaced body simply returns to its equilibrium position in an exponential fashion (fig. 15c). The damping for the limit case of $\zeta = 1$ is called critical damping.

Since the solution eqn. (7) dies out with time on account of damping, only the steady-state solution eqn. (8) is of general interest. Its frequency is equal to the forcing frequency $f_s = \omega_s/2\pi$ and the amplitude $x_0$ is obtained as:

$$x_0 = \frac{F_0/C}{\sqrt{[1 - (\omega_s/\omega_n)^2]^2 + (2\zeta\omega_s/\omega_n)^2}}, \tag{12}$$

The phase angle $\varphi$ by which the response $x$ lags the exciting force $F$ is as follows:

$$\tan\varphi = \frac{2\zeta\omega_s/\omega_n}{1 - (\omega_s/\omega_n)^2}, \tag{13}$$

where $\omega_n = \sqrt{C/m + m_a}$ is the natural frequency of the undamped system.

For a body with one torsional degree of freedom, eqn. (5) takes the form:

$$I_\theta \ddot{\theta} + B_\theta \dot{\theta} + C_\theta \theta = M(t) \tag{14}$$

where $I_\theta$ is the mass moment of inertia of the body, $M(t)$ is the exciting moment or torque, $B_\theta \dot{\theta}$ is the damping moment, $C_\theta \theta$ is the restoring moment, $\zeta_\theta = B_\theta / 2 I_\theta \omega_n$ is the damping ratio and $\omega_{\theta n} = \sqrt{C_\theta / I_\theta}$ is the undamped circular natural frequency. The response to a harmonic exciting moment $M(t) = M_0 \cos \omega_s t$ is:

$$\theta = \theta_0 \cos(\omega_s t - \varphi) \tag{15}$$

where $\theta_0$ and $\varphi$ are the amplitude of torsional vibration and the phase angle respectively.

### 3.2 Dynamic response in resonance conditions

In the study of flow-induced vibrations a condition of particular interest is the resonance phenomenon when the vortex shedding frequency is close to the natural frequency of the structure. This occurs because the resonance phenomenon generates critical conditions in the structures in terms of stability and structural stress corresponding to potential collapsing of the structures themselves.

Another case of relevant importance is one in which the frequency of the body oscillations matches the frequency of the wake vortex. In such cases the processes are outlined as lock-in or synchronization. The body tends to "pulse" presenting large amplitude and the system, even if it does not assume the resonance condition, is subject to relevant stress. In such conditions the oscillation amplitudes, transversal to the fluid flow (y-direction), are always found to be much larger than streamwise motions (x-direction).

Studies on the analysis of the vibrating structures nearing the conditions of resonance in bounded and free surface flows, have highlighted the existence of a strong dependence of the maximum transverse amplitude $A^*_{max}$ on some non-dimensional groups, as shown below:

$$A^*_{max} = A^* \left( \frac{U_0}{f_n D}; S_G; \frac{h}{D} \right) \tag{16}$$

where $A^*_{max} = y_0 / D$ is the ratio between the maximum transverse amplitude and the characteristic dimension $D$ of the body, the ratio $\frac{U_0}{f_n D}$ is the reduced velocity $U^*$, $h/D$ is the ratio between the water depth and the characteristic size of the body and $S_G$ is the Skop-Griffin parameter defined as follows:

$$S_G = 2\pi^3 \mathrm{Sh}^2 (m^* \zeta), \tag{17}$$

where Sh is the Strouhal number $\frac{U_0}{fD}$.

Concerning the influence of the body shape, the $S_G$ parameter takes into account both boundary conditions of the oscillating body, through $m^*$ (ratio between the structural mass $m$ and the added mass $m_a$) and $\zeta$, as well as the flow induced force

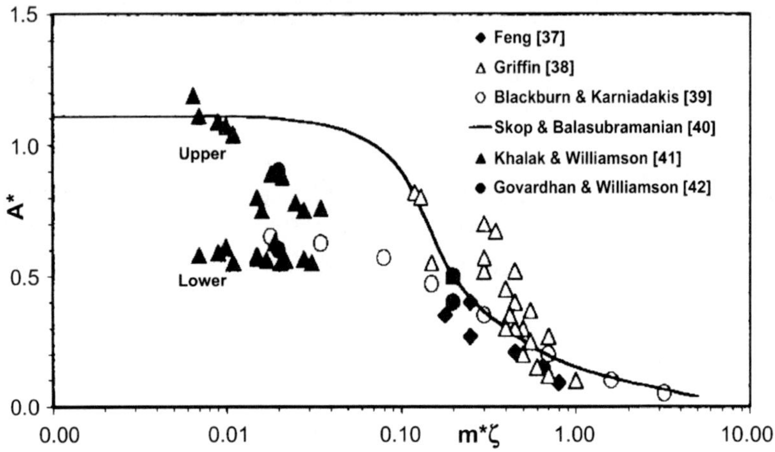


Figure 16: Griffin plot showing maximum amplitude observed in different experiments versus the combined mass-damping parameter (Khalak & Williamson [41]).

through Sh. Under resonance conditions, the Strouhal number is assumed constant, thus the maximum transverse amplitude, $A^*_{max}$, depends on the Skop-Griffin parameter and, as seen in eqn. (17), on the combined mass-damping parameter $m^*\zeta$.

Figure 16 summarizes the results of several experiments for different values of $m^*$ in terms of maximum transverse amplitude, $A^*_{max}$, versus $m^*\zeta$. From this figure, it does not seem possible to make a "singular" curve of $A^*_{max}$ versus $m^*\zeta$. Sarpkaya [1] originally stated that a simple observation of the motion equation immediately shows that the response of the system is independently governed by mass and damping. By analyzing three pairs of low-amplitude response data, each pair of them at similar values of $m^*\zeta$ but different $m^*$ values, he observed a large influence of mass ratio on $A^*_{max}$. In fact Sarpkaya [43] states: one should use the combined parameter $m^*\zeta$ only for $m^*\zeta > 0.40$ while for $m^*\zeta < 0.40$ the dynamic response of system is governed by $m^*$ and $\zeta$ independently.

Khalak & Williamson [41] carried on a set of experiments over a wide range of $m^*$ ($m^* = 1 \div 20$) under the same experimental conditions showing that even for low $m^*$ of the order 2 and very low mass-damping down to the value $m^*\zeta \sim 0.006$, the use of a single combined mass-damping parameter collapses peak amplitude data very well, even for a wide independent variation of parameters $m^*$ and $\zeta$ (fig. 16). In this way they extended the value of $m^*\zeta$ proposed by Sarpkaya by two orders of magnitude.

Furthermore, in the case of elastically mounted systems, they observed two different types of response depending on the high or low combined mass-damping parameter $m^*\zeta$. In fact for low $m^*\zeta$ values, there are three different branches of response: the initial, the upper and the lower ones which present two jumps in

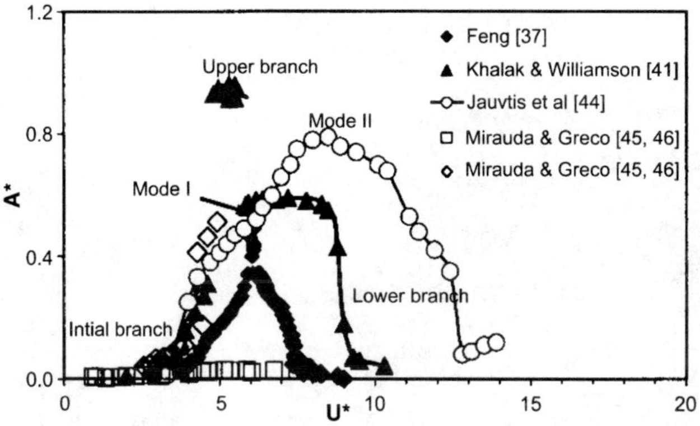


Figure 17: Maximum amplitude versus reduced velocity for different bodies: Khalak & Williamson [41] and Feng [37] on the cylinders; Jauvtis *et al* [44] and Miraudа & Greco [45, 46] on the spheres.

the magnitude of oscillating displacement (fig. 17). They found that the transition between the "initial" and "upper" branch was hysteretic, while the transition from the "upper" to "lower" branch involved an intermittent switching.

On the contrary, for high values of combined mass-damping parameter $m^*\zeta$, Feng [37] observed only two branches of response: the initial branch and the lower one. The passage between the two branches, as can be seen in fig. 17, occurs with a jump and the body reaches conditions of resonance.

Furthermore, Govardhan & Williamson [42], by visualization techniques (Digital Particle Image Velocimetry), showed that the change from the initial branch to the upper one, depends on the jump in the angle phase between the force induced by the shedding of the main vortex and the displacement of the body (fig. 18). This jump is characterized by a change in the form of the vortex wake downstream of the body by a mode "2S", indicating 2 single vortices shed per cycle, to mode "2P", meaning 2 pairs of vortices per cycle (fig. 18). Under this condition the value of the body oscillating frequency, $f$, passes across the natural frequency in water generating a resonance phenomenon. On the other hand, the passage from the upper branch to the lower one is characterized by the presence of a phase-difference between the total fluid force and the displacement of the body which tends to go toward a periodic uniform trend. In such cases no change in the form of the wake is observed.

For high values of $m^*\zeta$ the passage from the initial branch to the lower one depends on the jump of a phase between both the force components, the total force and the force induced by the vortex and such jump is related to a change in the form of the wake.

Referring to fig. 17, the behavior found for three-dimensional structures, with elementary geometrical forms (ex. spheres), is sensitively different from that observed for two-dimensional structures. In fact, the data of Jauvtis *et al* [44] relating to the oscillations of a sphere, indicate the presence of two distinct modes of

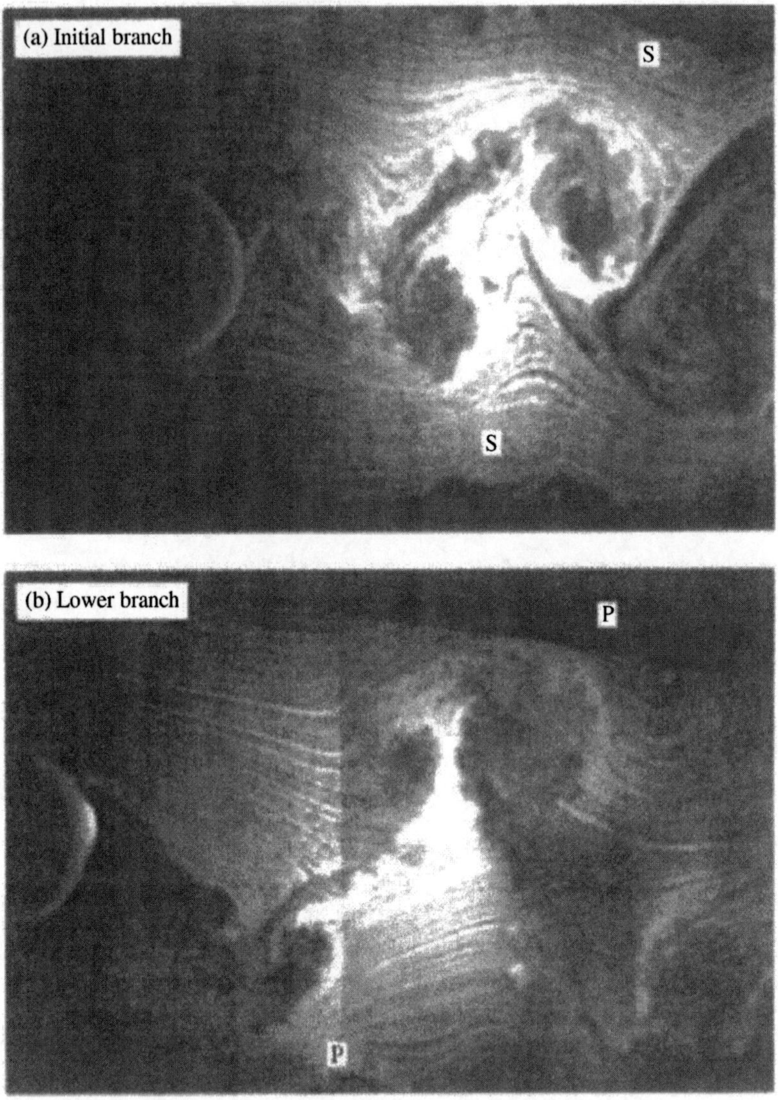


Figure 18: Example of flow visualization of the initial branch (2S mode) and the lower branch (2P mode) (Khalak & Williamson [41]).

response. The first mode of response (Mode I) is manifested in the presence of resonance conditions, when the frequency of the shedding of the vortex is close to the natural frequency of the body, and a synchronization regime is observed between the force and the response. When the average velocity of the flow increases, the system shows the presence of periodic oscillations characterized by high values of displacement that represent the second mode of response (Mode II).

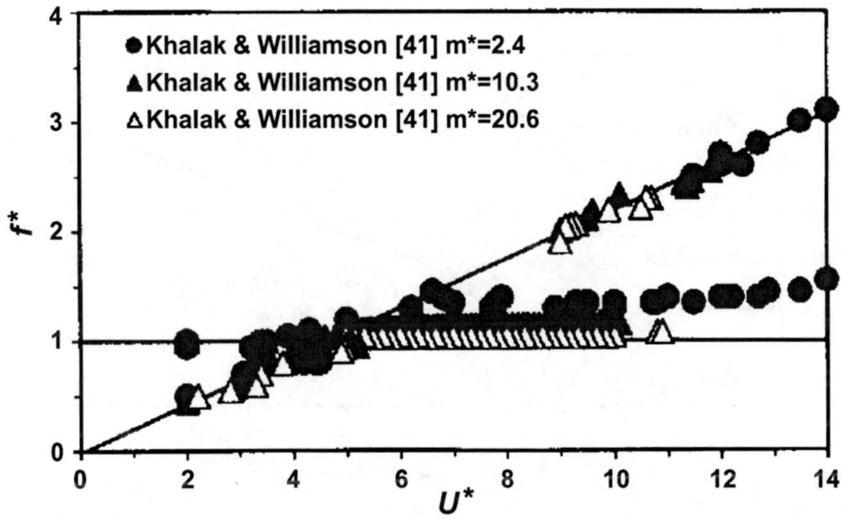


Figure 19: Frequency ratio versus reduced velocity for vibrating cylinder.

In fig. 17, data from Mirauda & Greco [45, 46] are also reported. The first set (squares), referring to a steel sphere in a free surface flow with a high value in the combined mass-damping parameter, is characterized by low oscillations and show only the initial branch without a jump in amplitude and, therefore, they do not exhibit hysteresis phenomena. The second series (triangles), characterized by values of $m^*\zeta$ lower than the previous ones, are close to the first mode of response. It outlines how the system tends to reach the resonance conditions where vortex-shedding frequency is equal to the natural frequency.

The results reported in fig. 17 can be better outlined by referring to figs. 19 and 20 which report the values of the $f^*$, ratio between the body oscillating frequency $f$ and the body natural frequency $f_n$, versus the reduced velocity $U^*$. In particular fig. 19 shows the data observed by Khalak & Williamson [41] for vibrating circular cylinders with mass ratio equal to 2.4, 10.3, and 20.6 and fig. 20 the data observed by Jauvtis *et al* [44] and Mirauda & Greco [45, 46] for vibrating spheres with a mass ratio equal to 80, 7.9 and 1.14, respectively.

In the figures, the horizontal line represents the condition in which the oscillating frequency $f$ is equal to the natural frquency and the diagonal line is the condition in which $f$ is equal to the vortex-shedding frequency for the static cylinder.

It has been observed that for low mass ratios, oscillation frequency starts from the natural frequency as the velocity $U^*$ increases and this transition is characterized by the presence of hysteresis.

On the contrary, in the case of high mass ratios the synchronization regime decreases and the value of $f^*$ remains close to the unity for all values of $U^*$.

This is true both for the two-dimensional structures (cylinders) as well as for three-dimensional structures (spheres).

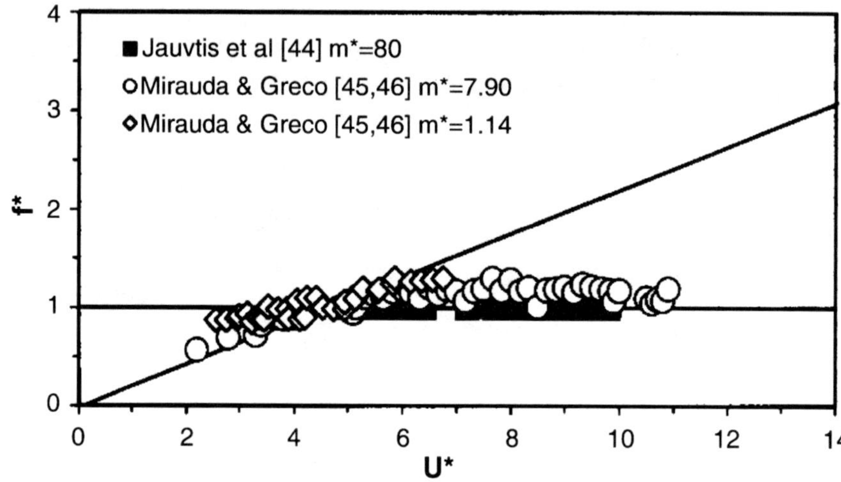


Figure 20: Frequency ratio versus reduced velocity for vibrating spheres.

In the case of free surface flow, the dynamic response is also conditioned by the parameter h/D. In fact, fig. 21 shows the experimental results of Miraudа & Greco [45, 46] for different values of relative submergence and for a limited range of reduced velocity ($U^* = 0.98 \div 8$).

In this range, it is possible to observe how the relative submergence influences the dynamic response of the system, the frequency ratio $f^*$ increases with h/D.

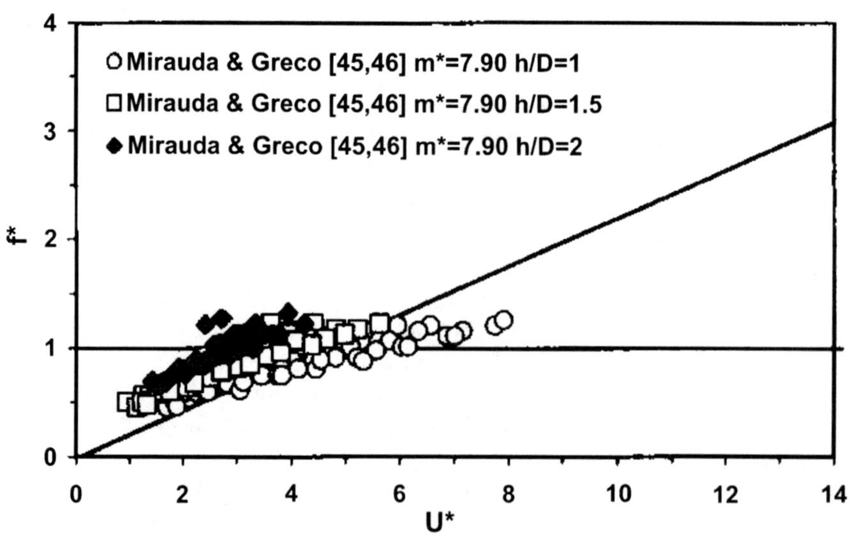


Figure 21: The influence of relative submergence h/D for vibrating spheres.

This behavior can be shown through the effect that the deformation of the free surface has on the oscillations of the sphere. In fact, for values of h/D = 1 the free surface deforms and the vortex layer is generated between the free surface and the upper obstacle surface. This layer gives rise to a near-wake conditioning the frequency body response. Vortex generation selects frequency ranges, which can include the "proper" obstacle frequency and can involve typical aspects related to the locking-in effects. Values of h/D > 1, on the other hand, pull the system away from the condition of lock-in and synchronization.

## 4 Conclusion

Flow induced excitations on structures represent a relevant and topic related to several modern theoretical and practical engineering problems. The aim of this chapter was to provide updated information about findings concerning two aspects related to the excitation on vibrating structures. In fact, the approach followed takes into account two main points of view on the processes: firstly, the flow field and the effect due to the turbulent features of the wake have been discussed as the source of the vibration on the structure. Secondly, the interaction between flow and structure has been proposed in terms of dynamic response of the obstacle. The process of flow induced excitations focuses on the framework of cause-effect referring to basic-shape structures like circular, and/or rectangular cylinders and spheres.

## 5 Acknowledgments

The authors gratefully acknowledge the financial support of the Italian Ministry of Scientific and Technology Research, for the project PRIN 2002 entitled "Influence of vorticity and turbulence in interactions of water bodies with their boundary elements and effect on hydraulic design".

Further, sincere thanks to the professors Naudascher & Rockwell for the use of some of their figures, plots and information as well as the anonymous reviewers for their kind and precious support during the writing of the chapter.

### List of symbols

$B$ = coefficient of mechanical damping $[MT^{-1}]$
$B_\theta$ = torsional damping coefficient $[ML^2T^{-1}]$
$C$ = spring constant coefficient $[MT^{-2}]$
$C'$ = added stiffness $[MT^{-2}]$
$C_A$ = potential added mass coefficient [/]
$C_D$ = drag coefficients [/]
$C_L$ = lift coefficient [/]
$C_\theta$ = torsional spring constant $[ML^2T^{-2}]$

| | | |
|---|---|---|
| $D$ | = | characteristic dimension of body $[L]$ |
| $f$ | = | oscillating frequency of the body $[T^{-1}]$ |
| $f^*$ | = | frequency ratio $(f/f_n)$ [/] |
| $f_n$ | = | natural frequency of body oscillator $[T^{-1}]$ |
| $f_0$ | = | frequency of vortex shedding $[T^{-1}]$ |
| $f_s$ | = | forcing frequency $[T^{-1}]$ |
| $F$ | = | fluid force on an obstacle $[MLT^{-2}]$ |
| $F_0$ | = | force amplitude $[MLT^{-2}]$ |
| $F_D$ | = | unit per length drag force $[MT^{-2}]$ |
| $Fr_s$ | = | Froude number referred to the cylinder thickness [/] |
| $G$ | = | elevation of the obstacle by the wall $[L]$ |
| $h$ | = | depth of water $[L]$ |
| $h^*$ | = | dimensionless parameter of obstacle submersion [/] |
| $h/D$ | = | relative submergence [/] |
| $I_\theta$ | = | mass moment of inertia $[ML^2]$ |
| $L$ | = | length of body along the flow direction $[L]$ |
| $m$ | = | mass of body $[M]$ |
| $m^*$ | = | mass ratio [/] |
| $m_a$ | = | added mass $[M]$ |
| $M$ | = | moment $[ML^2T^{-2}]$ |
| $M_0$ | = | moment amplitude $[ML^2T^{-2}]$ |
| Re | = | Reynolds number [/] |
| $S_G$ | = | Skop-Griffin parameter [/] |
| Sh | = | Strouhal number [/] |
| $t$ | = | time $[T]$ |
| Tu | = | free stream turbulent level [/] |
| $u'_{rms}$ | = | standard deviation of $U_0$ on inflow main direction $[LT^{-1}]$ |
| $U^*$ | = | normalized velocity [/] |
| $U_0$ | = | mean velocity of the incoming flow $[LT^{-1}]$ |
| $U_w$ | = | average velocity in the wake $[LT^{-1}]$ |
| $x, y$ | = | stream-wise, transverse displacement $[L]$ |
| $x_0, y_0$ | = | vibration amplitude in x,y direction $[L]$ |
| $y$ | = | transverse displacement $[L]$ |
| $y_0$ | = | vibration amplitude in y direction $[L]$ |
| $\delta$ | = | boundary layer thickness $[L]$ |
| $\gamma_b$ | = | blockage ratio [/] |

$\mu$ = fluid dynamic viscosity $[ML^{-1}T^{-1}]$
$\nu$ = fluid kinematic viscosity $[L^2T^{-1}]$
$\rho$ = fluid density $[ML^{-3}]$
$\zeta$ = damping ratio [/]
$\zeta_\theta$ = torsional damping ratio [/]
$\theta$ = angular or torsional displacement [/]
$\theta_\theta$ = amplitude of torsional vibration [/]
$\varphi$ = phase angle [/]
$\phi$ = logarithmic decrement of mechanical damping [/]
$\omega$ = circular frequency $[T^{-1}]$
$\omega_d$ = circular frequency of damped oscillator $[T^{-1}]$
$\omega_n$ = circular natural frequency $[T^{-1}]$
$\omega_s$ = circular forcing frequency $[T^{-1}]$
$\omega_{\theta n}$ = circular natural frequency of torsional vibration $[T^{-1}]$

## References


[1] Sarpkaya, T., Vortex-induced oscillations, *Journal of Applied Mechanics*, **46**, pp. 241–258, 1979.

[2] Ramberg, S.E. & Griffin, O.M., Hydroelastic response of marine cables and risers, In *Hydrodynamics in Ocean Engineering*, Norwegian Institute of Technology, Trondheim, Norway, pp. 1223–1245, 1981.

[3] Bearman, P.W., Vortex shedding from oscillating bluff bodies, *Annual Review of Fluid Mechanics*, **16**, pp. 195–222, 1984.

[4] Blevins, R.D., *Flow-induced vibrations,* New York: Van Nostrand Reinhold, 1990.

[5] Naudascher, E. & Rockwell, D., *Flow induced vibration: an engineering guide*, Rotterdam, Balkema, 1993.

[6] Bearman, P.W., Graham, J.M.R. & Obasaju, E.D., A study of forces, circulation and vortex patterns around a circular cylinder in oscillating flow, *Journal of Fluid Mechanics*, **196**, pp. 467–494, 1988.

[7] Keulegan, G.M. & Carpenter, L.H., Forces on cylinders and plates in an oscillanting fluid, *Journal of Research of the National Bureau of Standards*, **60(5)**, pp. 423–440, 1958.

[8] Blackburn, H.M. & Henderson, R.D., A study of two-dimensional flow past an oscillating cylinder, *Journal of Fluid Mechanics*, **385**, pp. 255–286, 1999.

[9] Lin, J.C., Vorobieff, P. & Rockwell, D., 3-dimensional patterns of streamwise vorticity in the turbulent near-wake of a cylinder, *Journal of Fluids and Structures*, **9**, pp. 231–234, 1995.

[10] Sheridan, J., Lin, J.C. & Rockwell, D., Flow past a cylinder close to a free surface, *Journal of Fluid Mechanics*, **300**, pp. 1–30, 1997.

[11] Naudascher, E., *AIRH Design Manual: Hydrodynamic forces*. Rotterdam: A.A. Balkema Publishers, 1991.

[12] Guilmineau, E. & Queutey, P., A numerical simulation of vortex shedding from an oscillating circular cylinder, *Journal of Fluids and Structures*, **16(6)**, pp. 773–794, 2002.

[13] Lam, K.M. & Dai, G.Q., Formation of vortex street and vortex pair from a circular cylinder oscillating in water, *Experimental Thermal and Fluid Science*, **26**, pp. 901–915, 1998.

[14] Yan, B., Oscillatory flow beneath a free surface, *Fluid Dynamic Research*, **22**, pp. 1–23, 1998.

[15] Munson, B.R., Young, D.F. & Okiishi, T.H., *Fundamentals of Fluid Mechanics* (third edition), John Wiley & Sons, Inc., 1998.

[16] Yu, D. & Kareem, A., Parametric study of flow around rectangular prisms using LES, *Journal Wind Engineering and Industrial Aerodynamics*, **78**, pp. 653–662, 1998.

[17] Naudasher, E. & Wang, Y., Flow-induced vibrations of prismatic bodies and grids of prisms, *Journal of Fluids and Structures*, **7**, pp. 341–373, 1993.

[18] Deniz, S. & Staubli, Th., Oscillating rectangular and octagonal profiles: interaction of leading- and trailing-edge vortex formation, *Journal of Fluids and Structures*, **11(1)**, pp. 3–31, 1997.

[19] Bearman, P.W. & Trueman, D.M., An investigation of the flow around rectangular cylinder, *The Aeronautical Quarterly*, **23**, pp. 229–237, 1972.

[20] Shimada, K. & Ishiara, T., Application of modified k-e model to the prediction of aerodynamic characteristics of rectangular cross section cylinders, *Journal of Fluids and Structures*, **16(4)**, pp. 465–485, 2002.

[21] Okajima, A., Strouhal numbers of rectangular cylinders, *Journal of Fluids Mechanics*, **123**, pp. 379–398, 1982.

[22] Haan, F.L., Kareem, A. & Szewczyk, A.A., The effects of turbulence on the pressure distribution around a rectangular prism, *Journal of Wind Engineering and Industrial Aerodynamics*, **78**, pp. 381–392, 1998.

[23] Lin, J.C. & Melbourne, W.H., Turbulence effects on surface pressure of rectangular cylinders, *Wind and Structures*, **2(4)**, pp. 253–266, 1999.

[24] Noda, H. & Nakayama, A., Free-stream turbulence effects on the instantaneous pressure and forces on cylinders of rectangular cross section, *Experiments in Fluids*, **34**, pp. 332–344, 2003.

[25] Saathoff, P.J. & Melbourne, W.H., Effects of free-stream turbulence on surface pressure fluctuations in a separation bubble, *Journal of Fluids Mechanics*, **337**, pp. 1–24, 1997.

[26] Nakamura, Y., Ohia, Y., Ozono, S. & Nakayama, R., Experimental and numerical analysis of vortex shedding from elongated rectangular cylinders at low Reynolds numbers 200–1000, *Journal of Wind Engineering and Industrial Aerodynamics*, **65**, pp. 301–308, 1996.

[27] Price, S.J., Sumner, D., Smith, J.G., Leong, K. & Paig Doussis, M.P., Flow visualization around a circular cylinder near to a plane wall, *Journal of Fluids and Structures*, **16(2)**, pp. 175–191, 2002.

[28] Angrilli, F., Bergamaschi, S. & Cossalter, V., Investigation of wall induced modifications to vortex shedding from a circular cylinder, *ASME Journal of Fluids Engineering*, **104**, pp. 518–522, 1982.

[29] Bearman, P.W. & Zdravkovich, M.M., Flow around a circular cylinder near a plane boundary, *Journal of Fluid Mechanics*, **89**, pp. 33–47, 1978.

[30] Buresti, G. & Lanciotti, A., Vortex shedding from smooth and roughened cylinders in cross-flow near a plane surface, *The Aeronautical Quarterly*, **30**, pp. 305–321, 1979.

[31] Taniguchi, S. & Miyakoshi, K., Fluctuating fluid forces acting on a circular cylinder and interference with a plane wall, *Experiments in Fluids*, **9**, pp. 197–204, 1990.

[32] Lei, C., Cheng, L. & Kavanagh, K., Re-examination of the effect of a plane boundary on force and vortex shedding of a circular cylinder, *Journal of Wind Engineering and Industrial Aerodynamics*, **80**, pp. 263–286, 1999.

[33] Cigada, A., Malavasi, S. & Vanali, M., Experimental studies on the boundary condition effects on the flow around a rectangular cylinder, *Fluid Structure Interaction 2003*, Cadiz, Spain, 24–26 June, 2003.

[34] Malavasi, S. & Guadagnini, A., Hydrodynamic loading on river bridges, *Journal Hydraulic Engeneering (ASCE)*, **129(11)**, November 2003, pp. 854–861, 2003.

[35] Denson, K.H., Steady-state drag, lift, and rolling-moment coefficients for inundated inland bridges, *Rep. No. MSHD-RD-82-077*, reproduced by National Technical Information Service, Springfield, Virg., pp. 1–23, 1982.

[36] Malavasi, S., Franzetti, S. & Blois, G., PIV Investigation of Flow Around Submerged River Bridge, *Proc. of River Flow 2004*, Napoli (Italy), June 23–25, 2004.

[37] Feng, C.C., The measurement of vortex-induced effects in flow past a stationary and oscillating circular and D-section cylinders, *Master's Thesis*, Univ. of British Columbia, Vancouver, B.C., Canada, 1968.

[38] Griffin, O.M., Vortex-excited cross-flow vibrations of a single cylindrical tube, *ASME Journal of Pressure Vessel Technology*, **102**, pp. 158–166, 1980.

[39] Blackburn, H., & Karniadakis, G.E., Two and Three dimensional simulations of vortex-induced vibration of a circular cylinder, In *3rd International Offshore and Polar Engineering Conference*, **3**, pp. 715–720, 1993.

[40] Skop, R.A. & Balasubramanian, S., A new twist on an old model for vortex-excited vibrations, *Journal of Fluids and Structures*, **11**, pp. 395–412, 1997.

[41] Khalak, A., & Williamson, C.H.K., Motion, forces and mode transitions in vortex-induced vibrations at low mass-damping, *Journal of Fluids and Structures*, **13**, pp. 813–851, 1999.

[42] Govardhan, R. & Williamson, C.H.K., Modes of vortex formation and frequency response of a freely vibrating cylinder, *Journal of Fluid Mechanics*, **420**, pp. 85–130, 2000.

[43] Sarpkaya, T., Hydrodynamics damping, flow-induced oscillations, and biharmonic response, *ASME Journal of Offshore Mechanics and Artic Engineering*, **117**, pp. 232–238, 1995.

[44] Jauvtis, N., Govardhan, R. & Williamson, C.H.K., Multiple modes of vortex-induced vibration of a sphere, *Journal of Fluids and Structures*, **15**, pp. 555–563, 2001.

[45] Mirauda, D. & Greco, M., Transverse vibrations of an sphere at high combined mass-damping parameter, *Shallow Flows* – Jirka and Uijittewaal (eds) Balkema Publisher, Taylor and Francis Group, London, ISBN 90 5809 700 5, pp. 111–116, 2004.

[46] Mirauda, D. & Greco, M., Flow-induced vibration of an elastically sphere at high combined mass-damping parameter, *Journal of IASME Transactions on Mechanical Engineering*, **1**, pp. 486–491, 2004.

# CHAPTER 7

# Air entrainment in vertical dropshafts with an orifice

P. Gualtieri & G. Pulci Doria
*Hydraulic and Environmental Engineering Dept.,*
*University of Naples, Naples, Italy.*

## Abstract

In the last decades, air entrainment by plunging liquid jets has been studied in international literature pointing out the various aspects of the phenomenon, in particular the involved variables and mechanisms.

An accurate experimental observation and a deep theoretical analysis of air entrainment in dropshafts allowed the authors to propose, also on the ground of the existing theories, a complete innovative model, that represents the phenomenon even in working conditions of the dropshaft not previously taken into consideration.

This model accepts, as input values, two characteristic experimental lengths of the plant (suitably made nondimensional ones) and returns, as output value, the ratio $\beta$ between entrained air and falling water volumetric flow-rates. This model holds four experimentally obtainable parameters.

The model has been tested and its parameters have been estimated by least squares method through a wide series of experimental tests (159) performed on a physical model on a big scale, organised in order to eliminate secondary effects due to viscosity and surface tension of the liquid, and also to the generally existing jet turbulence.

## 1 Introduction, state of art, goal of the paper

The phenomenon of air entrainment by a liquid jet plunging into a pool of the same liquid has been studied for a long time, either experimentally and theoretically. In fact, it belongs to a group of problems concerning the mechanical interactions between two-phases flows, that are frequent in industrial as well as in environmental situations.

In spite of its seeming simplicity, the understanding of this process is not complete at all. From studies of many authors [1–13], it clearly appears that air entrainment by a plunging jet is a very complex process, as many factors affect it. The primary variables are the jet diameter, the jet velocity and length, the jet turbulence generated by the nozzle, the nozzle geometry and the physical properties of the involved fluids. Moreover, different mechanisms of air entrainment complicate a quantitative prediction of the entrained air flow-rate.

Referring to the entrained air flow-rate measurements, since air entrainment by a plunging jet occurs as a localized phenomenon at the plunging point, basically two groups of different methods have been developed [1]: (i) catching air after it has been entrained into the liquid pool; (ii) measuring the removal of air after it has been entrained into the plunging point. In the second group, the gaseous space above the pool in the vicinity of the plunging point is separated from the ambient, and a supplementary air is let into this space through an appropriate flow-rate device (orifice, anemometer, volume and time readings).

By means of a careful analysis of the state of art, it has been realized that a large part of the experimental studies was carried out through plants in which the jet was produced through nozzles with a very small diameter (less than 25mm). In these conditions, the entrained air flow-rate depended also on the viscosity, on the surface tension of the liquid, and on the turbulence generated by the nozzle, as well as on the jet velocity and length. Therefore, the results of these experimental studies, especially with regard to the amount of the entrained air flow-rate, cannot easily be extrapolated from a working system. In relation to this last circumstance, recent detailed experiments, performed by Chanson *et al* [14] on small scale laboratory plants of plunging jets, show that scale effects can explain the discrepancies by factors of three, or more, different plants, shown by Bin [1] in his detailed review.

A clear example of the air entrainment problem, that cannot be studied through small scale plants, concerns the air entrainment by plunging jets falling down within a dropshaft. In fact, the only experimental studies about the air entrainment in dropshafts had been performed on big scale plants. They had been described in [15–17] that refer to dropshafts with vortex inlet. Figure 1, in particular, drawn by Viparelli [17] with simplifications, reports the experimental results of M. Viparelli, and Laushey & Mavis, concerning the case of some vortex dropshafts. The meaning of this sketch will be better explained in the subsequent paragraph. It is obvious that, in these plants, the second group of entrained air flow-rate measurers appears to be suitable.

As far as the air-entraining mechanisms in a dropshaft are concerned, in a previous work of one of the authors [18], a theoretical frame of five possible air-entraining mechanisms (drawn by previous literature) was presented. Some are common to free plunging jets, whereas others are peculiar to plunging jets in dropshafts.

The first mechanism (fig. 2) depends on the fact that a plunging jet directly entrains air along its surface: this mechanism acts only when the bottom of the dropshaft is directly open to the atmosphere.

The second mechanism (fig. 3) depends on the fact that the plunging jet could entrap some air bubbles within itself, due to different causes as, for instance, some

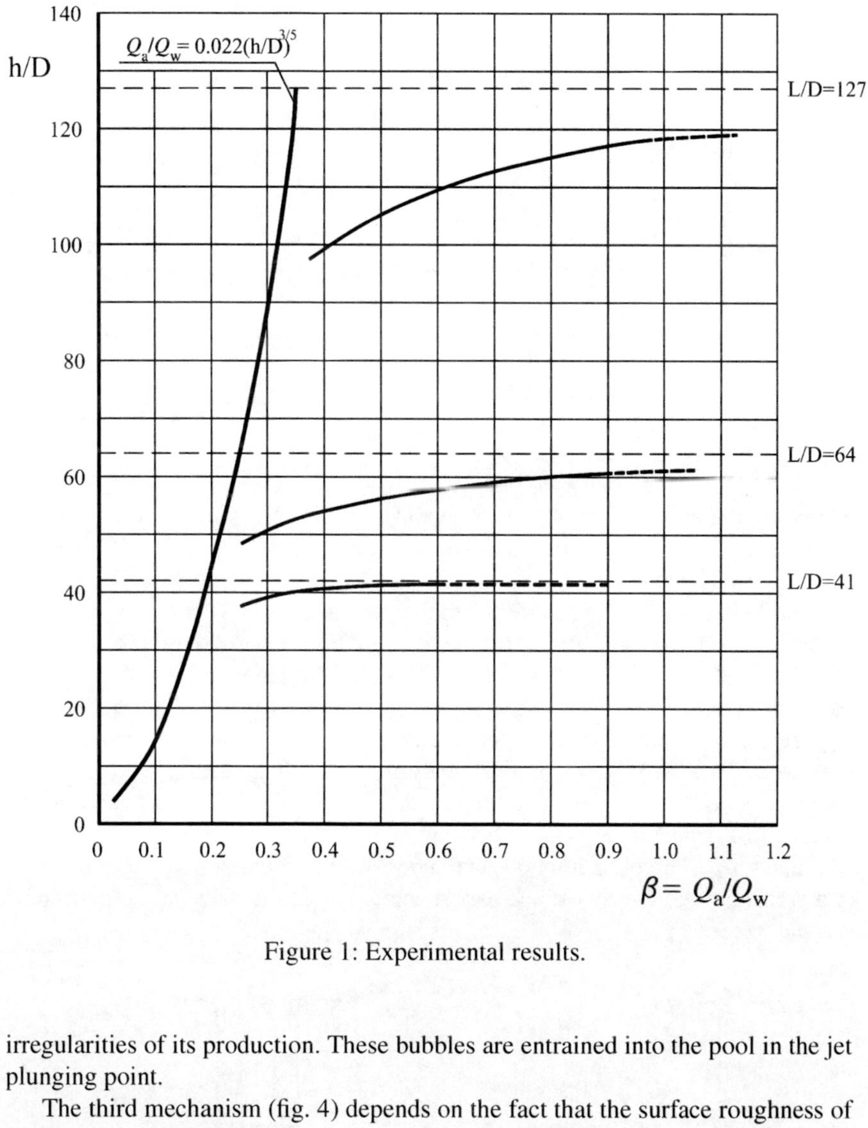


Figure 1: Experimental results.

irregularities of its production. These bubbles are entrained into the pool in the jet plunging point.

The third mechanism (fig. 4) depends on the fact that the surface roughness of the plunging jet could entrap air pockets, because of the presence of turbulence and inertial forces in the jet itself. Also these pockets are entrained into the pool in the jet plunging point.

The fourth mechanism (fig. 5) acts directly where the jet impinges on the water: a thin torus shaped air-cushion is born between the jet and the surrounding water and the air entrapped is pushed by the plunging jet velocity to enter the water.

Finally, the fifth mechanism (fig. 6) is a very particular one. The impingement of the plunging jet on the water column, partially filling the dropshaft, generates a foamy or bubbly zone, just like a hydraulic jump, where the air exchanges between atmosphere and water column arise. They can be either from air to water (so

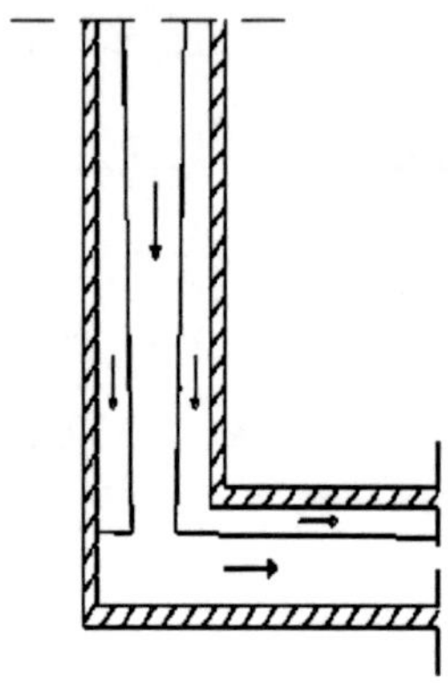

Figure 2: First mechanism.

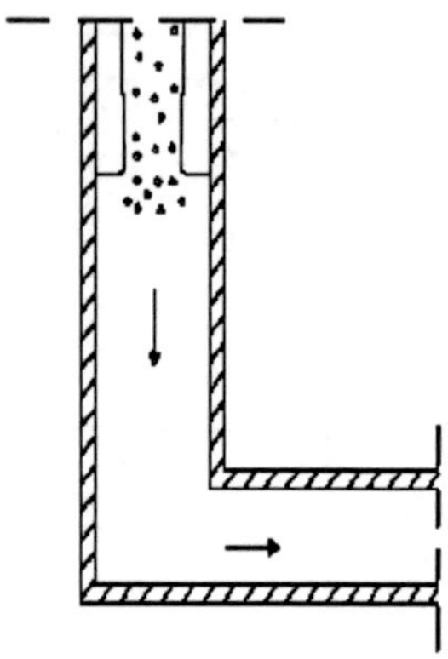

Figure 3: Second mechanism.

increasing the air entrainment), or from water to air (so lowering the air entrainment). Therefore this fifth air-entraining mechanism can be considered either as a positive or a negative one.

That being stated, the authors of the paper have studied for a long time the air entrainment phenomenon of a dropshaft with an orifice, both theoretically or experimentally on a big scale plant. The scale of the plant and the absence of a nozzle at its inlet made it possible to neglect the liquid viscosity, the surface tension and the turbulence effects, in order to extrapolate the experimental results to a prototype or simplify the theoretical analysis of the phenomenon. The plant had a fixed geometry, but it was investigated in all its possible working conditions to obtain a full sketch of its physical behaviors.

Therefore, a complete fully original model of the air entrainment phenomenon in a dropshaft was conceived and experimental laboratory data were produced to test and calibrate it.

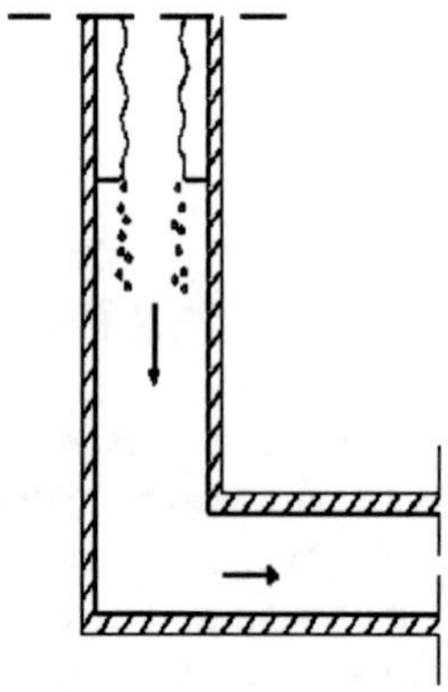

Figure 4: Third mechanism.

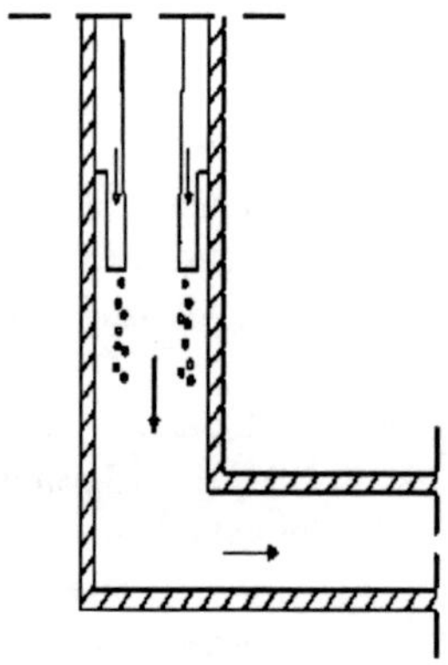

Figure 5: Fourth mechanism.

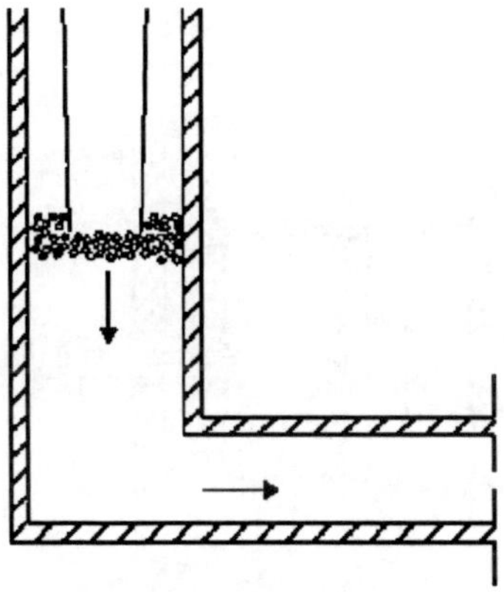

Figure 6: Fifth mechanism.

## 2 Experimental plant and paper development description

Figure 7 shows the experimental plant employed in Naples. The water flows from a large suitably fed tank into the dropshaft through a sharp-edged orifice, whose diameter ($D_o$) is 100mm large. The plunging jet is a central one. The water surface level (h) in the tank is measured through a lateral piezometer and the water flow-rate arriving to the tank is measured through a calibrated pressure orifice. The dropshaft is made of plexiglas, its diameter ($D_s$) is 200mm large, and its geometrical length is 6.25m. It is joined to an horizontal pipe through a 90° curve in which the center of the 45° cross section lies 0.25cm under the entering mouth, so that the equivalent dropshaft length (L) can be considered equal to 6.50m. The horizontal pipe, through a final gate valve, discharges in a recipient tank where a weir, higher than the axis of the horizontal pipe, keeps the same pipe under pressure. In steady conditions, the dropshaft is full of water up to a certain height, which can be changed by suitable regulation of the final gate valve.

In the steady working conditions, the water jet coming out from the orifice impinges on the water column in the dropshaft and generates a visible and measurable (through a graduated bar) foamy or bubbly zone, where the air-entraining mechanisms arise. The entrained air is continuously replaced by new air through a first air bleeder, laterally connected to the dropshaft. Moreover, this air bleeder allows also the entrained air flow-rate measurement. Finally, in the horizontal pipe, just after the curve, a second air bleeder has been inserted, to let most of the air, dragged by the jet, slip away.

In literature, three different working conditions for a dropshaft are described [18] (fig. 8).

They differ from one another by first of all referring to the way in which the water jet falls down in the dropshaft and impinges, or does not impinge, the water column in it. They finally depend either on the value of the water flow-rate or the opening degree of the final gate valve.

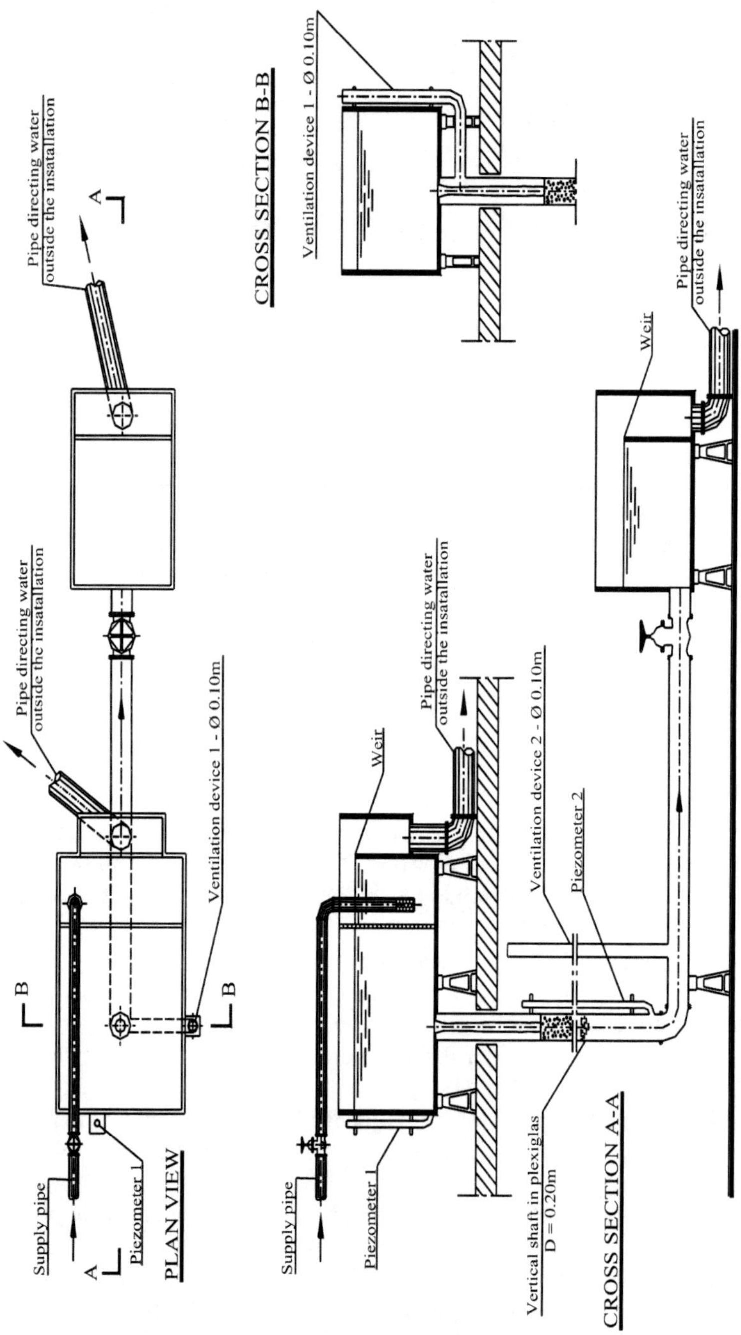


Figure 7: Experimental plant employed in Naples.

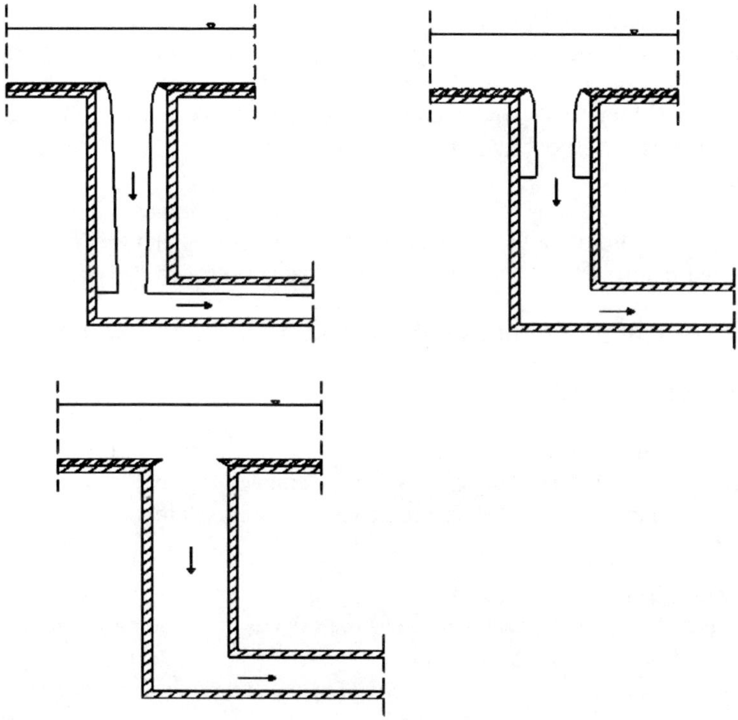

Figure 8: Working conditions in a dropshaft.

In particular:

- the dropshaft is said to work in Region I if the valve is fully opened, the water jet falls down along the whole length of the dropshaft with a free surface so that the air dragged can arrive directly to the second air bleeder;
- the dropshaft is said to work in Region II if the valve is about half opened, the water jet falls down along the first part of the whole length of the dropshaft with a free surface, and afterwards, at a certain distance H from the orifice, it begins to fall down in pressure;
- the dropshaft is said to work in Region III if the valve is only partially opened and the water jet falls down along the whole length of the dropshaft in pressure.

As in Region III there is no air entrainment, this phenomenon develops in Region I, Region II, and in Transition between Region I and Region II.

Figure 1 refers to experimental plants of vortex dropshafts characterized by different lengths L and different diameters $D_s$ (in the original figure $D_s$ is called simply $D$). The experimental points show working conditions relative to different free fall heights H (in the original figure called h) and different water flow-rates $Q_w$ (in the original figure called simply $Q$). The diagram gives the values of air flow-rates $Q_a$ in different working conditions. In particular, the diagram represents, in the ordinate, the ratios between free fall heights and diameter of the dropshaft and, in the abscissa, the ratios between entrained and water flow-rates. The sub-vertical part of the diagram corresponds to the Region II, whereas the sub-horizontal part of the diagrams corresponds to a Transition between Region II and Region I, where air entrainment phenomenon is higher.

In this paper new theoretical models of the air entrainment phenomenon in Region I, in Region II and in Transition between Region II and Region I are presented and compared to experimental data obtained through the plant in fig. 2.

## 3 Theoretical model of the air entrainment phenomenon

### 3.1 Theoretical model in Region I

The first part of the air entrainment model will deal with Region I. A first idea of this part of the model is linked to M. Viparelli's model [16], but his idea has been deeply revised through Fluid Mechanics considerations. In order to state a suitable model the starting hypotheses are the following ones:

- the dropshaft can be considered as a vertical pipe;
- the Region I can be schematized as the working condition of a vertical pipe completely open on the bottom;
- the water jet falling within the vertical pipe can be considered as a taper full cylinder, whose length corresponds to the length of the vertical pipe, dragging the neighbouring air through the first air bleeder.

This model was presented for the first time in [19].

In M. Viparelli's analysis, related to a vortex dropshaft, the dragged air is internal to the jet and attains the jet velocity itself, if the dropshaft is sufficiently long. In central jet conditions, on the contrary, the dragged air is external to the jet and its velocity varies from zero at the dropshaft wall up to the jet velocity itself at the jet surface.

In order to obtain the air mean velocity value, firstly an equilibrium has been supposed to be present, at any cross section of the dropshaft, between the pipe unit wall resistance to the air flow $\tau_s$ (upward) and the unit air-entraining force of the jet $\tau_j$ (downward). This hypothesis leads to the following relation:

$$\tau_s D_s = \tau_j D_j \tag{1}$$

Yet, introducing the friction factor and following the dimensional analysis, it is assumed that:

$$\tau_s = \left(\frac{\lambda_s}{8}\right) \rho_a V_a^2 \tag{2}$$

$$\tau_j = \left(\frac{\lambda_j}{8}\right) \rho_a \left(V_j - V_a\right)^2 \tag{3}$$

Finally, a modified version of Blasius law is thought to hold in this situation:

$$\lambda_s = Const \left[ \rho_a \left( D_s - D_j \right) \frac{V_a}{\mu_a} \right]^{-1/4} \tag{4}$$

$$\lambda_s = Const \left[ \rho_a \left( D_s - D_j \right) \left( V_j - V_a \right) \frac{1}{\mu_a} \right]^{-1/4} \tag{5}$$

On the basis of previous equations, the ratio between air velocity and jet velocity can be expressed as:

$$\frac{V_a}{V_j} = \frac{1}{\left[ 1 + \left( \frac{D_s^2}{D_j^2} \right)^{2/7} \right]} \tag{6}$$

In order to deduce eqn. (6), it is first of all necessary to obtain expressions of $V_a$ and $(V_j - V_a)$ from eqns. (2) and (3), respectively, and express their ratio; in the so obtained formula, the new ratios $\tau_s/\tau_j$ and $\lambda_j/\lambda_s$ appear: the first one can be replaced by $D_j/D_s$ employing eqn. (1) and the second one can be replaced by a further ratio obtainable by eqns. (4) and (5). The final expression after these computations can be considered as an equation whose unknown is the ratio $(V_j - V_a)/V_a$, which can be obtained. Once the ratio $(V_j - V_a)/V_a, = V_j/V_a - 1$ has been got, also the ratios $V_j/V_a$ or $V_a/V_j$ can be easily and finally obtained.

This formula is the true core of the model. The value of the ratio $V_a/V_j$ changes along the jet because of the continuity equation. The temporal mean value of $(D_s^2/D_j^2) = D^*$, calculated on the basis of a water particle falling down from the vena contracta to the bottom opening of the dropshaft, can be evaluated as:

$$D^*_{mean} = \left( \frac{D_s^2}{C D_o^2} \right) \left[ 1 + \left( \frac{1}{2} \right) \sqrt{\frac{L_{eff}}{h_{eqtot}} + 1} - 1 \right] \tag{7}$$

In this equation, $C$ is the contraction coefficient, $L_{eff} = (L - D_o/2)$ and $h_{eqtot} = (h + D_o/2)$ with $h$ the water head into the feeding tank. Details for obtaining this temporal mean value are reported in [19]. Therefore, the eqn. (6) can be rearranged in the following way:

$$\left( \frac{V_a}{V_j} \right)_{mean} \cong \frac{1}{1 + (D*_{mean})^{2/7}} \tag{8}$$

Until now, it has been supposed that the air enters the dropshaft through an orifice made in the wall of the dropshaft itself. The air enters the dropshaft through the first air bleeder, so that it is necessary to take into account the distributed and localized head losses in it too. In fact, these losses affect the air velocity; in particular, the ratio between the air velocity value $V_a^*$ that takes into account the head losses, and the previous air velocity value (without taking into account the head losses) $V_a$, can be expressed in the following way:

$$\frac{V_a^*}{V_a} = \frac{1 + (D*_{mean})^{2/7}}{1 + (D*_{mean})^{2/7} \left[1 + M\left(1 - \frac{1}{D*_{mean}}\right)^{7/4}\right]^{4/7}} \tag{9}$$

In this expression, the term $M$ means $M = (L_{lat}/L)(D_s^2/D_{lat}^2)$, where $L_{lat}$ is the whole virtual length of the first air bleeder, already suitably increased in order to take into account also the localized head losses and $D_{lat}$ is its diameter. It is obvious that the specific influence of the first air bleeder is included in the term in curly brackets. It is also clear that if $L_{lat}$ is zero, then the aforementioned ratio will be equal to unity. Details for obtaining this formula are not only given in [19] but also in [23]. In particular, in the experimental plant in fig. 7, the term M can be considered equal to 12.5.

After all, if the two ratios $V_a/V_j$ and $V_a^*/V_a$ are multiplied, the first one by the second one, to obtain the final ratio $V_a^*/V_j$, and if the air flow cross section and the water flow cross section are taken into account (always in their mean values) to transform a velocities ratio in a flow-rates ratio, the final expression of the air entrainment coefficient $\beta$ (which is defined as the nondimensional ratio $Q_a/Q_w$) in Region I will become the following one:

$$\beta = \frac{D^*_{mean} - 1}{1 + (D^*_{mean})^{2/7} \left[1 + M\left(1 - \frac{1}{(D^*)_{mean}}\right)^{7/4}\right]^{4/7}} \tag{10}$$

This expression represents the air entrainment model in Region I.

In the experimental plant, as all the plant parameters are kept constant, it is possible to state that the $\beta$ value depends only on the $D^*$ value, and therefore, due to eqn. (7), on the $h_{eqtot}$ value. Moreover, it is possible to express this dependence in a much easier way, as a function of the nondimensional ratio $h_{eqtot}/D_s$, by approximating the true law by a second order law, valid within the possible values of $h_{eqtot}/D_s$ (from 1.5 to 4.5). This simpler expression of $\beta$ is the following one:

$$\beta = \beta_0 + \beta_1\left(\frac{h_{eqtot}}{D_s}\right) + \beta_2\left(\frac{h_{eqtot}}{D_s}\right)^2 \tag{11}$$

The exact values to be given to $\beta_0$, $\beta_1$ and $\beta_2$ can be practically obtained on the basis of the true dimensions of an experimental plant, as it is shown in Section 4.1, where their actual values, relative to our specific plant, will be computed too.

## 3.2 Theoretical model in Region II

### 3.2.1 Choice of entraining-air involved mechanisms

In the "Introduction, state of art, goal of the paper" section, the five possible air-entraining mechanisms were described.

The experimental plant in fig. 7 is said to work in Region II if the water jet falls down along the first part of the whole dropshaft length with a free surface, and afterwards, at a certain height, it begins to fall down in pressure.

In this plant, in Region II, the first mechanism cannot develop because the dropshaft is not open to the atmosphere in the bottom, but the existence of a sub-mechanism is possible. The second mechanism does not appear as in a jet from an orifice bubble formation is avoided. The third mechanism does not appear because in a jet from an orifice turbulence is almost totally avoided. The fourth mechanism, on the contrary, can be strongly present. And, finally, also the fifth mechanism is clearly present, with a foaming zone many decimeters high.

Consequently, in the here presented theoretical model, it is assumed that the main air-entraining mechanism is the fourth one, but the arising air-cushion is generated within a very foaming zone. Moreover, the first mechanism is thought to act in any case, giving to the air, nearer to the jet, a vertical downward momentum, which causes a rise of pressure at the inlet of the air-cushion.

It is also important to make a distinction between entrapped air and entrained air. The entrapped air is put into the water by the action of the fourth mechanism, aided by the first one, and it feeds the rise of the foaming zone. The fifth mechanism acts as a negative one, letting a very high percentage of the entrapped air come back, so that the value of the truly entrained air is the small percentage of the entrapped air that cannot return back upward. This model was presented for the first time very recently in [20].

### 3.2.2 The momentum equation into the air-cushion

In the model, the jet comes down vertically from a circular orifice of diameter $D_o$. After the vena contracta (of diameter $D_c$, and placed $D_o/2$ downward with respect to the orifice), the jet falls down with a cross section always smaller due to the increasing falling velocity. When the jet plunges into the water, at a distance H from the orifice, its diameter becomes $D_j$, and its velocity $V_j$ depends on falling length H and tank head h, following the expression $V_j = [2g(H+h)]^{1/2}$. A torus shaped air volume is born around the jet in the surrounding foaming water. The cross section of this torus is supposed to be rectangular, wide (d) and high ($\zeta$). Within this torus, there is a downward flux of air coming from the upper region and going downward into the water, generating the entrapped air; in the torus, air velocities are constant along vertical directions and their values linearly vary from $V_i$, near the jet, to 0 near the foaming water. The momentum equation, applied to the air moving into the tore, as the motion has the aforementioned characteristics, coincides with the equilibrium equation. Therefore, the sum of the four forces applied to the four boundaries of the tore is zero:

$$(\tau' - \tau)\,\sigma_1 + (p_b - p_i)\,\sigma_b = 0 \tag{12}$$

$\tau_i$ is the unit entraining force of the jet at the impingement, which coincides with the viscous Newtonian stress on the flowing jet surface = $\mu_a\, V_j/d$. $d$ is the thickness of the torus.

$\tau'$ is the unit upward resistance to air flow of the foaming water, which coincides with the Newtonian stress on the foaming water surface. It is assumed that this Newtonian stress is reduced in the ratio $\rho_{rel}$ (called relative density, which is the ratio between density of foamy water and density of water) because the contact surface with water is diminished due to foam, so that $\tau' = \rho_{rel}\mu_a V_j/d$.

$p_b$ is the pressure on the bottom of the torus. This pressure is evaluated through Stevin's law applied to the foam on the basis of a density $\rho_{rel}\ \rho_w$, so that $p_b = \rho_{rel}\ \rho_w\ g\zeta$.

$p_i$ is the mean pressure on the top of the torus. This pressure is due to the air impact velocity. This air changes its velocity, as already said, linearly from $V_j$ to 0 just within the torus, so that the mean impact pressure is $p_i = \rho_a\ V_j^2/3$.

$\sigma_l$ is the lateral surface of the torus = $\pi D_j\zeta$; $\sigma_b$ is the bottom surface of the torus = $\pi D_j d$; $\mu_a$ is the viscosity of air; $\rho_a$ is the density of air; $\rho_w$ is the density of water.

With the aforementioned positions, eqn. (12) gives the following expression of $d$:

$$d = \sqrt{\left[\frac{\mu_a V_j\,(1-\rho_{rel})}{\rho_{rel}\rho_w g\varsigma - \frac{\rho_a V_j^2}{3}}\right]} \tag{13}$$

### 3.2.3 Modelling relative density of foamy water and computing entrapped air flow-rate $Q_{entr}$

The relative density $\rho_{rel}$ is modelled through the assumption that in the foamy zone, whose height $\xi$ is some decimeters, its value varies linearly from 0 at the top of the foamy zone to 1 at the bottom of the same zone. With this assumption, the mean relative density along the lateral surface of the torus $\rho_{rel}$, necessary to compute the pressure $p_b$, can be considered to be equal to the ratio $\zeta/2\xi$. Therefore, eqn. (13), taking into account also that this ratio is very small in comparison with unity, becomes:

$$d = \sqrt{\left[\frac{(\mu_a V_j\zeta)}{\left(\frac{\rho_w g\zeta^2}{2\xi} - \frac{\rho_a V_j^2}{3}\right)}\right]} \tag{14}$$

Once $d$ has been modelled by eqn. (14), the entrapped air flow-rate $Q_{entr}$ can be expressed as:

$$Q_{entr} = \sigma_b\left(\frac{V_j}{2}\right) = (\pi D_j d)\left(\frac{V_j}{2}\right) = \left(\frac{\pi D_j V_j}{2}\right)\sqrt{\left[\frac{(\mu_a V_j\zeta)}{\left(\frac{\rho_w g\zeta^2}{2\xi} - \frac{\rho_a V_j^2}{3}\right)}\right]} \tag{15}$$

### 3.2.4 Modelling the foamy zone height $\xi$

The foamy zone height $\xi$ is modelled through the assumption that it can be considered proportional to the entrapped air flow-rate $Q_{entr}$ through a coefficient $K_1$, whose dimensions are [$L^{-2}T$]. This is a consequence of the fact that the jet is falling within a dropshaft of a fixed diameter: therefore, the larger $Q_{entr}$ is, the larger the air bubble presence, and consequently so is the volume they occupy. And, as the diameter of the dropshaft is fixed, this larger volume can expand only in height. It is clear that, if the diameter of the dropshaft changes, the value of the constant $K_1$ will change; but, in this paper, the dependence of the air-entraining laws on the dropshaft diameter will not be investigated, so that the $K_1$ parameter will always be considered as a constant. With the previous assumptions, eqn. (15) can be further expressed and rearranged as follows:

$$Q_{entr}^2 - \left(\frac{3\rho_w g\zeta^2}{2K_1\rho_a V_j^2}\right) Q_{entr} + \left(\frac{3\pi^2\mu_a D_g^2 V_j\zeta}{4\rho_a}\right) = 0 \qquad (16)$$

Equation (16) appears as a second degree algebraic equation whose unknown is just $Q_{entr}$. This equation can be solved in a standard way; but, in fact, in eqn. (16), it results that the third term is very small in comparison with the two previous ones. With this assumption and discarding solution $Q_{entr} = 0$, it results:

$$Q_{entr} = \frac{3\rho_w g\zeta^2}{2K_1\rho_a V_j^2} \qquad (17)$$

This is a very simple and explicit expression, where the torus height $\zeta$ yet appears as an unknown: this height will be afterwards modelled.

### 3.2.5 Relation between entrapped air and entrained air flow-rates and consequent $\beta$ expression

After having obtained the entrapped air flow-rate $Q_{entr}$ through eqn. (17), now it is necessary to obtain the entrained air flow-rate, called $Q_E$. As already stated, due to the fifth mechanism, $Q_E$ is a fraction of $Q_{entr}$. The main assumption is that the $Q_E$ is proportional to $Q_{entr}$ through a not constant proportionality parameter, but proportional, in its turn, to the water flow-rate $Q_w$.

This assumption is in part connected with the matter that the jet is plunging in a fixed diameter dropshaft. The entrapped bubbles are entrained by the downward velocity of the jet in the dropshaft: the small bubbles are entrained more easily, and the big ones with more difficulty, due to the different re-ascending velocity. Therefore, it is clear that the greater the entrained air flow-rate is, the greater the vertical downward velocity is, and, consequently, due to the fixed value of the dropshaft diameter, the greater the water flow-rate is. Furthermore, if the water flow-rate is zero, the entrapped bubbles will all re-ascend toward the air overlying the water column and the entrained air flow-rate will be zero too. On the contrary, if the water flow rate increases (towards infinity), the entrapped air bubbles will be all entrained downward and the proportionality parameter will be the unity. As it appears from experimental data, the percent of entrapped air flow-rate which is entrained is always a small one, and it is therefore possible to state, as first approximation level, a linear proportionality, so that it definitively results:

$$Q_E = (K_2 Q_w)\, Q_{entr} \tag{18}$$

with $K_2$ suitable parameter with dimensions [L$^{-3}$T] and generally depending, as also $K_1$, from the diameter of the dropshaft. Also $K_2$ parameter will be here considered as a constant. With these assumptions eqn. (17) can easily give, for the air entrainment coefficient $\beta$ in Region II, the following expression:

$$\beta = \frac{Q_E}{Q_w} = K_2 \left( \frac{3\rho_w g \zeta^2}{2K_1 \rho_a V_j^2} \right) \tag{19}$$

### 3.2.6 Modelling the air tore height $\zeta$ and consequent $\beta$ expression

In order to model the air tore height $\zeta$, it is necessary to find out on which parameters it depends. It is assumed that they are the following ones.

The first one is the impact jet velocity $V_j$, as a greater velocity clearly gives rise, through its strength, to a greater torus height. The second one is the water flow-rate. In fact, the increasing of water flow-rate causes, as it has been experimentally shown, the increasing of the entrapped air and, consequently, of the bubble presence. The water flow-rate will indeed be here substituted by the velocity in the vena contracta $V_c$. Generally speaking, the parameters of this model $\zeta$ depend on both diameters of orifice and dropshaft, but obviously they will be here considered constant too, as previous ones.

The aforementioned dependence of $\zeta$ on $V_j$ and $V_c$ can be expressed through a power law:

$$\xi = K_3 V_g^{\alpha'} V_c^{\gamma'} \tag{20}$$

where the dimensions of $K_3$ depend on $\alpha'$ and $\gamma'$ values.

Due to eqn. (20), eqn. (19) becomes:

$$\beta = K_2 \frac{\left[ 3\rho_w g \left( K_3 V_g^{\alpha'} V_c^{\gamma'} \right)^2 \right]}{\left( 2K_1 \rho_a V_g^2 \right)} \tag{21}$$

### 3.2.7 Final transformations of the model

The last transformations are conceived in order to obtain for $\beta$ an expression simpler than eqn. (21), depending on clearly nondimensional parameters measurable during an experiment. To obtain it, the two characteristic velocities $V_j$ and $V_c$ are transformed through the heights:

H = the jet falling height from the orifice to the impact point;
h = the water head in the tank over the orifice.

The previously considered velocities can be rewritten as:

$$V_j = \sqrt{2g\,(H + h))} = \sqrt{2gH_{tot}} \tag{22}$$

$$V_c = \sqrt{2g\left(h + \frac{D_0}{2}\right)} = \sqrt{2gh_{eqtot}} \tag{23}$$

where the meaning of $h_{eqtot}$ is well known and, of course:

$$H_{tot} = H + h_{eqtot} - \frac{D_o}{2} \tag{24}$$

$H_{tot}$ can be called the whole fall length.

Equation (21) therefore can be rewritten inserting the previous assumptions and rearranging:

$$\beta = K' H_{tot}^{\alpha} h_{eqtot}^{\gamma} \tag{25}$$

where $K'$ contains $\rho_a$, $\rho_w$, $g$, and $K_1$, $K_2$, $K_3$; while $\alpha$ and $\gamma$ depend on both $\alpha'$ and $\gamma'$. Obviously, as $K'$ contains $K_1$, $K_2$, $K_3$, it depends on the diameters of orifice and dropshaft: but, if they are fixed, also $K'$ must be considered as a true constant.

Finally, multiplying and dividing in eqn. (25) by $D_s^{\alpha+\gamma}$, and yet rearranging:

$$\beta = K \left(\frac{H_{tot}}{D_s}\right)^{\alpha} \left(\frac{h_{eqtot}}{D_s}\right)^{\gamma} \tag{26}$$

This expression depends on nondimensional parameters that are very simple to be measured during an experiment; in particular, the second one was already used in eqn. (11). Also $K$, $\alpha$ and $\gamma$ are nondimensional parameters that are constant in a plant with fixed characteristics (*i.e.* diameters of the dropshaft and of the orifice). Their values will be established on the basis of suitable experimental surveys.

### 3.3 Theoretical model in Transition between Region I and Region II

At this point, two nondimensional expressions of coefficient $\beta$, namely expression (11) which is valid in Region I, and expression (26) which is valid in Region II, have been stated. Here, a new expression of coefficient $\beta$ valid in Transition between Region II and Region I will be defined.

First of all, the reason why there is a Transition Region will be explained.

In fact, the experimental plant always presents, all through its working time, an oscillatory behavior of the water column. Moreover, as it has been already shown, this water column has a foamy and bubbly top zone. Therefore, when this foamy zone, due to oscillations, arrives up to the downstream curve of the shaft, a less resistant way for the air to attain the second air bleeder springs up.

If the fall length of the jet is $H$ in the mean, the instantaneous fall length oscillates between two values, $H_{min} = H - \Delta H$ and $H_{max} = H + \Delta H$. When $H_{max} > L$, then a direct exit of the air from the dropshaft to the second air bleeder springs up so that Region I conditions arise during short time periods. When $H_{max} < L$, the horizontal pipe and the lower part of the dropshaft are always under pressure so that Region II conditions are always present. The $H$ limit value of Region II is called $H_{\text{II}}$ and its value is $H_{\text{II}} = L - \Delta H$. In this condition, the whole fall length $H_{tot}$ is worth:

$$H_{totII} = H_{II} + \left(h_{eqtot} - \frac{D_o}{2}\right) = L - \Delta H + \left(h_{eqtot} - \frac{D_o}{2}\right) \tag{27}$$

Experimental results show that the $H_{totII}$ value is substantially constant and can be obtained just from experiments. The eqn. (27) shows it is possible to evaluate $\Delta H$ as:

$$\Delta H = (L - H_{totII}) + \left(h_{eqtot} - \frac{D_o}{2}\right) = \Delta H_0 + \left(h_{eqtot} - \frac{D_o}{2}\right) \tag{28}$$

where $\Delta H_0$ has the constant value obtainable just as:

$$\Delta H_0 = L - H_{totII} \tag{29}$$

and consequently also $H_{II}$ as:

$$H_{II} = L - \left[\Delta H_0 + \left(h_{eqtot} - \frac{D_o}{2}\right)\right] \tag{30}$$

and $H_{totII}$ as:

$$H_{totII} = L - \Delta H_0 \tag{31}$$

and, in non-dimensional form:

$$\frac{H_{totII}}{D_s} = \frac{(L - \Delta H_0)}{D_s} \tag{32}$$

The correspondent $\beta$ expression will be called henceforth $\beta_{II}$:

$$\beta_{II} = K\left(\frac{H_{totII}}{D_s}\right)^{\alpha}\left(\frac{h_{tot}}{D_s}\right)^{\gamma} \tag{33}$$

Now, if $H$ grows beyond the $H_{II}$ value, supposing that an oscillation behavior between $H - \Delta H$ and (virtually) $H + \Delta H$ always exists with the already calculated $\Delta H$ value, it is evident that the higher $H$ is, the longer the time during which the fall length is greater than L and the plant works in Region I. In particular, following previous assumptions, if $H - \Delta H$ is greater than L, the plant will work perpetually in Region I. Therefore, a second limit virtual value $H_I$ exists, over which the plant works completely in Region I. It is given by:

$$H_I - \Delta H = L \tag{34}$$

Namely:

$$H_I = L + \Delta H = L + \Delta H_0 + \left(h_{eqtot} - \frac{D_o}{2}\right) \tag{35}$$

Also the correspondent $H_{totI}$ will be worthy:

$$H_{totI} = H_I + \left(h_{eqtot} - \frac{D_o}{2}\right) = L + \Delta H_0 + 2\left(h_{eqtot} - \frac{D_o}{2}\right) \tag{36}$$

As a consequence of the previous assumptions, and reasoning henceforth in nondimensional shape, it is possible to state that, once the $h_{tot}/D_s$ value has been fixed, two fundamental data can be obtained.

The first one is the $H_{totI}/D_s$ ratio, from:

$$\frac{H_{totI}}{D_s} = \frac{L + \Delta H_0}{D_s} + 2\frac{h_{eqtot} - \frac{D_o}{2}}{D_s} \tag{37}$$

This is the $H_{totI}/D_s$ value which lets the plants' $\beta$ attain the characteristic value $\beta_I$ of Region I.

The second one is just the actual $\beta$ value of Region I, which obviously can be obtained from eqn. (11), and will be henceforth called $\beta_I$:

$$\beta_I = \beta_0 + \beta_1\frac{h_{eqtot}}{D_s} + \beta_2\left(\frac{h_{eqtot}}{D_s}\right)^2 \tag{38}$$

On the basis of all aforementioned statements, it is possible to state a believable entraining air law $\beta = \beta(H_{tot}/D_s, h_{tot}/D_s)$ in Transition Region.

First of all, it is possible to state that the Region II field is defined by: $H_{tot}/D_s \leq H_{totII}/D_s$, and the Transition Region field by: $H_{tot}/D_s \geq H_{totII}/D_s$, and $H_{tot}/D_s \leq H_{totI}/D_s$. Within the Transition Region, the entraining air law $\beta = \beta(H_{tot}/D_s, h_{tot}/D_s)$ can be established on the basis of continuity criteria.

Therefore, in Transition Region, the working curve, relative to a well defined $h_{tot}/D_s$ value, must start from the point $[H_{totII}/D_s, \beta_{II}]$ with a derivative:

$$D = K\alpha\left(\frac{H_{totII}}{D_s}\right)^{\alpha-1}\left(\frac{h_{tot}}{D_s}\right)^{\gamma} \tag{39}$$

and tends to the final point $[H_{totI}/D_s, \beta_I]$. Moreover, the $\beta(H_{tot}/D_s, h_{tot}/D_s)$ law always increases with $(H_{tot}/D_s)$ and $(h_{tot}/D_s)$. A sufficiently simple law that meets these requirements can be the following one:

$$\eta = D\chi + \left[\frac{(\eta_1 - D\chi_1)}{\chi_1^{\varphi}}\right]\chi^{\varphi} \tag{40}$$

where:

$$\chi = \frac{H_{tot}}{D_s} - \frac{H_{totII}}{D_s} \tag{41}$$

$$\eta = \beta - \beta_{II} \tag{42}$$

$$\chi_1 = \frac{H_{totI}}{D_s} - \frac{H_{totII}}{D_s} = 2\frac{\left[\Delta H_0 + \left(h_{eqtot} - \frac{D_o}{2}\right)\right]}{D_s} \tag{43}$$

$$\eta_1 = \beta_I - \beta_{II} \tag{44}$$

In eqn. (40), the parameter $\varphi$ is a free one, and can be modelled on the basis of the experimental data.

# 4 Experimental calibration and control of the model

## 4.1 The Region I plant working

On the basis of the theoretical analysis and the actual dimensions of the plant, it is possible to obtain the numerical values for coefficients $\beta_0$, $\beta_1$, and $\beta_1$ of eqn. (11).

The first choice to carry out is the suitable value of the whole virtual length of the first air bleeder, in order to take into account also the localized head losses, that has been called $L_{lat}$. The real length is 1.20m. The additional equivalent length $L_{eq}$ has been computed on the basis of the Darcy–Weisbach equation:

$$\frac{\lambda L_{eq}}{D_{lat}}\frac{V^2}{2g} = \sum \varepsilon_i \frac{V^2}{2g} \tag{45}$$

and therefore:

$$L_{eq} = \sum \varepsilon_i \frac{D_{lat}}{\lambda} \tag{46}$$

$\Sigma\varepsilon_i$ is the sum of all local head losses coefficients in our plant and can be put equal to 2.5. It is evident that the true value of $L_{eq}$ changes with hydrodynamic conditions (namely with $h_{tot}/D_s$ values), due to variability of friction factor $\lambda$ with Reynolds number.

Once the correct $L_{eq}$ has been calculated in each different hydrodynamic condition, it is possible to employ the already presented algorithm to evaluate the $\beta$ value. In this evaluation, it is necessary to use a trial and error method between $L_{eq}$ and $\beta$.

Four different values of $h_{tot}/D_s$ ratio, chosen within the values allowed by the plant ($h_{tot}/D_s$ = 1.75, 2.85, 3.45, 4.55) have been tested. For each one of these $h_{tot}/D_s$ ratios, the trial and error method has been used, and the correspondent $\beta$ value in Region I has been computed, obtaining $\beta$ = 2.124, 1.847, 1.754, 1.633, respectively. It is noteworthy that in these four cases the $L_{eq}$ values result 9.88m, 10.20m, 10.29m, 10.59m, respectively, namely length values which appeared not to be very different from one another.

On the basis of the couples of $h_{tot}/D_s$ and $\beta$ values, a 2nd order curve has been found imposing its transit on the first couple, the last couple and an intermediate couple between the two central ones. The final result for the $\beta_0$, $\beta_1$, and $\beta_2$ coefficients has been: $\beta_0 = 2.768$, $\beta_1 = -0.444$, $\beta_2 = 0.0406$, so that eqn. (11) finally becomes:

$$\beta = \beta_I = 2.768 - 0.444\frac{h_{tot}}{D_s} + 0.0406\left(\frac{h_{tot}}{D_s}\right)^2 \tag{47}$$

## 4.2 The Region II and the Transition

Whereas the parameters of Region I have been obtained in a fully theoretic way, a similar procedure is impossible in Region II and in Transition Region. In particular, eqn. (26) holds three unknown parameters (K, $\alpha$, $\gamma$) and eqn. (40) holds one more unknown parameter ($\varphi$). These parameters can only be obtained through suitable experimental surveys. In the following paragraphs, the experimental surveys and the subsequent numerical processes, to obtain the unknown parameters values, will be described.

Generally speaking, to obtain the unknown parameters, the least square method will be performed, on the basis of a sufficiently significant number of experimental data. The first objective is therefore to obtain good experimental values of $\beta$ parameter correspondent to different working conditions in Region II and in Transition between Region II and Region I. Different working conditions mean different values of $(H_{tot}/D_s)$ and $(h_{tot}/D_s)$.

The experimental ways to obtain the $\beta$ values corresponding to different $(H_{tot}/D_s)$ and $(h_{tot}/D_s)$ values have been obtained through two different methods called, respectively, Anemometric Method and Volumetric Method. These methods will be explained in Section 4.2.1 and 4.2.2.

### 4.2.1 Anemometric Method

The Anemometric Method has been the first one presented by authors in [21–23].

The two main quantities that must be evaluated to compute $\beta$ values are $Q_w$ and $Q_a$, whose ratio gives the $\beta$ value.

The $Q_w$ flow-rate has been evaluated through a double method: (1) through a pipe orifice inserted within the feeding circuit of the plant; (2) on the ground of the $h_{eqtot}$ value which allowed to compute the flow-rate using the fluid mechanics basic rules. The two methods gave $Q_w$ values sufficiently close to one another.

The core of the Anemometric Method consists indeed in evaluating $Q_a$, through direct measurements performed within the first air bleeder. A hot wire anemometer was placed on its axis, within a branch of this air bleeder 1m long, to avoid local perturbations. To obtain $Q_a$, the velocity obtained through the hot wire was directly multiplied by the cross-section of the first air bleeder, bearing the small error consequent to leaving out the actual velocity distribution (having also performed a control of this procedure as it will later be referred in this section). The velocity values, that were read by the hot wire anemometer, were always digitally acquired in a PC.

The main problem during those measurements was the already remembered oscillatory phenomenon of the free surface within the dropshaft. This phenomenon is due to oscillations, characterized by their own time period, of the total water mass included into the dropshaft, its final curve towards the horizontal pipe and the second air bleeder. The characteristic period of this oscillation could be either experimentally detected or theoretically calculated [22]. Due to this oscillation, the $Q_a$ within the first air bleeder presented an oscillatory trend too (in the shape of an oscillatory flux caused by the action of the liquid free surface in the dropshaft superimposed to the continuous flux due to air entrainment). The oscillations, in some conditions, were so strong that they caused even a reversal of velocities: but, in this case, the anemometer signal did not change sign, due to the characteristics of the anemometer itself. Therefore, the experimental data had to be suitably processed to obtain the true mean value of velocities. To perform this process, a graph of the direct observation of the recorded signal was carried out, and this observation, together with either the theoretical or experimental knowledge of the oscillation period, allowed us to fix on the graph the exact instants of air velocity reversals and, consequently, change sign to the velocity diagram in the suitable time periods. Finally, the mean value of the velocity was computed through the corrected diagrams.

An accurate measurements statistical errors control was carried out in [21] to decide the acquisition frequency and duration, and the final decision was to use a 1000Hz sampling frequency and a 10-min time period for each experimental point. Through the aforementioned control, it has been possible to state that with these choices the residual statistical error in mean velocities had to be considered of the order of 1.2%.

A control of the velocity profile flatness into the first air bleeder was also performed repeating, in some preliminary tests, the aforementioned procedure in different points of the diameter, and drawing the conclusion that the error due to neglecting of velocity profile's curvature was not higher than the statistical one.

In these experimental conditions, the complete Anemometric Method assures by itself an error in $Q_a$ measurement of very few percent units. The final control of the method was in any case entrusted to a direct comparison with the results of the second (volumetric) experimental method that has been employed afterwards.

### 4.2.2 Volumetric Method

Two different implementation ways of the Volumetric Method will be here described. The first one is the basic simpler implementation and allows us to comprehend the main characteristics of the method: it was presented in [24]. The second one is a more advanced implementation through which some approximations of the first implementation can be avoided: it was presented in [25].

**4.2.2.1 Simpler implementation of Volumetric Method** In the experimental plant description it was stressed that the entrained air is fed through the first air bleeder. If this air bleeder is instantaneously closed, a depression $\Delta p$ (considered henceforth as a positive number) arises within the dropshaft because of the air entrainment phenomenon, and the longer the experiment lasts, the greater this depression becomes. First of all, the depression acts as an additional head $\Delta p/\gamma_w$ in the tank and increases more and more the water flow-rate: therefore, as the arriving flow-rate does not change, the water height in the feeding tank becomes lower and lower, but no more than a few centimeters during the whole test.

Moreover, in the dropshaft the water column rises more and more, with respect to the previous steady level, of a quantity which holds two terms: the first one is necessary to allow the increased water flow-rate to stream from the final gate valve and the second one is necessary to compensate the air depression. The experiment is stopped after a certain time, when the water column rise attains some dozens of centimeters: this time is generally comprised of between a few seconds and some dozens of seconds. In these conditions, the final value of the depression is always worth no more than some percent units of atmospheric pressure, so that, in this simpler implementation of the method, air is considered as an incompressible fluid. To complete the description of the method, it is suitable to recall that, due to the adopted closure system for the first air bleeder, a small value of depression is present also at the beginning of the test.

The main output of the described experiment is the possibility of computing directly the mean value of $Q_a$ during the test time as the ratio between the volume $\tau_a$ of the entrained air (considered equal to the volume previously occupied by air and at the end of the test occupied by the water raised in the shaft) and the time duration $t$ of the experiment. To completely perform the whole experiment, it is necessary to obtain values of $\beta$, $h_{eqtot}$, $H_{tot}$ during the test. Now it is clear that, during the test, these values are all changing from initial to final values. The fundamental hypothesis is that it is possible to refer to mean values of all quantities, through a linearization procedure. This hypothesis can be considered valid until the water column rise is not too high.

To write correct expressions of the mean values of $\beta$, $h_{eqtot}$, $H_{tot}$, it is firstly necessary to take into account the circumstance that the additional head $\Delta p/\gamma_w$ must also be taken into account, so that the simple head $h$ must be replaced by the equivalent head:

$$h_{eq} = h + \frac{\Delta p}{\gamma_w} \tag{48}$$

and:

$$h_{eqtot} = h + \frac{D_0}{2} + \frac{\Delta p}{\gamma_w} \tag{49}$$

At this point, with the aid of "in" and "fin" symbols, the following substitutions are made in order to refer to mean values:

$$h_{eqtot} = \frac{h_{in} + h_{fin} + \frac{\Delta p_{fin}}{\gamma_w} + D_o}{2} \tag{50}$$

$$H_{tot} = \frac{H_{in} + H_{fin}}{2} + h_{eqtot} - \frac{D_o}{2} \tag{51}$$

Finally, $\beta$ can be computed directly in mean as the ratio $\tau_a/\tau_w$ between the already mentioned volume of entrained air and the volume of flowed water during the experiment duration time.

In particular, $\tau_a$ is computed as the air volume which is initially held in the dropshaft between the levels $H_{in}$ and $H_{fin}$ (volume that has been entrained by the jet, that does not hold the central jet volume and that can be calculated through fluid mechanics laws), and $\tau_w$ is the sum of the volume which flowed during the time of the experiment due to the steady (initial) water flow-rate $Q_{win}$ (externally measured) and the further water volume detracted from the tank (that can be computed through the values $h_{in}$ e $h_{fin}$). Final water flow-rate $Q_{wfin}$ can be computed through the knowledge of the initial value of $Q_{win}$, the detracted water volume and the experiment duration time *t* (always through a linear model). $\Delta p_{in}$ and $\Delta p_{fin}$ are obtained from $Q_{win}$ and $Q_{wfin}$ through the simple orifice fluid mechanics laws.

In such a way, all experimental parameters are known, and the Volumetric Method can give the same information ($\beta$) as the Anemometric Method directly correspondent to the $h_{eqtot}$ and $H_{tot}$ values.

More particulars, especially in calculating the volume $\tau_a$ that is one of the most difficult tasks of the computation, are given in [24].

**4.2.2.2 Advanced implementation of Volumetric Method** A more advanced implementation of the Volumetric Method holds also the consequences of the air compressibility. The fundamental feature of this advanced implementation lies in the fact that the volume $\tau_a$ of the entrained air is no longer computed in a direct way as in the simpler Volumetric Method, but before calculating the entrained mass and therefore dividing this mass by the density of the air in conditions of pressure drop $\Delta p$ presence. More particulars about this advanced method, which is simple as an idea but not very simple for implementation, are given in [25].

This implementation gives final values of $\beta$ greater than those furnished by the simpler method with the same experimental data (fig. 9). The causes of this increasing are two. The first one is linked to the circumstance that the entrained air volume, stressed by the ascent of the water column, grows during the experiment due to the pressure drop. The second one is linked to the circumstance that the air expansion, due to the pressure drop, produces a further entraining also of a fraction of the air held in the whole dropshaft and yet in the first air bleeder, and it is necessary to compute this air too. More particulars about these two causes are given in [26].

**4.2.3 Comparisons between Anemometric Method and Volumetric Method**
In order to control the performances of the Anemometric Method and the advanced Volumetric Method, some tests have been performed.

In particular, two sets of $\beta$ measurements, the first one through the Anemometric Method and the second one through the advanced Volumetric Method, were compared. In the first set the $h_{eqtot}$ value has been kept constant and equal to 0.42m, but the $H_{tot}$ values differ from one another. In the second set, because of the chosen methodology, it was not possible to decide *a priori* the value of $h_{eqtot}$:

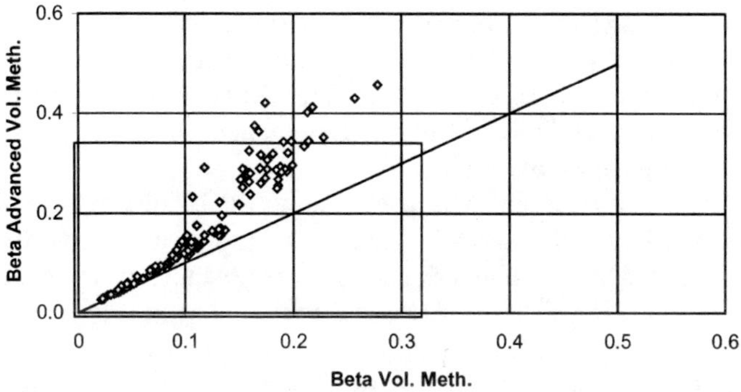


Figure 9: Volumetric method: Comparison between simple method and advanced method $\beta$ values.

consequently, the measurements have been arranged in order to have an $h_{eqtot}$ value as near as possible to 0.42m. The chosen values of H/$D_s$ were held between 19 and 30: this interval was large enough to be suitable for the test and, in particular, it corresponded to Transition conditions between Region II and Region I: these conditions are indeed the heaviest conditions as regards to either the air-entraining phenomenon or the measuring possibilities.

Figure 10 reports the results of these two test series: white squares refer to Anemometric Method and black circles refer to Volumetric Method. It is evident that the series of points overlap very well one another on average. Moreover, it is evident too that black circles show a greater spreading than white squares: this is the obvious and natural consequence of the circumstance that the $h_{eqtot}$ values of the different tests are not alike, and the spreading of $h_{eqtot}$ values causes a consequent spreading in the correspondent $\beta$ values.

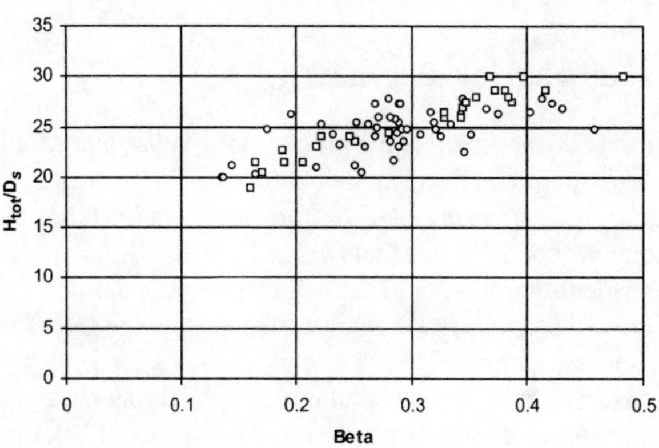


Figure 10: Comparison between volumetric method and anemometric method $\beta$ values.

### 4.2.4 Performed tests and obtained calibration results

In order to calibrate eqns. (26) and (40) parameters, a considerable number of experimental tests have been performed. In each test, either it was carried out through Anemometric Method or through Volumetric Method, the three fundamental values $H_{tot}$, $h_{eqtot}$, $\beta$, have been obtained.

A first insight into the obtained results showed that the $H_{totII}$ value, which keeps Region II tests and Transition Region tests apart, can be considered to be worthy 4.4 m, so that the nondimensional $H_{totII}/D_s$ value is worth 22. Therefore, $\Delta H_0 = 2.1$m and $\Delta H_0/D_s$ value is worth 10.5.

The performed tests can be divided into tests in which $H_{tot}/D_s$ is worth less than 22, and tests in which $H_{tot}/D_s$ is worth more than 22. The first ones have been 87 tests and their data have all been obtained through the Volumetric Method, which is the only one that is worthy in these conditions. The second ones have been 72 tests and their data have been obtained either through the Volumetric Method (48 cases) or through the Anemometric Method (24 cases).

In both situations (Region II or Transition Region), a least squares method has been implemented to obtain the parameters values ($K, \alpha, \gamma$ in the first case and $\varphi$ in the second case) that correspond to the best fit between experimental and modelled ($\beta_{exp}$ and $\beta_{mod}$, respectively) values.

The final obtained values are the following ones: $K = 0.00425$; $\alpha = 0.81$; $\gamma = 0.83$; $\varphi = 1.87$.

In order to ascertain the value of the mean error associated with these types of measurements, the r.m.s. value of the $(\beta_{exp} - \beta_{mod})$ difference has been evaluated, either in Region II or in Transition Region. The results have been 0.0228 in the case of Region II working and 0.0617 in the case of Transition Region. But, in any case, it is noteworthy that the maximum $\beta_{exp}$ obtained value has been 0.152 in the first case and 0.482 in the second case. Therefore, the r.m.s. values of the correspondent sets relative to maximum values have been 0.150 in the first case and 0.128 in the second case. These data show that the relative accuracy of measurements can be considered quite constant.

## 5 Control and behavior diagrams

In order to visualize the accuracy of the proposed formulas, in fig. 11 a comparison is shown between experimental and modelled $\beta$ values. In particular, in abscissa experimental values $\beta_{exp}$ and in ordinate the correspondent modelled (either through eqn. (26) for Region II or through eqn. (40) for Transition Region) values $\beta_{mod}$ are reported. It is evident that the points lie sufficiently near the line 45° inclined. It is also evident that the $\beta$ values are higher, the absolute spreading of points with respect to this line is wider.

Moreover, also as a simple and useful tool for predicting behavior of the dropshafts with regard to air entrainment phenomenon, in fig. 12 the proposed laws eqns. (26) and (40) with the already obtained correspondent $K$, $\alpha$, $\gamma$, $\varphi$ values are presented in a single diagram. In this figure also the borderlines of Transition Region are presented. The four presented lines represent $\beta$ versus $H_{tot}/D_s$ laws

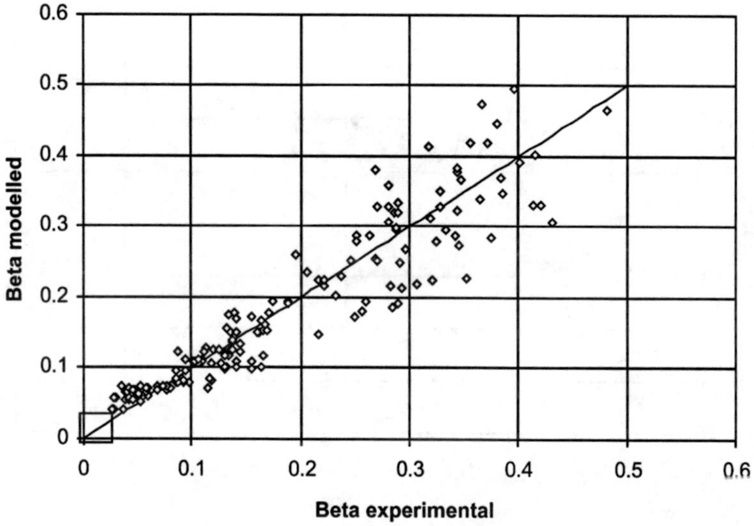


Figure 11: Comparison between experimental and modelled $\beta$ values.

relative to $h_{eqtot}/D_s$ values, respectively, of 1.5, 2.5, 3.5, 4.5 (respective lines are thicker and thicker).

It is noteworthy that eqn. (26) diagrams present clear intersections from one another in a zone of the diagram near $H_{tot}/D_s = 30$ and $\beta = 0.3$ values. This is

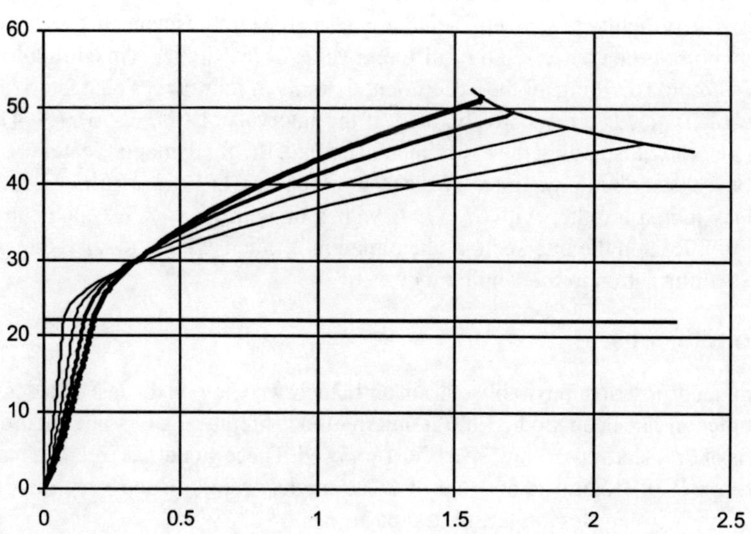


Figure 12: Sketch of proposed laws for predicting behavior of dropshafts with regard to air entrainment phenomenon (line thickness grows with $h_{eqtot}/D_s$).

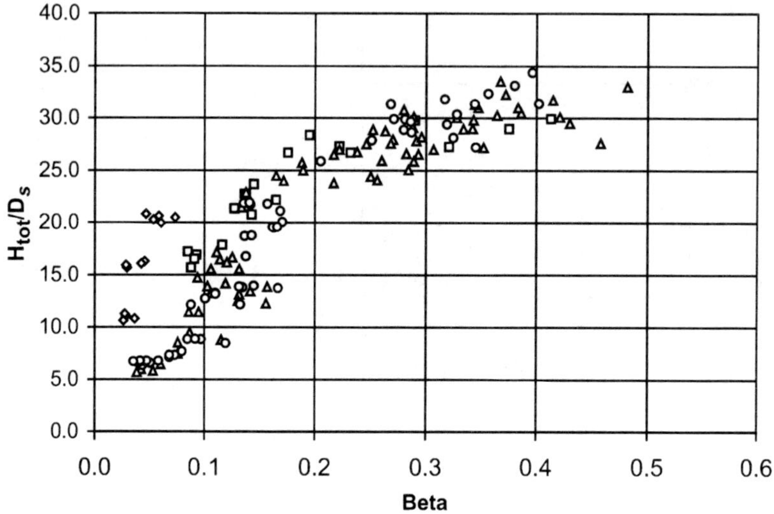


Figure 13: Experimental points: Rhombs: $1 \leq h_{eqtot}/D_s < 2$ Squares: $2 \leq h_{eqtot}/D_s < 3$ Triangles: $3 \leq h_{eqtot}/D_s < 4$ Circles: $4 \leq h_{eqtot}/D_s < 5.5$.

due to the circumstance that the $h_{eqtot}/D_s$ values are higher, the correspondent $\beta_I$ values are lower. This behavior has been firstly stressed by the author in [23] and is probably one of the causes of the $\beta$ measuring difficulty in Transition Region.

Finally, fig. 13 reports the present 159 experimental points. As elsewhere, in the abscissa there are always $\beta$ values and in the ordinate there are always $H_{tot}/D_s$ values. The difficulty of this representation lies in the circumstance that each experimental point is characterized by a different value of $h_{eqtot}/D_s$. This difficulty has been overcome dividing all the experimental points in four ranges characterized by values of $h_{eqtot}/D_s$, respectively, held in the intervals $(1 \div 2)$, $(2 \div 3)$, $(3 \div 4)$ and $(4 \div 5.5)$. Notwithstanding the experimental spread, the fundamental features of the previously described laws are very clear. In particular, in Region II it is clear that $\beta$ values increase either with $H_{tot}/D_s$ values or with $h_{eqtot}/D_s$ values; and in Transition Region the intersection phenomenon is much evident too. All that gives a good confirmation to the whole model.

## 6 Conclusions

Central jet dropshafts inevitably entrain air in their working. In the last decades, this phenomenon has been studied in the international literature and some theoretical models of dropshafts working have been proposed. These models are relative mainly to Region II or Region I; not one of these models refers to the very important Transition between Region II and Region I.

An accurate experimental observation and a deep theoretical analysis of the phenomenon, allowed the authors to propose, on the ground of the aforementioned already existing theories, improved models either of Region II or Region I.

Both models accept as input values two characteristic experimental lengths called $H_{tot}$ and $h_{eqtot}$ (which are transformed in nondimensional ones through the shaft diameter $D_s$) and return as output value the nondimensional ratio $\beta$ between entrained air and falling water flow-rates. The Region II model is a simple power law model with three experimentally obtainable parameters; the Region I model is a very complex one and can be transformed in a simpler second degree polynomial model with three theoretically obtainable parameters.

The same experimental observation and theoretical analysis allowed the authors to produce a model of the Transition Region, which is fully original with respect to international literature. This new model accepts and returns the same input and output values of the previous ones and holds only one more experimentally obtainable parameter.

Consequently, it is possible to state that a fully complete working condition's model of the air entrainment in a central jet dropshaft has been carried out for the first time in international literature: this model accepts $H_{tot}/D_s$ and $h_{eqtot}/D_s$ nondimensional lengths as input values and return as output value the ratio $\beta$ between entrained air and falling water flow-rates. This model, moreover, holds four experimentally obtainable parameters.

This model was tested and its parameters were obtained by least squares method through a wide series of experimental tests (159 experimental tests) relative to so big a prototype to eliminate errors caused by viscosity and other secondary effects in air entrainment phenomenon.

This model refers to a single value either of the dropshaft diameter (0.2m) or the orifice diameter (0.1m). Therefore, the theoretically and experimentally obtained parameters refer only to a shaft of these dimensions. But the whole theory can be applied also to other dropshaft dimensions, obviously computing over again the relative theoretical and experimental parameters.

## 7 Acknowledgments

The helpful comments of the anonymous referees are acknowledged.

The present paper has been carried out as a part of Italian MURST PRIN 2002 granted Research "*Influence of vorticity and turbulence in interactions of water bodies with their boundary elements and effects on hydraulic design*".

## References

[1] Bin, A.K., Gas entrainment by plunging liquid jets. *Chemical Engineering Science*, **48**, 1993.

[2] Chanson, H., Aoki, S. & Hoque, A., Similitude of air entrainment at vertical circular plunging jets. *ASME 2002 Fluids Engineering Division Summer Meeting*, Montreal, 2002.

[3] Chanson, H., *Air Bubble Entrainment Upon Liquid Impact*. Academic Press: London, 1996.

[4] Chanson, H., *Air bubble Entrainment in Free Surface Turbulent Shear Flows*. Academic Press: London, 1997.

[5] Ervine, D.A. & Ahmed, A.A., A scaling relationship for two dimensional vertical dropshaft. *Proc. B.H.R.A. Conf. On Hydr. Struct. Paper E1*, Coventry, U.K., 1982.

[6] Ervine, D.A., *A Review of Bubbly and Air Pockets Flows in Civil Engineering Hydraulic Structures*. Lecture to United States Bureau of Reclamation, Denver, 1985.

[7] Lin, T.J. & Donnelly, H.G., Gas bubble entrainment by plunging laminar liquid jets. *A.I. Ch.E. Journal*, **12**, pp. 563–571, 1969.

[8] Oguz, H.N., Prosperetti, A. & Lezzi, A.M., Examples of air-entrainment flows. *Physics of Fluids*, **4**, 1992.

[9] Oguz, H.N. & Prosperetti, A., Mechanics of air entrainment by a falling liquid. *ASME*, **187**, 1994.

[10] Oguz, H.N., Prosperetti, A. & Kolaini, A.R., Air entrainent by falling water mass. *Journal of Fluid Mechanics*, **294**, 1995.

[11] Oguz, H.N. & Prosperetti, A., Air entrainment upon liquid impact. *Phil. Trans. R. Soc. London*, 1997.

[12] Rajaratnam, N. & Kwan, A.P., Air entrainment at drops. *Journal of Hydraulic Research*, **5**, 1996.

[13] Zhu, Y., Oguz H.N. & Prosperetti, A., On mechanism of air entrainment by liquid jets a free surface. *Journal of Fluid Mechanics*, **404**, 2000.

[14] Chanson, H., Aoki, S. & Hoque, A., Similitude of air entrainment at vertical cirular plunging jets. *ASME Fluid Engineering Division Summer Meeting*, Montreal, 2002.

[15] Laushey, L.M. & Mavis, F.T., Air entrainment by water flowing down vertical shafts. *$5^{th}$ Congress I.A.H.R. Minneapolis*, 1953.

[16] Viparelli, M., *Air Entrainment by water jets in closed conduits*. (in Italian) L'Energia Elettrica Fasc. XI, XXXI 1954.

[17] Viparelli, M., Air and water currents in vertical shafts. *Le Houille Blanche*, **6**, pp. 857–869, 1961.

[18] Pulci Doria, G., Scuderi, M. & Viparelli, R., Air entrainment in vertical dropshaft. (in italian), *XXII National Conference of Hydraulic Engineering*, Cosenza, 1990.

[19] Gualtieri, P., Pulci Doria, G. & Viparelli, R., Air entrainment model in a dropshaft working in region I experimental control. (in Italian) *XXVI National Hydraulics and Waterworks Conference*, Catania – Vol III pp. 541–552, 1998.

[20] Gualtieri, P. & Pulci Doria, G., Air entrainment in vertical dropshafts: a new hydrodynamic model. *$6^{th}$ Int. Conf. On Hydroscience and Engineering ICHE* 2004.

[21] Gualtieri, P. & Viparelli, R., Hydraulic plant in order to study air entraining in presence of unsteady pressure motions. (in Italian) *XXV National Conference of Hydraulic Engineering*, Torino, pp. 33–44, 1996.

[22] Gualtieri, P., Unsteady pressure motion generated by jets in dropshafts. (in Italian) *XXV National Conference of Hydraulic Engineering*, Torino, pp. 45–56, 1996.

[23] Gualtieri, P., Pulci Doria, G. & Viparelli R., Air entrainment in a dropshaft working in transition between the so-called Region I and Region II. *XIV National Conference of the Italian Association of Theoretical and Applied Mechanics A.I.ME.TA.*, Como. 1999.

[24] Gualtieri, P. & Pulci Doria, G., Volumetric methodology to measure air entrainment in vertical dropshaft. (in Italian) *XXVII National Conference of Hydraulic Engineering*, Genova, 2000.

[25] Gualtieri, P. & Pulci Doria, G., An improvement of volumetric methodology to measure air entrainment in vertical dropshaft. (in Italian) *XXVIII National Conference of Hydraulic Engineering*, Potenzav, 2002.

[26] Gualtieri, P., Volumetric method in order to measure air entraining in central jet vertical dropshafts: influence of air compressibility. (in Italian) *Symposium in Prof. Taglialatela'a honour*. Naples, 2002.

# CHAPTER 8

# Variational methods in sloshing problems

M. La Rocca, G. Sciortino, P. Mele & M. Morganti
*Department of Civil Engineering Sciences,*
*Roma TRE University, Italy.*

## Abstract

In this work a variational approach to sloshing problems is presented. Such an approach is based on the definition of a functional of the fluid motion which becomes stationary in correspondence with the motions of the fluid system. In the present case the fluid is assumed to be inviscid and attention is focused on the Hargneaves–Luke formulation, which consists of defining the functional as the work done by the pressure inside the fluid domain. Such a formulation allows the determination of all the most interesting field variables, namely the velocity, the pressure and the instantaneous and local shape of the moving surface.

## 1 Introduction

Sloshing is the motion of a body of fluid, partly limited by free boundaries, in a moving vessel. Sloshing constitutes a problem of the highest complexity and scientific-technical interest. The scientific interest concerns the variety of physical phenomena – mostly nonlinear – which can be investigated even in very simple sloshing configurations and which can easily be realized experimentally. The technical interest lies in the several practical situations in which sloshing occurs. The transportation of free-surface fluids by ship or rail and the sloshing of free-surface fluids in the fuel tanks of airplanes and rockets represent only two of many examples of sloshing, but their clear relation to the safety of human life highlights the importance of investigating this phenomenon. For this reason the first studies on sloshing were stimulated by the necessity of evaluating the stability of rockets and airplanes when their fuel tanks were filled with free-surface fluids – the works of Abramson [1] and Moiseiev & Rumyantsev [2] can be considered milestones.

Among the methods that can be applied to the investigation of sloshing, variational methods have recently received increasing attention. A variational method, as applied to sloshing, is a method based on the definition of a functional of the fluid motion. The motions of the fluid system make the functional stationary and such motions are then found by imposing the stationarity of the functional. Moiseiev and Rumyantsev's book is the first systematic scientific work on the application of variational methods to sloshing problems whose results are of theoretical rather than practical importance. The work of Miles [3] is perhaps the first to undertake an exhaustive and wide-ranging investigation of the application of several variational approaches to the sloshing problem. Miles and Whitham have also applied variational methods to the investigation of water waves [4–7].

Different definitions can be adopted for the functional and these give rise to different variational approaches. The definition of the functional as the work done by the pressure on the fluid domain [8] is in any case simple and effective. Luke [9] showed that the classical problem of waves in an inviscid free-surface liquid is equivalent to the variational formulation obtained by adopting such a definition. In the following, attention will be focused on such a definition of the functional and on its consequences for the investigation of sloshing. It is interesting to observe that very recently this approach has received increasing attention [10–18] due to its utility in calculating all the interesting quantities concerning sloshing – the shape and motion of the free surface and the pressure and velocity distributions inside the fluid domain.

## 2 The mathematical model

### 2.1 Statement of the problem

Let us consider a rigid container of arbitrary cylindrical shape, with transverse section A, filled with an incompressible, quiescent liquid up to height H. A rigid motion is imposed on the container and can be decomposed into the superposition of a translation and a rotation. Translation and rotation are identified by the vectors $\mathbf{V}_O(t)$ and $\mathbf{\Omega}(t)$, respectively, the translation and rotation velocity vectors. Figure 1 is a schematic representation of the container, showing the frame of reference used and the characteristic elements of the imposed motion.

The velocity of a point $P$ attached to the container, assuming the frame of reference with origin in $O$ is given by:

$$\mathbf{V}_P = \mathbf{V}_O(t) + \mathbf{\Omega}(t) \times \mathbf{OP} \tag{1}$$

Let $\mathbf{v}$ and $\mathbf{u}$ indicate the velocity of a fluid particle, respectively, in the absolute and relative frame of reference with respect to the container. They are related by:

$$\mathbf{v} = \mathbf{V}_O(t) + \mathbf{\Omega}(t) \times \mathbf{OP} + \mathbf{u} \tag{2}$$

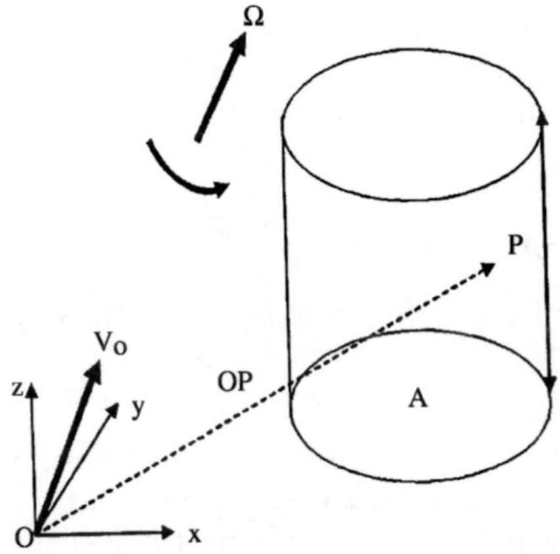


Figure 1: Schematic representation of the problem.

In the absolute frame of reference, if viscous forces are neglected, no forces exist that could give rise to vorticity. So the absolute velocity vector field can be assumed as:

$$\mathbf{v} = \boldsymbol{\nabla}\varphi \tag{3}$$

The inviscid hypothesis is quite reasonable in sloshing problems, as it is well known that viscous forces are important only in relation to the boundary layers of the container wall [19, 20]. Moreover, the ratio of the boundary lengthscale on a container's characteristic dimension $D$ is given by:

$$\delta = \frac{1}{D}\sqrt{\frac{2\nu}{\omega}} \tag{4}$$

$\nu$ being the kinematic viscosity of the liquid and $\omega$ a characteristic pulsation for the motion imposed on the container. Such a ratio is in general very small: *e.g.* $\delta = \mathcal{O}\left(10^{-3}\right)$ for water in a container with $D = 1m$, $\omega = 1\frac{rad}{s}$. This fact suggests that the viscous forces should be omitted from consideration, at least at this stage of the formulation.

Now let us consider the Euler equation in the relative frame of reference:

$$\begin{aligned} &\frac{\partial \mathbf{u}}{\partial t} + \frac{1}{2}\nabla\left(\mathbf{u}\cdot\mathbf{u}\right) + \mathbf{w}\times\mathbf{u} + \frac{d\mathbf{V}_O}{dt} \\ &+2\boldsymbol{\Omega}\times\mathbf{u} + \frac{d\boldsymbol{\Omega}}{dt}\times\mathbf{OP} + \Omega\times\left(\boldsymbol{\Omega}\times\mathbf{OP}\right) + \frac{1}{\rho}\nabla p + g\nabla\Psi = 0 \end{aligned} \tag{5}$$

where $\rho$ is the fluid density, $\mathbf{w} = -2\boldsymbol{\Omega}$ is the relative vorticity field and $\boldsymbol{\Psi}$ is the potential function of the body force (the weight), which depends on the imposed motion and on the spatial coordinates. Due to the rigid motion imposed on the container, the apparent forces are accounted for as well. Substituting the expression for the relative velocity obtained from eqn. (2) – in which definition eqn. (3) has been used – in Euler eqn. (5), the pressure field is obtained:

$$p = -\rho\left(\frac{\partial\varphi}{\partial t} + \frac{\nabla\varphi\cdot\nabla\varphi}{2} - \nabla\varphi\cdot\mathbf{V}_O - \nabla\varphi\cdot\boldsymbol{\Omega}\times\mathbf{OP} + \mathbf{V}_O\cdot\boldsymbol{\Omega}\times\mathbf{OP} + g\Psi\right) \tag{6}$$

which is determined only by inertia and body forces.

The classic formulation for a sloshing problem consists of finding the scalar function $\varphi$, given the vectors $\mathbf{V}_O, \boldsymbol{\Omega}$, the initial conditions and the boundary conditions. For this purpose, the mass conservation equation has to be used:

$$\nabla^2\varphi = 0 \tag{7}$$

in the fluid domain $\mathcal{D}_f$, defined as:

$$\mathcal{D}_f \equiv \{(x, y, z)|\,(x, y) \in A, z \in [-H, \eta\,(x, y, t)]\} \tag{8}$$

$z = \eta(x, y, t)$ being the instantaneous and local elevation of the free surface of the liquid over the quiescent level $H$, during the sloshing motion. The boundary conditions must be imposed on the container walls and state that the absolute velocity of the fluid along the direction normal to the wall must be equal to the wall velocity along the same direction:

$$\nabla\varphi\cdot\mathbf{n} = (\mathbf{V}_O + \boldsymbol{\Omega}\times\mathbf{OP})\cdot\mathbf{n} \tag{9}$$

Moreover, on the free surface $z = \eta(x, y, t)$, two other boundary conditions must be fulfilled: the first is the so-called kinematic boundary condition and expresses the fact that the free surface is a material surface:

$$\frac{\partial\eta}{\partial t} + \left(u\frac{\partial\eta}{\partial x} + v\frac{\partial\eta}{\partial y} - w\right)\Bigg|_{z=\eta(x,y,t)} = 0 \tag{10}$$

$u, v, w$ being the components of the relative velocity. The second is the so-called dynamic boundary condition and expresses the fact that on the free surface during the motion the pressure must remain constant and equal to zero:

$$\frac{\partial\varphi}{\partial t} + \frac{\nabla\varphi\cdot\nabla\varphi}{2} - \nabla\varphi\cdot\mathbf{V}_O - \nabla\varphi\cdot\boldsymbol{\Omega}\times\mathbf{OP} + \mathbf{V}_O\cdot\boldsymbol{\Omega}\times\mathbf{OP} + g\Psi\Bigg|_{z=\eta(x,y,t)} = 0 \tag{11}$$

Boundary conditions eqns. (10) and (11) are expressed in terms of the unknown functions $\varphi, \eta$. So the classic problem consists of finding the functions $\varphi, \eta$ which satisfy eqn. (7) with boundary conditions eqns. (9), (10), (11) and initial conditions like:

$$\varphi(x,y,z,0) = \phi_0(x,y,z)$$
$$\eta(x,y,z,0) = \eta_0(x,y,z) \tag{12}$$

The main difficulty of such a problem concerns boundary conditions eqns. (10) and (11), which are not only strongly nonlinear, but also present a very complicated coupling between $\varphi$ and $\eta$, $\varphi$ in eqns. (10) and (11) being calculated for $z=\eta(x,y,t)$. A common approach in solving the classic problem is to expand the unknown dependent variables in perturbative series of a small parameter, which usually coincides with a non-dimensional motion amplitude (*e.g.* the amplitude of the imposed sloshing motion). Such an approach is rather complicated, however, and requires a tremendous amount of calculation even if a low number of terms is considered in the series expansion [21–24].

## 2.2 The variational formulation for sloshing

Let us consider the functional [8, 9]:

$$\mathcal{F} = -\int_{t_1}^{t_2}\int_A\int_{-H}^{\eta(x,y,t)} p dz dA dt \tag{13}$$

The functions $\varphi, \eta$, solutions of the classical problem constituted by eqn. (7) with boundary conditions eqns. (9), (10) and (11), and initial conditions eqn. (12), make the functional stationary. In other words the first variation of $\mathcal{F}$, $\delta\mathcal{F}$, calculated with respect to $\varphi, \eta$, remains equal to zero in correspondence with the solutions of the classical problem. Such solutions can then be found by calculating the first variation of $\mathcal{F}$ with respect to $\varphi, \eta$ and imposing on it a value of zero. On the other hand, it is possible, starting from expression (13), to demonstrate the equivalence of the variational formulation to the classical formulation. To do this, following a well known line of thinking [3, 9], it is necessary to substitute the expression for the pressure eqn. (6) in the definition of the functional eqn. (13), and then apply the operator $\delta$ to the resulting expression. Such an operator acts on $\mathcal{F}$ like the total functional derivative with respect to $\varphi, \eta$ [25]. In other words:

$$\delta\mathcal{F} = \delta_\eta\mathcal{F} + \delta_\varphi\mathcal{F} \tag{14}$$

The calculation of $\delta_\eta\mathcal{F}$ is rather simple – it is like the calculation of the derivative $\frac{d}{dx}\int_{x_0}^{x} f(t)\,dt$:

$$\delta_\eta\mathcal{F} = \int_{t_1}^{t_2}\int_A \rho\bigg(\frac{\partial\varphi}{\partial t} + \frac{\nabla\varphi\cdot\nabla\varphi}{2} - \nabla\varphi\cdot\mathbf{V}_O - \nabla\varphi\cdot\mathbf{\Omega}\times\mathbf{OP}$$
$$+\mathbf{V}_O\cdot\mathbf{\Omega}\times\mathbf{OP} + g\Psi\bigg)\bigg|_{z=\eta(x,y,t)} \delta\eta dA dt \tag{15}$$

The calculation of $\delta_\varphi \mathcal{F}$ is more complicated due to the presence of time and space derivatives and nonlinear terms. Applying the Leibniz rule and the divergence theorem, it may be shown that:

$$\begin{aligned}\delta_\varphi \mathcal{F} = &-\int_{t_1}^{t_2}\int_A \rho\left(\frac{\partial\eta}{\partial t}+u\frac{\partial\eta}{\partial x}+v\frac{\partial\eta}{\partial y}-w\right)\delta\varphi|_{z=\eta}\,dAd\\ &-\int_{t_1}^{t_2}\int_{\Sigma_w}\rho(\nabla\varphi-\mathbf{V}_O-\mathbf{\Omega}\times\mathbf{OP})\cdot\mathbf{n}\delta\varphi d\Sigma dt\\ &-\int_{t_1}^{t_2}\int_A\int_{-H}^{\eta(x,y,t)}\rho\nabla^2\varphi\delta\varphi dzdAdt\\ &+\int_A\int_{-H}^{\eta(x,y,t)}\rho\left(\delta\varphi|_{t=t_2}-\delta\varphi|_{t=t_1}\right)dzdA \qquad (16)\end{aligned}$$

where $\Sigma_w$ is the surface of the fluid domain coinciding with the rigid container walls. For $\delta\mathcal{F}=0$, it is necessary that the integrands in eqns. (15) and (16) are identically equal to zero. It is straightforward to see that these requirements coincide with the requirement that eqn. (7) and boundary conditions eqns. (9), (10) and (11) are satisfied. Moreover, the conditions $\delta\varphi|_{t=t_2}=\delta\varphi|_{t=t_1}=0$ must be fulfilled – they have no counterpart in the classical formulation, but state the fact that the variations with respect to the actual solution are zero at the initial and final instant of motion.

At this point the equivalence between the classical and variational formulations for sloshing has been demonstrated. The variational formulation of sloshing can now be stated as: find the functions $\varphi, \eta$ which make the functional $\mathcal{F}$ stationary and satisfy initial condition eqn. (12).

## 2.3 The Lagrange equations

Let us consider the following series expansion of the unknown functions $\varphi, \eta$:

$$\begin{aligned}\varphi &= \varphi_p(x,y,z,t)+\sum_{nm}A_{nm}(t)\,c_{nm}(x,y)\,d_{nm}(z)\\ \eta &= \eta_p(x,y,t)+\sum_{nm}q_{nm}(t)\,c_{nm}(x,y) \qquad (17)\end{aligned}$$

Where $c_{nm}(x,y)$ and $d_{nm}(z)$ are known functions of $x, y$ and $z$, $A_{nm}(t)$ and $q_{nm}(t)$ are unknown functions of $t$, $\eta_p(x,y,z,t)$ and $\varphi_p(x,y,z,t)$ are known functions, the latter of which satisfies eqn. (7) and boundary conditions eqn. (9). The functions $c_{nm}(x,y), d_{nm}(z)$ are the elements of a functional space with suitable properties. They derive from the separation of variables applied to eqn. (7) with homogeneous boundary conditions. The series expansions for $\varphi, \eta$, though complete in the unperturbed fluid domain, nevertheless become incomplete in the perturbed domain. However, as noted by Faltinsen & Timokha [15], their use is justified when the time varying domain is close to its mean shape.

Moreover, let us introduce the following definition for the lagrangian of the fluid motion:

$$\mathcal{L} = -\int_A \int_{-H}^{\eta(x,y,t)} p dz dA \tag{18}$$

so that, the functional (13) assumes the following expression:

$$\mathcal{F} = \int_{t_1}^{t_2} \mathcal{L} dt \tag{19}$$

Substituting expansions (17) in (19) and integrating, at least formally, with respect to the space variables, functional (19) becomes a functional of $\mathcal{A}_{nm}(t)$, $q_{nm}(t)$ and $\frac{d\mathcal{A}_{nm}(t)}{dt}$. Regarding $\mathcal{A}_{nm}(t)$,$q_{nm}(t)$,$\frac{d\mathcal{A}_{nm}(t)}{dt}$ as the components of the vectors $\mathbf{A}(t)$,$\mathbf{q}(t)$,$\mathbf{A}'(t)$, a more compact notation can be adopted and expression (19) can be written as:

$$\mathcal{F} = \int_{t_1}^{t_2} \mathcal{L}\left(\mathbf{A}(t), \mathbf{q}(t), \mathbf{A}'(t), t\right) dt \tag{20}$$

Now for the functional is stationary, it is necessary and sufficient that the unknown time dependent coefficients satisfy the Lagrange equations [25].

$$\begin{cases} \dfrac{d}{dt}\dfrac{\partial \mathcal{L}}{\partial \mathbf{A}'} - \dfrac{\partial \mathcal{L}}{\partial \mathbf{A}} = 0 \\ -\dfrac{\partial \mathcal{L}}{\partial \mathbf{q}} = 0 \end{cases} \tag{21}$$

It is interesting to observe that with this line of thinking, the unknown time dependent coefficients $\mathcal{A}_{nm}(t)$, $q_{nm}(t)$ can be considered as nothing but the generalized coordinates of the motion. Moreover, the term $\frac{d}{dt}\frac{\partial \mathcal{L}}{\partial \mathbf{q}'}$ does not appear in eqn. (21) because the adopted definition of the lagrangian coincides with the work done by the pressure over the whole fluid domain, and the expression of this latter eqn. (6) contains the time derivative of the velocity potential explicitly, but not that of the free surface $\eta$. Equation (21) constitute an infinite set of nonlinear, first order, ordinary differential equations for $\mathcal{A}_{nm}(t)$, $q_{nm}(t)$.

### 2.4 Computational considerations

The Lagrange eqns. (21) are very difficult to handle, as the coupling between $\mathcal{A}_{nm}(t)$, $q_{nm}(t)$ occurs by means of the functions $d_{nm}(z)$. Let us consider the substitution of the expansions eqns. (17) in the definition of the lagrangian eqns. (18) and perform the integrations. It comes out that:

$$\mathcal{L} = \sum_{nm} \mathcal{A}'_{nm} \langle M_{nm} (\eta (\mathbf{q},t), x, y) \rangle + \langle N (\mathbf{A},\eta (\mathbf{q},t), x, y, t) \rangle \tag{22}$$

in which the operator $\langle \cdot \rangle$ has been defined as: $\langle \cdot \rangle \equiv \iint_A \cdot dxdy$ and $M_{nm}, N$ are known functions of $\mathbf{A},\eta(\mathbf{q},t), x, y, t$. Then, at least formally, the Lagrange equations assume the form:

$$\begin{cases} \sum_{lp} \left\langle \dfrac{\partial M_{nm}}{\partial \eta} \dfrac{\partial \eta}{\partial q_{lp}} \right\rangle q'_{lp} + \left\langle \dfrac{\partial M_{nm}}{\partial \eta} \dfrac{\partial \eta}{\partial t} \right\rangle - \left\langle \dfrac{\partial N}{\partial \mathcal{A}_{nm}} \right\rangle = 0 \\ \sum_{lp} \left\langle \dfrac{\partial M_{lp}}{\partial \eta} \dfrac{\partial \eta}{\partial q_{nm}} \right\rangle \mathcal{A}'_{lp} + \left\langle \dfrac{\partial N}{\partial \eta} \dfrac{\partial \eta}{\partial q_{nm}} \right\rangle = 0 \end{cases} \tag{23}$$

where the coefficients of the equations appear as integrals of functions of $\eta$, which is, in turn, unknown.

In matrix form, eqns. (23) assume the following expression:

$$\begin{cases} \dfrac{\partial \mathbf{M}}{\partial \mathbf{q}} \mathbf{q}' + \mathbf{P} - \dfrac{\partial N}{\partial \mathbf{A}} = 0 \\ \dfrac{\partial \mathbf{M}}{\partial \mathbf{q}}^{\mathbf{T}} \mathbf{A}' + \dfrac{\partial N}{\partial \mathbf{q}} = 0 \end{cases} \tag{24}$$

where the following definitions have been used: $\frac{\partial \mathbf{M}}{\partial \mathbf{q}} \equiv (((\langle \frac{\partial M_{nm}}{\partial \eta} \frac{\partial \eta}{\partial q_{lp}} \rangle))$, $\frac{\partial \mathbf{M}}{\partial \mathbf{q}}^{\mathbf{T}} \equiv (((\langle \frac{\partial M_{lp}}{\partial \eta} \frac{\partial \eta}{\partial q_{nm}} \rangle))$, $\mathbf{P} \equiv ((\langle \frac{\partial M_{nm}}{\partial \eta} \frac{\partial \eta}{\partial t} \rangle)$, $\frac{\partial N}{\partial \mathbf{A}} \equiv ((\langle \frac{\partial N}{\partial \mathcal{A}_{nm}} \rangle)$, $\frac{\partial N}{\partial \mathbf{q}} \equiv ((\langle \frac{\partial N}{\partial \eta} \frac{\partial \eta}{\partial q_{nm}} \rangle)$. It is possible, at least formally, to invert the matrices $\frac{\partial \mathbf{M}}{\partial \mathbf{q}}, \frac{\partial \mathbf{M}}{\partial \mathbf{q}}^{\mathbf{T}}$ in order to put the system (24) in normal form. The result is:

$$\begin{cases} \mathbf{q}' = -\dfrac{\partial \mathbf{M}}{\partial \mathbf{q}}^{-1} \left( \mathbf{P} - \dfrac{\partial N}{\partial \mathbf{A}} \right) \\ \mathbf{A}' = -\left( \dfrac{\partial \mathbf{M}}{\partial \mathbf{q}}^{\mathbf{T}} \right)^{-1} \dfrac{\partial N}{\partial \mathbf{q}} \end{cases} \tag{25}$$

Equations (25) are the evolution equations for $\mathcal{A}_{nm}(t), q_{nm}(t)$ – it is important to observe that their coefficients are integrals of functions which depend on $\eta$ and cannot be calculated analytically. This means that to solve eqns. (25) two strategies are possible. The most common is to expand the functions of $\eta$ in Taylor series. Then the dependency of the equations coefficients on $\eta$ becomes explicit and the integrals can be calculated. The second strategy consists of integrating the differential system eqns. (25) numerically, calculating the integrals at every time step. Following the first strategy there is the evident advantage that the integrals are calculated only once, but their number increases enormously even if an $\mathcal{O}\left(\eta^3\right)$

expansion is adopted. On the other hand, the second strategy makes it possible to account for the nonlinearities as they are. A possible criticism of the second strategy is that it does not allow the application of analytical methods to the equations.

It should be noted that the first important point in solving system eqns. (25) is to define a finite set of integers $n, m$ – *i.e.* to choose a finite number of functions $c_{nm}(x, y)$ which are representative for the case in question. In other words, a truncation criterion must be adopted for the series representation eqns. (17), such that the leading modes – *i.e.* those whose evolution captures most of the fluid system's energy – are considered. Useful truncation criteria can be obtained by a preliminary analysis of the linearized eqns. (23). Experiments are also very useful in defining truncation criteria, particularly when suitable signal analysis is applied to the result.

## 3 Dissipative effects

Dissipative effects play an important role in the long time numerical simulation of sloshing dynamics. Experimental observations [11] show that after a long enough transient, the motion does not depend on the initial conditions. Moreover, the solutions of a mathematical model in which dissipative effects are correctly taken into account exhibit a weaker and weaker dependence on the initial conditions as the time increases. This dynamic behavior is typical for any weakly dissipative system.

In the framework of the variational formulation, the fluid motion is described using a potential formulation. Nevertheless, damping effects, due mainly to the boundary layers in correspondence with the container walls, can be introduced in eqns. (21) as generalized dissipative forces, defined as the derivatives of the dissipation function $\mathcal{G}(\mathbf{A}', \mathbf{q}')$ (*i.e.* the work done by the dissipative forces) with respect to the generalized velocities [25]:

$$\begin{cases} \dfrac{d}{dt}\dfrac{\partial \mathcal{L}}{\partial \mathbf{A}'} - \dfrac{\partial \mathcal{L}}{\partial \mathbf{A}} = \dfrac{\partial \mathcal{G}}{\partial \mathbf{A}'} \\ -\dfrac{\partial \mathcal{L}}{\partial \mathbf{q}} = \dfrac{\partial \mathcal{G}}{\partial \mathbf{q}'} \end{cases} \tag{26}$$

The dissipation function $\mathcal{G}(\mathbf{A}', \mathbf{q}')$ is defined by means of the logarithmic decrement coefficient of the amplitude of each mode [3] – the evaluation of such a coefficient then becomes a crucial point. For this purpose there are two main theoretical approaches. The first consists of defining a suitable linear eigenvalue problem which furnishes a dispersion relationship between the wave vector $\mathbf{k}$ and the corresponding complex pulsation $\Phi_{\mathbf{k}}$ [11, 26]. The second consists of calculating the ratio between the dissipated and the total energy of the fluid motion [12, 27, 28].

Let us analyze the first approach, assuming for the sake of simplicity that the wavevector $\mathbf{k}$ and the functions are respectively defined as $\mathbf{k} \equiv \left(\frac{n\pi}{B}, \frac{m\pi}{L}\right) = (k_n, \tau_m)$, $q_{nm}(t) \equiv e^{-I\Phi_{\mathbf{k}}t}$, $c_{nm}(x, y) \equiv \cos(k_n x)\cos(\tau_m y)$. Such definitions correspond to the case of a prismatic container whose section is a rectangle with sides $B$ and $L$. The dispersion relationship between the wave vector $\mathbf{k}$ and the complex pulsation $\Phi_{\mathbf{k}}$ can be obtained only in a linear regime of motion, namely in the case of small amplitude waves. Then, considering the velocity and pressure fields

$$\begin{aligned} u_{\mathbf{k}} &= U_{\mathbf{k}}(z)\sin(k_n x)\cos(\tau_m y)e^{-I\Phi_{\mathbf{k}}t} \\ v_{\mathbf{k}} &= V_{\mathbf{k}}(z)\cos(k_n x)\sin(\tau_m y)e^{-I\Phi_{\mathbf{k}}t} \\ w_{\mathbf{k}} &= W_{\mathbf{k}}(z)\cos(k_n x)\cos(\tau_m y)e^{-I\Phi_{\mathbf{k}}t} \\ p^d_{\mathbf{k}} &= P_{\mathbf{k}}(z)\cos(k_n x)\cos(\tau_m y)e^{-I\Phi_{\mathbf{k}}t} \end{aligned} \tag{27}$$

and substituting them in the linearized Navier–Stokes equations for an incompressible fluid in non-forced regime:

$$\begin{aligned} \frac{\partial \mathbf{v}}{\partial t} = -\frac{\nabla p}{\varrho} - \nabla(gz) + \frac{\mu}{\varrho}\nabla^2\mathbf{v} \\ \nabla\cdot\mathbf{v} = \mathbf{0} \end{aligned} \tag{28}$$

the following system of ordinary, linear, differential equations is obtained for the unknown functions $U_{\mathbf{k}}, V_{\mathbf{k}}, W_{\mathbf{k}}, P_{\mathbf{k}}$:

$$\begin{aligned} I\Phi_{\mathbf{k}}U_{\mathbf{k}} + \frac{1}{\varrho}k_n P_{\mathbf{k}} - \frac{\mu}{\varrho}\left(|\mathbf{k}|^2 - \frac{d^2}{dz^2}\right)U_{\mathbf{k}} &= 0 \\ I\Phi_{\mathbf{k}}V_{\mathbf{k}} + \frac{1}{\varrho}\tau_m P_{\mathbf{k}} - \frac{\mu}{\varrho}\left(|\mathbf{k}|^2 - \frac{d^2}{dz^2}\right)V_{\mathbf{k}} &= 0 \\ I\Phi_{\mathbf{k}}W_{\mathbf{k}} - \frac{1}{\varrho}\frac{dP_{\mathbf{k}}}{dz} - \frac{\mu}{\varrho}\left(|\mathbf{k}|^2 - \frac{d^2}{dz^2}\right)W_{\mathbf{k}} &= 0 \\ k_n U_{\mathbf{k}} + \tau_m V_{\mathbf{k}} + \frac{dV_{\mathbf{k}}}{dz} &= 0 \end{aligned} \tag{29}$$

These functions, together with the free surface $\eta_{\mathbf{k}} = q_{\mathbf{k}}\cos(k_n x)\cos(\tau_m y)$ $e^{-I\Phi_{\mathbf{k}}t}$, have to satisfy the following linear homogeneous boundary conditions:

$$\begin{aligned} u_{\mathbf{k}} = v_{\mathbf{k}} = w_{\mathbf{k}} &= 0, \quad z = -H \\ p^d_{\mathbf{k}} - \varrho g\eta - 2\mu\frac{\partial w_{\mathbf{k}}}{\partial z} &= 0, \quad z = 0 \\ \frac{\partial u_{\mathbf{k}}}{\partial z} + \frac{\partial w_{\mathbf{k}}}{\partial x} &= 0, \quad z = 0 \\ \frac{\partial v_{\mathbf{k}}}{\partial z} + \frac{\partial w_{\mathbf{k}}}{\partial y} &= 0, \quad z = 0 \\ \frac{\partial \eta_{\mathbf{k}}}{\partial t} &= w_{\mathbf{k}}, \quad z = 0 \end{aligned} \tag{30}$$

The first three conditions eqns. (30) represent the no-slip conditions on the bottom of the prismatic container, the last condition is simply the linearized kinematic condition on the free surface, and the other conditions represent the absence of normal and tangential stress on the free surface. The velocity and pressure fields eqns. (27) cannot satisfy the no slip conditions on the vertical walls of the container. In addition, the velocity and pressure fields eqns. (27) actually represent the exact linear motion for a fluid unbounded in $x$ and $y$ directions. This fact is equivalent to neglecting the effects of the container's lateral walls on the dissipation rate. As a consequence, this first approach generally furnishes damping rate values lower than those experimentally observed.

From the boundary conditions eqns. (30), the following six conditions are obtained:

$$
\begin{aligned}
U_{\mathbf{k}} = V_{\mathbf{k}} = W_{\mathbf{k}} = 0, \quad z = -H \\
-I\Phi_{\mathbf{k}}P_{\mathbf{k}} - \varrho g W_{\mathbf{k}} + 2\mu I\Phi_{\mathbf{k}}\frac{dW_{\mathbf{k}}}{dz} = 0, \quad z = 0 \\
\frac{dU_{\mathbf{k}}}{dz} - k_n W_{\mathbf{k}} = 0, \quad z = 0 \\
\frac{dV_{\mathbf{k}}}{dz} - \tau_m W_{\mathbf{k}} = 0, \quad z = 0
\end{aligned}
\tag{31}
$$

The general solution of eqns. (29) is:

$$
\begin{aligned}
W_{\mathbf{k}}(z) = a_1 e^{|\mathbf{k}|z} + a_2 e^{-|\mathbf{k}|z} + a_3 e^{\beta_{\mathbf{k}} z} + a_4 e^{-\beta_{\mathbf{k}} z} \\
V_{\mathbf{k}}(z) = -\tau_m \frac{a_1 e^{|\mathbf{k}|z} - a_2 e^{-|\mathbf{k}|z}}{|\mathbf{k}|} + a_5 e^{\beta_{\mathbf{k}} z} + a_6 e^{-\beta_{\mathbf{k}} z} \\
U_{\mathbf{k}}(z) = -\frac{1}{k_n}\left(\frac{dW_{\mathbf{k}}}{dz} + \tau_m V_{\mathbf{k}}\right) \\
P_{\mathbf{k}}(z) = I\rho\Phi_{\mathbf{k}} \frac{a_1 e^{|\mathbf{k}|z} - a_2 e^{-|\mathbf{k}|z}}{|\mathbf{k}|}
\end{aligned}
\tag{32}
$$

where $\beta_{\mathbf{k}} \equiv \sqrt{|\mathbf{k}|^2 - I\frac{\varrho\Phi_{\mathbf{k}}}{\mu}}$. Imposing the fulfillment of the six conditions eqns. (31), a linear homogeneous system for the unknowns $a_1, ..., a_6$ is obtained. The desired dispersion relationship is obtained by requiring that the $6x6$ determinant of the matrix coefficients be zero. Introducing the following dimensionless variables:

$$
\begin{aligned}
\Gamma_{\mathbf{k}} \equiv H|\mathbf{k}| \\
\phi_{\mathbf{k}} \equiv \frac{\varrho H^2 \Phi_{\mathbf{k}}}{\mu} \\
R_E = \frac{\varrho^2 g H^3}{\mu^2}
\end{aligned}
\tag{33}
$$

the dispersion relationship assumes the following implicit form:

$$
\Psi(\Gamma_{\mathbf{k}}, R_E, \phi_{\mathbf{k}}) = 0 \tag{34}
$$

In eqns. (34) $\Gamma_{\mathbf{k}}$ and $R_E$ have to be considered real parameters, while $\phi_{\mathbf{k}}$, defined as $\phi_{\mathbf{k}} = \phi_{R\mathbf{k}} + I\phi_{I\mathbf{k}}$, is a complex variable with the following meaning for its real

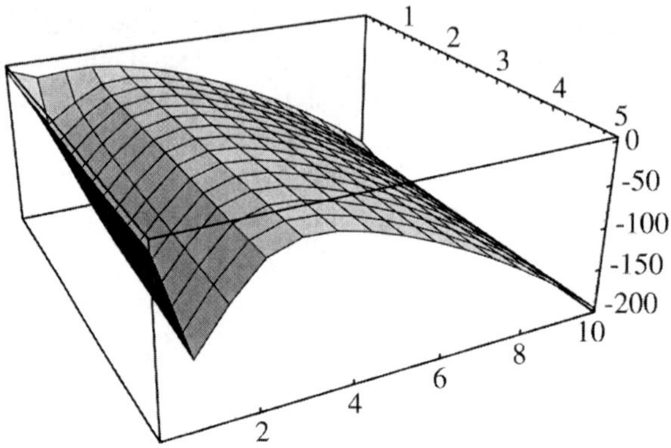


Figure 2: Plot of $\phi_{I\mathbf{k}}\,(R_E, \Gamma_{\mathbf{k}})$.

and imaginary parts: $\phi_{R\mathbf{k}} = 2\pi f_{\mathbf{k}}$, $\phi_{I\mathbf{k}} \;=\; -\,\gamma_{\mathbf{k}} \,<\, 0$ , $f_{\mathbf{k}}$ and $\gamma_{\mathbf{k}}$ being, respectively, the non-dimensional natural frequency and the non-dimensional logarithmic decrement of the mode whose wave vector is **k**.

Unfortunately, because of the transcendental nature of the function $\Psi$, eqns. (34) cannot be solved analytically with respect to $\phi_{\mathbf{k}}$. Applying Newton's method, the roots of eqn. (34) are evaluated as the limit of the following sequence:

$$\phi_{\mathbf{k}}^{n+1} = \phi_{\mathbf{k}}^{n} - \frac{\Psi(\Gamma_{\mathbf{k}}, R_E, \phi_{\mathbf{k}}^{n})}{\Psi_{\phi_{\mathbf{k}}}(\Gamma_{\mathbf{k}}, R_E, \phi_{\mathbf{k}}^{n})} \tag{35}$$

where $\Psi_{\phi_{\mathbf{k}}}(\Gamma_{\mathbf{k}}, R_E, \phi_{\mathbf{k}}^{n})$ is the complex derivative of $\Psi(\Gamma_{\mathbf{k}}, R_E, \phi_{\mathbf{k}})$ with respect to $\phi_{\mathbf{k}}$.

The first value $\phi_{\mathbf{k}}^{0}$ is assumed to be the non-dimensional, purely real value of the root of the inviscid dispersion relationship:

$$\phi_{\mathbf{k}}^{0} = \phi_{\mathbf{k}}^{inv}(R_E, \Gamma_{\mathbf{k}}) = \sqrt{R_E \Gamma_{\mathbf{k}} Tanh\,(\Gamma_{\mathbf{k}})} \tag{36}$$

Sequence eqns. (35) converges on the exact complex value $\phi_{\mathbf{k}} = \phi_{R\mathbf{k}}(R_E, \Gamma_{\mathbf{k}}) + I\phi_{I\mathbf{k}}(R_E, \Gamma_{\mathbf{k}})$ for a large range of values of $\Gamma_{\mathbf{k}}$ and $R_E$. The order of magnitude of $R_E$ is normally very high – *e.g.* $\mathcal{O}\left(10^{10}\right)$ as in the experimental configuration of La Rocca *et al* [11], where it assumes the value $\overline{R}_E = 2.467 \times 10^{10}$.

In figs. 2 and 3 the surfaces $\phi_{I\mathbf{k}}(R_E, \Gamma_{\mathbf{k}})$, $\frac{\phi_{R\mathbf{k}}(R_E,\Gamma_{\mathbf{k}})}{\phi_{\mathbf{k}}^{inv}(R_E,\Gamma_{\mathbf{k}})}$ , plotted versus $\Gamma_{\mathbf{k}}$ and $R_E/\,\overline{R}_E$, are shown in the range of values $10^{-3} \le \Gamma_{\mathbf{k}} \le 10, 10^{-3} \le R_E/\overline{R}_E \le 5$.

It is interesting to observe that the ratio $\frac{\phi_{R\mathbf{k}}(R_E,\Gamma_{\mathbf{k}})}{\phi_{\mathbf{k}}^{inv}(R_E,\Gamma_{\mathbf{k}})}$ is very close to the one in the examined parameters range, particularly for high $\Gamma_{\mathbf{k}}$. This means that the frequency of oscillations of an inviscid **k** mode is almost equal to the corresponding frequency of a viscous **k** mode. Moreover, fig. 2 highlights the fact that $\phi_{I\mathbf{k}}$ $(R_E, \Gamma_{\mathbf{k}})$ is always negative and for $\Gamma_{\mathbf{k}} > 4$ decreases monotonically as $\Gamma_{\mathbf{k}}$ increases (*i.e.* for high wave numbers).

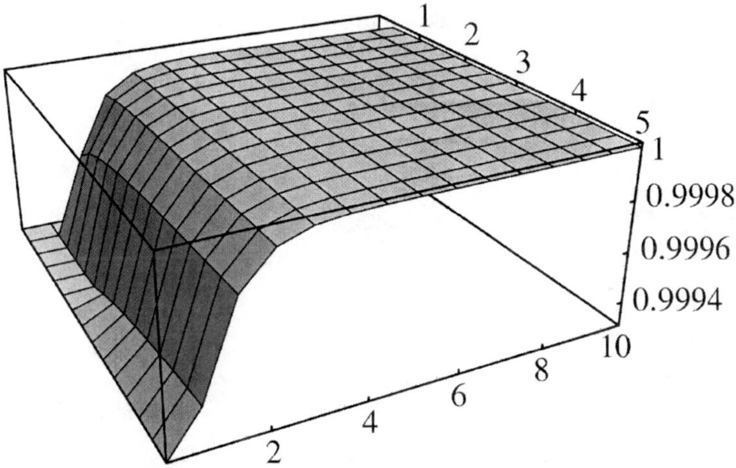


Figure 3: Plot of $\frac{\phi_{R\mathbf{k}}(R_E,\Gamma_{\mathbf{k}})}{\phi_{\mathbf{k}}^{inv}(R_E,\Gamma_{\mathbf{k}})}$.

In the second approach, the logarithmic decrement $\gamma_{\mathbf{k}}$ linked to the corresponding viscous $\mathbf{k}$ mode can be defined as [12, 27, 29]:

$$\gamma_{\mathbf{k}} = \frac{D_{\mathbf{k}}}{2E_{\mathbf{k}}} \tag{37}$$

where $D_{\mathbf{k}}$ and $E_{\mathbf{k}}$ are, respectively, the mean dissipation rate and the mean kinetic energy for the free oscillation of the $\mathbf{k}$ mode. The mean dissipation rate is due mainly to viscous dissipation at the rigid boundary of the container and viscous dissipation in the interior of the fluid domain, while dissipative effects on the free surface are negligible. As a consequence, $D_{\mathbf{k}}$ can be defined as the work done by viscous stresses in the whole fluid domain $D_f$ according to the following formula:

$$\begin{gathered} D_{\mathbf{k}} = \frac{\omega_{\mathbf{k}}}{2\pi} \int\limits_0^{2\pi/\omega_{\mathbf{k}}} dt \int\limits_{\mathcal{D}_f} 2\mu\, \mathfrak{D}_{\mathbf{k}} \colon \mathfrak{D}_{\mathbf{k}} dV \\ \mathfrak{D}_{\mathbf{k}} \colon \mathfrak{D}_{\mathbf{k}} \equiv \sum_{i,j=1}^{3} \mathfrak{D}_{\mathbf{k}i,j}^2 \end{gathered} \tag{38}$$

where $\omega_{\mathbf{k}} = \Phi_{\mathbf{k}}^{inv} = \sqrt{g\,|\mathbf{k}|\, Tanh\left(|\mathbf{k}|\, H\right)}$, $\mathfrak{D}_{\mathbf{k}}$ is the symmetrical part of the tensor $\nabla \mathbf{v}_{\mathbf{k}}$ and $\mathbf{v}_{\mathbf{k}}$ is the velocity field corresponding to the free oscillation of the $\mathbf{k}$ mode, with frequency $\frac{\omega_{\mathbf{k}}}{2\pi}$. The kinetic energy $E_{\mathbf{k}}$ is defined by the expression:

$$E_{\mathbf{k}} = \frac{\omega_{\mathbf{k}}}{2\pi} \int\limits_0^{2\pi/\omega_{\mathbf{k}}} dt \int\limits_{\mathcal{D}_f} \frac{1}{2} \varrho \mathbf{v}_{\mathbf{k}} \cdot \mathbf{v}_{\mathbf{k}} dV \tag{39}$$

The velocity field $\mathbf{v_k}$ can be decomposed into a sum of a rotational part $\mathbf{v}_\mathbf{k}^R$ (negligible outside the boundary layers) and an irrotational part $\nabla\varphi_\mathbf{k}$ :

$$\mathbf{v_k} = \mathbf{v}_\mathbf{k}^R + \nabla\varphi_\mathbf{k} \tag{40}$$

where:

$$\varphi_\mathbf{k} \equiv \mathrm{Re}(\widehat{\varphi_\mathbf{k}} e^{I\omega_\mathbf{k} t})$$
$$\widehat{\varphi_\mathbf{k}} \equiv A_\mathbf{k} \frac{\cosh\left(|\mathbf{k}|\,(z+H)\right)}{\cosh(|\mathbf{k}|\,(H))} \cos(k_n x)\cos(\tau_m y) \tag{41}$$

and $A_\mathbf{k}$ is a constant. In order to determine the velocity field in the boundary layers, it is necessary to substitute (40) into the linearized Navier–Stokes equation (28):

$$\nabla\left(\frac{\partial\varphi_\mathbf{k}}{\partial t} + \frac{p_\mathbf{k}}{\rho} + gz\right) + \frac{\partial \mathbf{v}_\mathbf{k}^R}{\partial t} = \nu\nabla^2\mathbf{v}_\mathbf{k}^R$$

Requiring that $\nabla(\frac{\partial\varphi_\mathbf{k}}{\partial t} + \frac{p_\mathbf{k}}{\rho} + gz) = 0$ it follows that the rotational part of the velocity field $\mathbf{v_k}$, has to satisfy the Stokes equation:

$$\frac{\partial \mathbf{v}_\mathbf{k}^R}{\partial t} = \nu\nabla^2\mathbf{v}_\mathbf{k}^R \tag{42}$$

Equation (42) must be solved in the boundary layer of each rigid wall of the container. To this purpose, let $x_1, x_2, x_3$ be a local system of Cartesian coordinates with the plane $x_3 = 0$ coinciding with the rigid wall ($x_3 > 0$ in the interior of fluid domain) and let $\varepsilon_1, \varepsilon_2, \varepsilon_3$ be the corresponding unit vectors. We further assume that in the boundary layer near a rigid wall the velocity is purely tangential to the wall, due to the small thickness of the boundary layer:

$$\mathbf{v}_\mathbf{k}^R \simeq u_\mathbf{k}^R \varepsilon_1 + v_\mathbf{k}^R \varepsilon_2$$

Equation (42) must be solved with the following boundary conditions:

$$u_\mathbf{k}^R = -\left.\frac{\partial\widehat{\varphi_\mathbf{k}}}{\partial x_1}\right|_{x_3=0} e^{I\omega_\mathbf{k} t},\, v_\mathbf{k}^R = -\left.\frac{\partial\widehat{\varphi_\mathbf{k}}}{\partial x_2}\right|_{x_3=0} e^{I\omega_\mathbf{k} t}$$
$$u_\mathbf{k}^R = v_\mathbf{k}^R = 0,\ x_3 \to +\infty \tag{43}$$

The first ones are the no-slip conditions on the rigid wall, while the conditions for $x_3 \to +\infty$ require that the rotational part of the velocity field vanish in the interior of the fluid domain where the motion is essentially irrotational.

The solution to eqns. (42) with boundary conditions eqns. (43) can be expressed in the form:

$$
\begin{aligned}
u_{\mathbf{k}}^R &= -\left.\frac{\partial\widehat{\varphi_{\mathbf{k}}}}{\partial x_1}\right|_{x_3=0} e^{-x_3/\delta}e^{I(\omega_{\mathbf{k}}t-x_3/\delta)} \\
v_{\mathbf{k}}^R &= -\left.\frac{\partial\widehat{\varphi_{\mathbf{k}}}}{\partial x_2}\right|_{x_3=0} e^{-x_3/\delta}e^{I(\omega_{\mathbf{k}}t-x_3/\delta)}
\end{aligned}
\tag{44}
$$

where $\delta \equiv \sqrt{\frac{2\nu}{\omega_{\mathbf{k}}}}$ is assumed as the boundary layer thickness due to the oscillation of the $\mathbf{k}$ mode [19, 20].

Now from the expression (41) for the potential $\varphi_{\mathbf{k}}$ and the expressions of $\mathbf{v}_{\mathbf{k}}^R$ in each boundary layer it is possible to obtain the velocity field $\mathbf{v}_{\mathbf{k}}$ at each point in the fluid domain and, in turn, the tensor $\mathbf{v}_{\mathbf{k}}$. At last it is possible to calculate the terms $D_{\mathbf{k}}$ and $E_{\mathbf{k}}$ according to definitions (38) and (39). Both $D_{\mathbf{k}}$ and $E_{\mathbf{k}}$depend on $A_{\mathbf{k}}^2$ (see eqns. 41), but the ratio $D_{\mathbf{k}}/E_{\mathbf{k}}$ does not and coincides with the desired logarithmic decrement $\gamma_{\mathbf{k}}$ for any assigned wave vector $\mathbf{k}$.

>From an experimental point of view, it is possible to determine the values of $\gamma_{\mathbf{k}}$ by means of direct measurements of free oscillations of the free surface. Two methods are worth remembering. In the first a least squared method is applied to fit experimental data of the decaying free surface with suitable analytical expressions to determine the coefficients $\gamma_{\mathbf{k}}$. In the second, by using the theory of wavelet transform, a direct determination of $\gamma_{\mathbf{k}}$ is obtained by using time histories of the decaying free surface.

Let us show the first method. For this purpose, let the free surface $\eta$ at the point $(\widetilde{x}, \widetilde{y})$ be represented by the following truncated series:

$$
\eta(\widetilde{x}, \widetilde{y}, t) = \sum_{n,m=1}^{N} q_{nm}(t)\cos(k_n\widetilde{x})\cos(\tau_m\widetilde{y}) \tag{45}
$$

where $N$ is the total number of considered modes. It follows that if $N$ measurements of time histories of the free surface are performed in $N$ distinct points $(\widetilde{x}, \widetilde{y})$, it is possible to define a linear system of equations which allows the single modal contributions $q_{nm}(t)$ to be determined and permits the calculation of the corresponding logarithmic decrement $\gamma_{\mathbf{k}}$. In order to show how the method works, let us consider a $2D$ case $(m=0)$ and choose the points $(\widetilde{x}, \widetilde{y}) = (0, B/2)$, $(B/2, B/2)$, $(B/6, B/2)$. Writing eqn. (45) in such points, the following linear system is obtained:

$$
\begin{aligned}
q_1 + q_2 + q_3 &= \eta(0, B/2, t) \\
q_2 &= -\eta(B/2, B/2, t) \\
q_1\cos(\pi/6) + q_2\cos(\pi/3) &= \eta(B/6, B/2, t)
\end{aligned}
\tag{46}
$$

The choice of these three measurement-points is due to these considerations: point $(B/2, B/2)$ is a nodal point for every $2D$ odd mode, while $(B/6, B/2)$ is a nodal point for the $2D$ odd mode $n=3, m=0$. Solving eqns. (46), it follows that:

$$
\begin{aligned}
q_1(t) &= \frac{\eta(B/2, B/2, t) + 2\eta(B/6, B/2, t)}{\sqrt{3}} \\
q_2(t) &= -\eta(B/2, B/2, t) \\
q_3(t) &= \frac{\sqrt{3}\eta(0, B/2, t) + (\sqrt{3}-1)\eta(B/2, B/2, t) - 2\eta(B/6, B/2, t)}{\sqrt{3}}
\end{aligned}
\tag{47}
$$

Now a least squared approximation technique is used in order to evaluate the damping coefficients $\gamma_{\mathbf{k}}$. In particular, supposing that the experimental time histories are discretized by using $N_{data}$ with $\Delta t$ as the time interval of sampling, it is possible to define the quantity:

$$
S_c(A_j, \gamma_j, \varkappa_j) \equiv \sum_{k=1}^{N_{data}} \left[q_j(k\Delta t) - A_j e^{-\gamma_j k\Delta t} \sin(\omega_j k\Delta t - \varkappa_j)\right]^2 \tag{48}
$$

where $\gamma_j$ is the damping coefficient linked to $q_j(t)$-mode and $A_j, \varkappa_j$, are, respectively, the amplitude and phase of $q_j(t)$. Stipulating that $S_c(A_j, \gamma_j, \varkappa_j)$ attains a minimum in correspondence with $A_j, \gamma_j, \varkappa_j$, the following system is obtained for the unknowns $A_j, \gamma_j, \varkappa_j$:

$$
\begin{aligned}
\frac{\partial S_c(A_j, \gamma_j, \varkappa_j)}{\partial A_j} &= 0 \\
\frac{\partial S_c(A_j, \gamma_j, \varkappa_j)}{\partial \gamma_j} &= 0 \\
\frac{\partial S_c(A_j, \gamma_j, \varkappa_j)}{\partial \varkappa_j} &= 0
\end{aligned}
\tag{49}
$$

The experimental free oscillations are brought about by putting the tank in oscillation with a given frequency and amplitude and suddenly arresting it in a horizontal position. When the amplitude of the free oscillations is sufficiently small to ensure the linear regime of motion, time histories of the decaying free surface are recorded.

This experimental technique, together with the second theoretical approach described in this section, was used to determine the damping coefficients $\gamma_1, \gamma_2, \gamma_3$ of the first three $2D$ modes for the sloshing of two immiscible liquid layers inside a closed square section tank [12]. The comparison between numerical and experimental values is shown in Table 1.

In some cases, the solution of system eqns. (49) is not easy due to badly conditioned Jacobian matrix problems caused by the phase $\varkappa_j$. In these cases the value of the phase $\varkappa_j$ has to be defined "empirically" by superimposing the graphics of

Table 1

| $\gamma$ | Numerical | Experimental |
|---|---|---|
| $\gamma_1$ | 0.038 | 0.040 |
| $\gamma_2$ | 0.052 | 0.058 |
| $\gamma_3$ | 0.066 | 0.069 |

curves $q_j(t)$ and $\sin(\omega_j t - \varkappa_j)$ and choosing $\varkappa_j$ in such a way as to minimize the phase difference. Then system (49) must be solved only with respect to the unknown $A_j, \gamma_j$ and generally does not exhibit bad conditioning problems.

To overcome these problems, an elegant procedure based on the property of the wavelet transform can be applied as follows. When the free surface decays in the linear regime of free oscillations, eqns. (45) becomes:

$$\eta(\widetilde{x}, \widetilde{y}, t) = \sum_{n,m=1}^{N} q_{nm} \sin(\omega_{nm} t - \varkappa_{nm}) e^{-\gamma_{nm} t} \cos(k_n \widetilde{x}) \cos(\tau_m \widetilde{y}) \qquad (50)$$

due to the absence of non linear interactions. Let us consider the wavelet transform [30] of the quantity $\eta(\widetilde{x}, \widetilde{y}, t)$ with respect to time $t$. Such a wavelet transform is defined by the formula:

$$W(t, f) = \sqrt{\frac{2\pi f}{k_s}} \int_{-\infty}^{+\infty} \eta(\widetilde{x}, \widetilde{y}, \tau) \widetilde{\Upsilon} \left( \frac{2\pi f}{k_s} (\tau - t) \right) d\tau \qquad (51)$$

where $f$ is the local frequency of the signal at instant $t$, $k_s$ a real parameter and $\widetilde{\Upsilon}$ the complex conjugate of the function $\Upsilon$. When $k_s \geq 6$ the admissibility of the function $\Upsilon$ is ensured [30]. $\Upsilon$ can be defined in several ways. One of the most adopted definitions is the so-called Morlet wavelet:

$$\Upsilon(\xi) \equiv e^{-\frac{\xi^2}{2} + Iks\xi} \qquad (52)$$

The term $|W(t, f)|^2 = W(t, f)\widetilde{W}(t, f)$ is proportional to the instantaneous energy wavelet spectrum. Assuming for $\eta$ the expression eqn. (50), that the frequencies $\omega_{nm}$ are well distinct and there is not reciprocal interaction among the spectral contributions of each linear mode, the analytical calculation of $\left|W(t, \frac{\omega_{nm}}{2\pi})\right|^2$ gives:

$$\begin{aligned} &\left| W\left(t, \frac{\omega_{nm}}{2\pi}\right) \right|^2 \\ &= \frac{k_s \pi e^{k_s^2 (\gamma_{nm}^2/\omega_{nm}^2 - 4)}}{2\omega_{nm}} e^{-2\gamma_{nm} t} \\ &\times \left[ 1 + e^{2k_s^2} \left( e^{2k_s^2} - 2\cos\left( 2\left( \omega_{nm} t - \frac{k_s^2 \gamma_{nm}}{\omega_{nm}} - \varkappa_{nm} \right) \right) \right) \right] \end{aligned} \qquad (53)$$

As $k_s \geq 6$, it follows that:

$$\left| \frac{2\cos\left(2\left(\omega_{nm}t - \frac{k_s^2\gamma_{nm}}{\omega_{nm}} - \varkappa_{nm}\right)\right)}{e^{2k_s^2}} \right| \leq \frac{2}{e^{2k_s^2}} < 10^{-31}$$

As a consequence, the numerical value of $\left|W(t, \frac{\omega_{nm}}{2\pi})\right|^2$ is almost equal to:

$$\left|W\left(t, \frac{\omega_{nm}}{2\pi}\right)\right|^2 = \frac{k_s\pi e^{k_s^2(\gamma_{nm}^2/\omega_{nm}^2-4)}(1+e^{4k_s^2})}{2\omega_{nm}}e^{-2\gamma_{nm}t} \tag{54}$$

The calculation of $\gamma_{nm}$ can, as a consequence, be obtained by the simple formula:

$$\gamma_{nm} = -\frac{1}{2}\frac{\partial}{\partial t}\ln\left[\frac{2\omega_{nm}\left|W(t, f_{nm})\right|^2}{k_s\pi e^{k_s^2(\gamma_{nm}^2/\omega_{nm}^2-4)}(1+e^{4k_s^2})}\right] \tag{55}$$

This technique also offers the possibility of controlling experimentally the linear regime hypothesis for the decaying motion. In fact, as highlighted in eqns. (54), if each modes $q_j(t)$ is defined by the law:

$$q_j(t) = A_{nm}\sin(\omega_{nm}t - \varkappa_{nm})e^{-\gamma_{nm}t}$$

then the logarithm of $\left|W(t, \frac{\omega_{nm}}{2\pi})\right|^2$ has to decay according to a linear law and this can be experimentally detected by the measurements of $\eta$ and applying the wavelet transform eqn. (51).

# 4 Results and discussion

## 4.1 Linear sloshing in a square section, prismatic tank

Let us consider a square section, prismatic tank, of side $B$ and height $H$. Let this tank be filled with quiescent fluid of density $\rho$, up to level $H$ above the bottom of the tank in horizontal position. The tank can rotate at an angle $\theta$ with respect to a horizontal axis of rotation, perpendicular to the front section of the tank and at a distance $R$ from the bottom of the tank. The law of motion:

$$\theta = \theta(t)$$

is given. A frame of reference $Oxyz$, attached to the tank, with origin on the axis of rotation is adopted.

All the above mentioned assumptions are schematically represented in fig. 4.

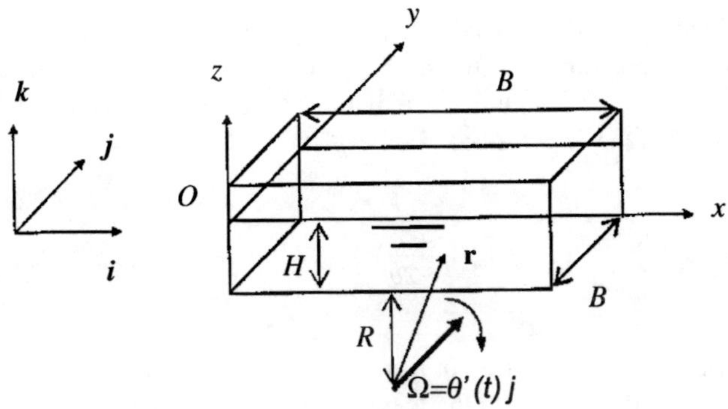


Figure 4: Schematic representation of the rotating tank.

Functions $\varphi_p(x,y,z,t), \eta_p(x,y,t)$ assume the form:

$$\begin{aligned}\varphi_p(x,y,z,t) &= (R+H)\left(x-\frac{B}{2}\right)\theta(t)\\&+\theta(t)\sum_k a_k\left\{\cosh(\lambda_k x)-\cosh\left[\lambda_k(x-B)\right]\right\}\sin(\lambda_k z)\\&-\theta(t)\sum_k b_k\sin\left[\mu_k\left(x-\frac{B}{2}\right)\right]\sinh(\mu_k z)\\\eta_p(x,y,t) &= \left(x-\frac{B}{2}+\left.\frac{\partial\varphi_p}{\partial z}\right|_{z=0}\right)\theta(t)\end{aligned} \tag{56}$$

where the definitions of $a_k, \lambda_k, b_k, \mu_k$ are omitted for the sake of simplicity (see La Rocca *et al* [10, 11]), while functions $c_{nm}(x,y)$ are defined as:

$$c_{nm}(x,y) = \cos\left(\frac{n\pi x}{B}\right)\cos\left(\frac{m\pi y}{B}\right) \tag{57}$$

Function $d_{nm}(z)$ has the expression:

$$d_{nm}(z) = \frac{\cosh\left[|\mathbf{k}|(z+H)\right]}{\cosh(|\mathbf{k}|h)}, |\mathbf{k}| = \frac{\pi\sqrt{n^2+m^2}}{B} \tag{58}$$

Functions $M_{nm}(\eta,x,y)$ are given by the expression:

$$M_{nm}(\eta,x,y) = \frac{\sinh\left[|\mathbf{k}|(\eta+H)\right]}{|\mathbf{k}|\cosh(|\mathbf{k}|H)}c_{nm}(x,y)$$

while the very complicate expression for $N(\eta,x,y,t)$ is omitted for the sake of simplicity [11]. If a Taylor series expansion respect to $\eta$, with initial point $\eta=0$, is performed on both $M_{nm}$ and $N$ up to first order and only: $\mathcal{O}(\mathcal{A}_{nm}(t)^2), \mathcal{O}(q_{nm}(t)^2)$, $\mathcal{O}(\mathcal{A}_{nm}(t)\,q_{nm}(t))$, $\mathcal{O}(\mathcal{A}_{nm}(t)\,\theta(t))$, $\mathcal{O}(q_{nm}(t)\;\theta(t))$ terms are retained, a *linearized* lagrangian $\mathcal{L}^*$ is obtained, in the sense that if Lagrange equations are applied to it, linear ordinary differential equations are obtained for the evolution of $\mathcal{A}_{nm}(t), q_{nm}(t)$. It is possible to see that, for this sloshing case, such a linearised lagrangian $\mathcal{L}^*$ has the expression:

$$\mathcal{L}^* = \rho B^2 \sum_{nm} \left[ \mathcal{A}'_{nm} q_{nm} + \frac{\omega_{nm}^2}{2g} \mathcal{A}_{nm}^2 + g \frac{q_{nm}^2}{2} - \mathcal{F}_{nm}(t)\, q_{nm} \right] \tag{59}$$

$\mathcal{F}_{nm}(t)$ being a known function of time, whose order of magnitude is $\mathcal{O}(|\theta(t)|)$, and $\omega_{nm}$ the eigenfrequency of the $\mathbf{k}$-mode, whose expression is:

$$\omega_{nm} = \sqrt{g\,|\mathbf{k}| \tanh(|\mathbf{k}|\, H)} \tag{60}$$

Applying Lagrange eqns. (21) to (59) and accounting for generalised dissipative forces [25], derived from the dissipation function [11]:

$$\mathcal{G} = \rho B^2 \sum_{nm} \frac{g \gamma_{nm}}{\omega_{nm}^2} \left( q'_{nm} \right)^2$$

the following equations are obtained:

$$\begin{cases} q'_{nm} - \dfrac{\omega_{nm}^2}{g} \mathcal{A}_{nm} = 0 \\ \mathcal{A}'_{nm} + g q_{nm} + \dfrac{2 g \gamma_{nm}}{\omega_{nm}^2} q'_{nm} = \mathcal{F}_{nm}(t) \end{cases} \tag{61}$$

It is interesting to note that such equations could be obtained directly from boundary conditions eqns. (10) and (11), linearized around the free surface at rest: $\eta = 0$ and adopting for $\varphi, \eta$ expansions eqns. (17). Moreover, eqns. (61) describe the evolution of the $nm$ mode without being influenced by the other modes: this is a consequence of linearization, because coupling among modes is due only to non-linearities. Eliminating $\mathcal{A}'_{nm}$ from eqns. (61), the following second order, linear differential equation is obtained for $q_{nm}$:

$$q''_{nm} + 2\gamma_{nm} q'_{nm} + \omega_{nm}^2 q_{nm} = \mathcal{F}_{nm}(t)$$

whose asymptotic solution is:

$$q_{nm} = \frac{\omega_{nm}^2}{g} \int_{t_0}^{t} \frac{e^{-\gamma_{nm}(t-\tau)}}{\sqrt{\omega_{nm}^2 - \gamma_{nm}^2}} \sin\left[ \sqrt{\omega_{nm}^2 - \gamma_{nm}^2}\, (t-\tau) \right] \mathcal{F}_{nm}(\tau)\, d\tau \tag{62}$$

valid for $\mathcal{O}(|\theta(t)|) \ll 1$ and far from resonance phenomena.

In fig. 5a comparison between the experimental and analytical time histories of the free surface is shown. The figure is concerned with a case of sloshing

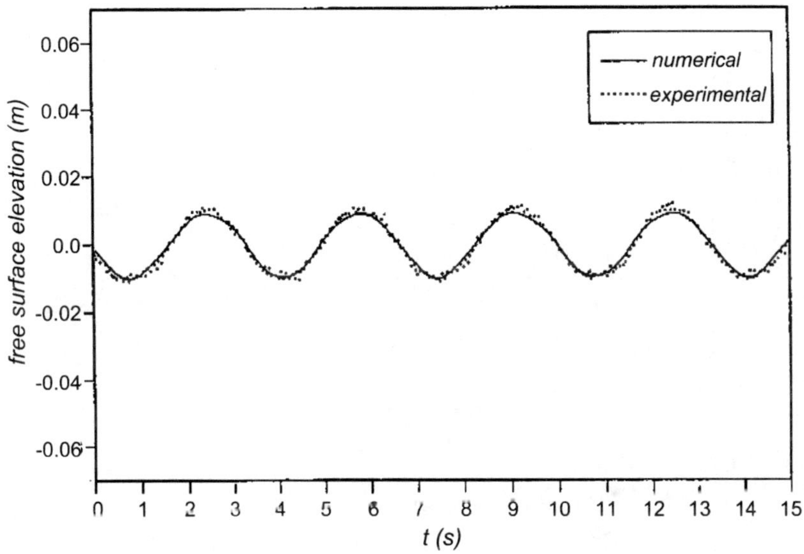


Figure 5: Experimental and numerical time history of the free surface.

with $B = 0.5m, \theta \simeq 0.087 \sin(2\pi f t), f = 0.3Hz, h = 0.136m$ – the oscillation imposed to the tank is harmonic, with a frequency value sufficiently far from the lowest resonance frequency value, which coincides with the resonance frequency of the first $2D$ odd mode $\left(\frac{\omega_{10}}{2\pi} = 1.04Hz\right)$. The time history is relative to the point $x = 0.15m, y = 0.25m$. The hypothesis for the validity of the analytical solution eqn. (62) are satisfied and, as a consequence, the agreement between the analytical and the experimental results is very good. An interesting confirmation of the validity of the analytical solution eqn. (62) is given by the spectrum of the time history of the experimental free surface at the center of the square section of the tank, shown in fig. 6, which reveals that at that point the motion of the free surface is negligible. The analytical solution predicts that the free surface at this point is always equal to zero in the linear approximation: this is due to the fact that only $2D$ odd modes (*i.e.* modes with $n$ odd and $m = 0$) are directly excited in eqns. (61) and such modes have a nodal line for $x = \frac{B}{2}$. Other modes ($2D$ even or $3D$ modes) can then be excited only by means of nonlinear interactions.

### 4.2 Nonlinear sloshing in a square section, prismatic tank. Importance of the dissipative terms. Further applications of the variational approach

When the characteristics of the imposed motion no longer satisfy the hypothesis for the validity of the linear analytical solution eqn. (62), the sloshing motion shows a wide variety of nonlinear behaviors. Nonlinear interactions among the modes, parametric resonance phenomena, and $2D - 3D$ transition are perhaps the most interesting and evident nonlinear behaviors.

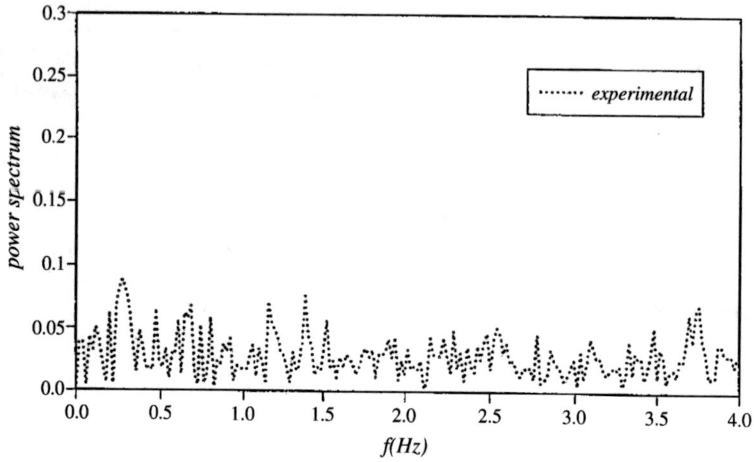


Figure 6: Experimental power spectrum.

The application of the variational formulation to nonlinear sloshing will yield results only if the eqns. (25) are solved numerically. For this purpose, a suitable truncation criterion must be adopted that will allow the consideration of a finite number of interacting modes. In particular, a preliminary direct perturbative expansion, performed on the eqns. (25), is able to determine the nonlinear resonance conditions, which are responsible for the nonlinear interactions among modes [10]. The processing of experimental data – for example, the application of proper orthogonal decomposition to sloshing data – is able to determine the leading modes – *i.e.* those whose evolution captures most of the energy of the motion. Interesting results are obtained in La Rocca *et al* [11]. Of particular interest is the investigation into the transition of sloshing from linear to nonlinear $2D$ and from nonlinear $2D$ to nonlinear $3D$ sloshing. Such transition phenomena are related to nonlinear interactions between the odd modes, directly excited by the imposed motion and the $2D$ even and $3D$ modes. A very interesting case of a nonlinear sloshing configuration with $2D$ to $3D$ transition is given in La Rocca *et al* [11], with $B=0.5m, \theta \simeq 0.087\sin(2\pi ft), f=0.7Hz, h=0.136m$.

In this case the exciting frequency is approximately 70% of the lowest resonance frequency. A frequency closer to the lowest resonance frequency could not be experimentally realized because of the amplitude of the resulting sloshing motion, which caused the outflowing of liquid. The exciting motion is then in quasi-resonance conditions. Experiments showed that this configuration is unstable – *i.e.* if the motion is started very carefully from a $2D$ free-surface configuration, the shape of this latter remains two-dimensional. If, on the contrary, a perturbation is suddenly imposed to the free-surface, this latter definitely evolves to a $3D$ configuration. Thus two kinds of numerical simulations can be performed: one with $2D$ modes only, and another with $2D$ and $3D$ modes. The results are shown in figs. 7 and 8, where a comparison of the spectra of the experimental and

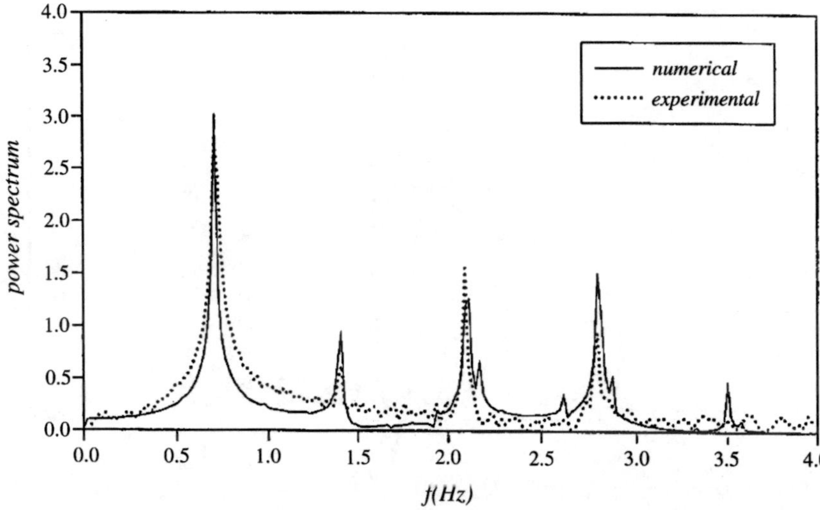


Figure 7: Experimental and numerical power spectrum of the free surface (2D case).

numerical time histories of the free surface in $x = 0.15m, y = 0.25m$ is shown, first when the free surface is two dimensional and then after the transition to the three dimensional configuration has occurred. The $3D$ case (fig. 8) is characterized by a transfer of energy from the frequency $f$ to its multiples $2f, 3f, 4f$. The largest spectral contribution occurs at the frequency $3f = 2.1Hz$ in the $3D$ case. The agreement between experimental and numerical results confirms that this contribution is due to the presence of the $3D$ mode with $n = 2, m = 2$ whose eigenfrequency $\frac{\omega_{22}}{2\pi} = 2.15Hz$ is very close to $3f$. Moreover, the frequency $1.7Hz$ coincides with the eigenfrequency of the mode $(2, 0)$, activated by nonlinear interactions. Such a free-surface transition from a $2D$ to a $3D$ shape is also observed in other sloshing configurations and confirmed by theoretical predictions based on perturbative analyses of lagrangian eqns. (25) [10]. In particular, in fig. 9 it is possible to see a $3D$ sloshing configuration in which the presence of the $3D$ mode with $n = 0, m = 4$ is evident. Such a sloshing configuration ($B = 0.5m, \theta \simeq 0.087 \sin(2\pi f t), f = 0.4Hz, h = 0.075m$) is the result of a transition from a previous $2D$ configuration, excited at a frequency which is 50% of the resonance frequency of the first odd mode $(n = 1, m = 0)$.

The importance of the dissipative terms must be highlighted. As a matter of fact, they are not important in linear sloshing, whose occurrence is in off-resonance conditions, but they become indispensable in nonlinear sloshing for at least two main reasons. First, they insure the independence of the solution from the initial conditions, and second, they limit the amplitude of the resulting sloshing motion in the proximity of resonance conditions, giving the numerical simulations robustness and reliability. It has been shown that, in the absence of such terms, numerical instability phenomena occur. It is interesting to see that although the first

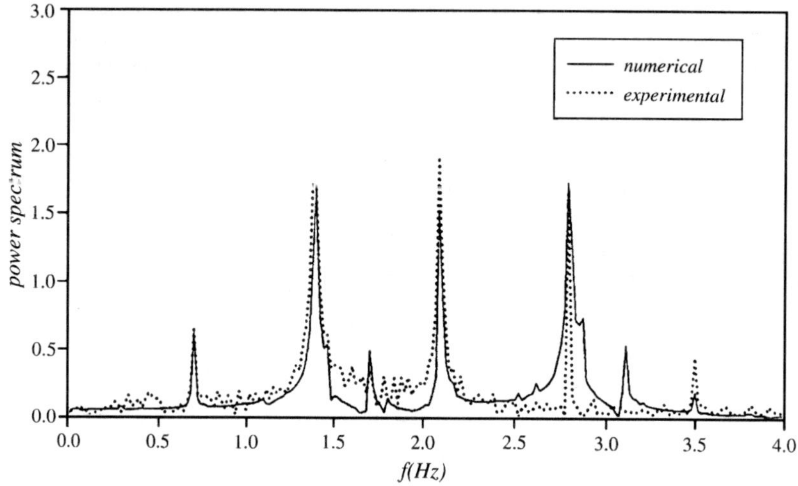


Figure 8: Experimental and numerical power spectrum of the free surface (3D) case.

assumption concerning fluid motion is the inviscid hypothesis, dissipative forces are then reintroduced as generalized dissipative forces, due to the lagrangian structure of the motion equations.

And last – but not least! – it is worth mentioning that the variational formulation can also be applied to sloshing in stratified fluids [12, 17, 18]. An interesting case is the sloshing of two-layer fluids (*i.e.* fluids composed of two layers of immiscible

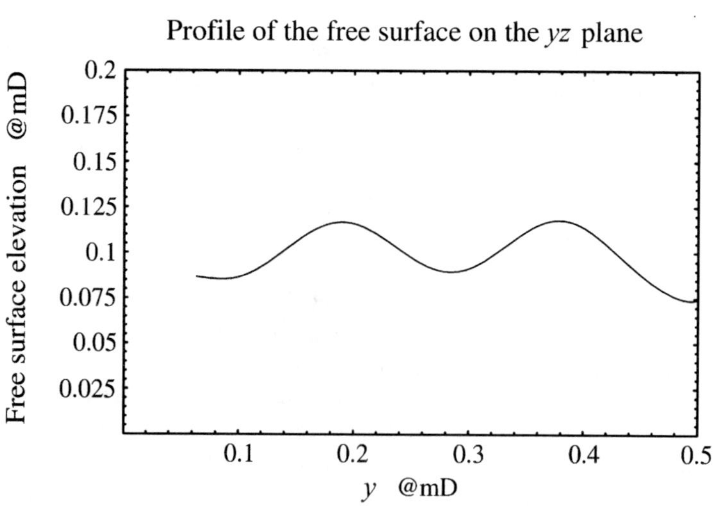


Figure 9: The 3D mode $n=0, m=4$ photographed on the $yz$ plane.

fluids of different densities), where the definition of the functional must account for the different pressure fields:

$$\begin{aligned}\mathcal{F} = &-\int_{t_1}^{t_2}\int_A\int_{-H_1}^{\eta_1(x,y,t)} p_1 dzdAdt \\ &-\int_{t_1}^{t_2}\int_A\int_{\eta_1(x,y,t)}^{\eta_2(x,y,t)} p_2 dzdAdt \\ &-\int_{t_1}^{t_2}\int_A \tau\sqrt{1+\left(\frac{\partial\eta_1}{\partial x}\right)^2+\left(\frac{\partial\eta_1}{\partial y}\right)^2}\,dAdt \end{aligned} \quad (63)$$

and where the surface tension $\tau$ on the interface $\eta_1(x, y, t)$ has been accounted for by the last term on the right-hand side of eqn. (63).

## 5 Acknowledgements

The authors wish to thank the referees for their useful suggestions and criticisms which permitted to improve their work.

This study has received financial support by the Italian Ministry of Scientific and Technology Research, project PRIN 2002 "Influence of vorticity and turbulence in interactions of water bodies with their boundary elements and effect on hydraulic design".

## References

[1] Abramson, H.N., *The dynamic behaviour of liquids in moving containers*, NASA Report, SP 106. 1966.

[2] Moiseiev, N.N. & Rumyantsev, V.V., *Dynamic stability of bodies containing fluids*, Springer Verlag: New York, 1968.

[3] Miles, J.W., Non-linear surface waves in closed basins, *J. Fluid Mech.*, **75**, 419–448, 1976.

[4] Whitham, G.B., Non-linear dispersion of water waves, *J. Fluid Mech.*, **27**, 399–412, 1967.

[5] Whitham, G.B., *Linear and nonlinear waves*, Interscience, 1974.

[6] Miles, J.W., Parametrically excited, standing cross-waves, *J. Fluid Mech.*, **186**, 119–127, 1988.

[7] Miles, J.W. & Becker, J., Parametrically excited, progressive cross-waves, *J. Fluid Mech.*, **186**, 129–146, 1988.

[8] Hargneaves, R., A pressure-integral as kinetic potential, *Phil. Mag*. **16**, 436–444, 1908.

[9] Luke, J.C., A variational principle for a fluid with a free surface, *J. Fluid Mech.*, **27**, 395–397, 1967.

[10] La Rocca, M., Mele, P. & Armenio, V., Variational approach to the problem of sloshing in a moving scontainer, *J. Theo. Applied Fluid Mech.*, **4**, 1997.

[11] La Rocca, M., Sciortino, G. & Boniforti, M.A., A fully nonlinear model for the sloshing in a rotating container, *Fluid Dyn. Res.*, **27**, pp. 23–52, 2000.

[12] La Rocca, M., Sciortino, G. & Boniforti, M.A., Interfacial gravity waves in a two-fluid system, *Fluid Dyn. Res.*, **30**, pp. 31–63, 2002.

[13] Faltinsen, O.M., Rognebakke, O.F., Lukovsky, I.A. & Timokha, A.N., Multidimensional modal analysis of nonlinear sloshing in a rectangular tank with finite water depth, *J. Fluid Mech.*, **407**, pp. 201–234, 2000.

[14] Faltinsen, O.M. & Timokha, A.N., Adaptive multimodal approach to nonlinear sloshing in a rectangular tank, *J. Fluid Mech.*, **432**, pp. 167–200, 2001.

[15] Faltinsen, O.M. & Timokha, A.N., Asymptotic modal approximation of nonlinear resonant sloshing in a rectangular tank with small fluid depth, *J. Fluid Mech.*, **470**, pp. 319–357, 2002.

[16] Faltinsen, O.M., Rognebakke, O.F. & Timokha, A.N., Resonant three-dimensional nonlinear sloshing in a square-base basin, *J. Fluid Mech.*, **487**, pp. 1–42, 2003.

[17] La Rocca, M., Sciortino, G. & Boniforti, M.A., Gravity-capillary waves in layered fluid, *Nonlinear Oscillation*, **6**, pp. 197–205, 2003.

[18] La Rocca, M., Sciortino, G., Adduce, C., Boniforti, M.A., Experimental and theoretical investigation on the sloshing of a two-liquid system with free-surface, *Physics of Fluids,* **17**, pp 1–17, 2005.

[19] Batchelor, G.K., *An introduction to fluid dynamics*, Cambridge University Press, 1987.

[20] Landau, L.D. & Lifshitz, E.M., *Fluid Mechanics*, Pergamon Press, 1989.

[21] Faltinsen, O.M., A nonlinear theory of sloshing in rectangular tanks, *J. Ship Res.*, **18**, 224–241, 1974.

[22] Flipse, J.E., Lou, Y.K. & Su, T.C., *A nonlinear analysis of liquid sloshing in rigid containers*, Texas A & M University, Report MA/RD/940/82046. 1980.

[23] Waterhouse, D.D., Resonant sloshing near a critical depth, *J. Fluid Mech.*, **281**, 313–318, 1994.

[24] Ockendon, J.R., Ockendon, H. & Waterhouse, D.D., Multi-mode resonances in fluids, *J. Fluid Mech.*, **315**, 317–344, 1996.

[25] Goldstein, H., *Meccanica Classica*, Zanichelli, Bologna, 1982.

[26] Martel, C., Nicolas, J.A. & Vega, J.M., Surface-wave damping in a brimful circular cylinder, *J. Fluid Mech.*, **360**, 213–228, 1998.

[27] Miles, J.W., Surface-wave damping in closed basins, *Proc. R. Soc.* Lond. A297, 459–473, 1967.

[28] Henderson, D.M. & Miles, J.W., Surface wave damping in a circular cylinder with a fixed contact line, *J. Fluid Mech.*, **275**, 285–299, 1994.

[29] Miles, J.W. & Henderson, D.M., A note on interior vs. boundary layer damping of surface waves in a circular cylinder, *J. Fluid Mech.*, **364**, 319–323, 1998.

[30] Farge, M., Wavelet transform and their application to turbulence, *Ann. Rev. Fluid Mech.*, **24**, pp. 395–457, 1992.

# CHAPTER 9

# Turbulence, friction, and energy dissipation in transient pipe flow

G. Pezzinga[1] & B. Brunone[2]
[1] *Department of Civil and Environmental Engineering, Catania University, Italy.*
[2] *Department of Civil and Environmental Engineering, Perugia University, Italy.*

## Abstract

The paper analyzes the energy dissipation in transient pipe flow focusing on friction and its relation with turbulence. Other important mechanisms of energy dissipation are also analyzed, in particular the viscoelastic behavior of polymeric pipes. Different models for the evaluation of friction forces and their limitations are examined by comparison with new experimental data.

## 1 Introduction

The analysis of turbulence in transient pipe flows has both a theoretical interest, with regard to velocity profiles and turbulence, and a practical one, for the evaluations of pressure oscillations and their damping. In fact, the velocity profiles in unsteady flow conditions may show greater gradients, and thus greater shear stresses, than the corresponding values in steady flow. On the other hand, the underestimation of energy dissipation in transient pipe flow can give rise to an overestimation of the maximum oscillation that can take place in pipe networks, or in transient cavitating flows.

The evaluation of energy dissipation due to friction can be carried out by two-dimensional models, in which the variation of the longitudinal component of velocity along the radial coordinate is considered. Different formulations of turbulent stress models, were studied in the two-dimensional flow schematization [1–5]. However, the validation of this kind of formulations was carried out only with experimental data of pressure oscillations for simple pipes and pipe networks.

The analysis carried out by one-dimensional models with steady or quasi-steady resistance formulas gives rise to underestimation of friction forces and damping

[6, 7]. In one-dimensional models, it is possible to adopt unsteady resistance, usually with dissipation terms to be added to quasi-steady resistance terms. However, in these models the evaluation of parameters is not general and rigorous [5].

In some conditions, different energy dissipation mechanisms can play an important role. Transient cavitating flows are an example of such conditions. Transient flows in polymeric pipes constitute cases in which a considerable part of energy dissipation is due to viscoelastic pipe behavior, and energy dissipation due to friction becomes secondary.

## 2 Turbulence

### 2.1 Quasi two-dimensional models

#### 2.1.1 Formulation of the equations

Quasi two-dimensional models are based on the continuity and momentum equations written for an elastic pipe with circular cross-section with axial symmetry. The continuity and momentum equations in cylindrical coordinates are:

$$\frac{\partial \rho}{\partial t}+\frac{\partial (\rho u)}{\partial x}+\frac{1}{r}\frac{\partial (\rho r v)}{\partial r}=0 \tag{1}$$

$$\frac{\partial u}{\partial t}+u\frac{\partial u}{\partial x}+v\frac{\partial u}{\partial r}=-g\frac{\partial H}{\partial x}-\frac{1}{\rho}\frac{\partial \sigma_x}{\partial x}-\frac{1}{\rho r}\frac{\partial (r\tau)}{\partial r} \tag{2}$$

$$\frac{\partial v}{\partial t}+u\frac{\partial v}{\partial x}+v\frac{\partial v}{\partial r}=-g\frac{\partial H}{\partial r}-\frac{1}{\rho}\frac{\partial \tau}{\partial x}-\frac{1}{\rho r}\frac{\partial (r\sigma_r)}{\partial r}+\frac{\sigma_\theta}{\rho r} \tag{3}$$

where $x$ = distance along pipe; $r$ = distance from the axis; $t$ = time; $H$ = pressure head; $u$ and $v$ = velocity components in the longitudinal and radial directions, respectively; $\rho$ = density of the liquid; $g$ = gravitational acceleration; $\sigma_x$, $\sigma_r$, and $\sigma_\theta$ = deviation of normal stresses from pressure in longitudinal, radial and angular direction; and $\tau$ = shear stress. These equations determine $H$, $u$ and $v$, as function of $x$, $r$ and $t$, if the relations between stresses and strain velocities, the boundary conditions and the initial conditions are defined. For expedience, some simplifications are acceptable.

First, the velocity component $v$ and its derivatives are neglected in both momentum eqns. (2) and (3). In fact, the numerical results of Vardy & Hwang [1] show, for both laminar and turbulent unsteady flow, maximum values of the radial velocities of the order of 10–20 $\mu$m/s. In the momentum equation in the longitudinal direction the normal stress is assumed to be equal to the pressure and the residual convective term is neglected, as usually done also in one-dimensional models. Furthermore, the stress terms in the radial momentum equation are neglected implying

$$\frac{\partial H}{\partial r}=0 \tag{4}$$

Thus, a single instantaneous value of the pressure head exists in each section. For this reason the model is quasi two-dimensional, the dependent variable $u$ is a function of $x$, $r$ and $t$, while the dependent variable $H$ is a function only of $x$ and $t$. These hypotheses, with the other usual simplification of neglecting the spatial derivative of $\rho$ in the continuity equation, allow one to rewrite eqns. (1) and (2) in the form

$$\frac{\partial H}{\partial t}+\frac{a^2}{gA_0}\frac{\partial Q}{\partial x}=0 \tag{5}$$

$$\frac{\partial u}{\partial t}+g\frac{\partial H}{\partial x}+\frac{1}{\rho r}\frac{\partial (r\tau)}{\partial r}=0 \tag{6}$$

where $A_0$ is the total cross sectional area of the pipe, $a$ is the pressure wave speed and $Q$ is the discharge, calculated by integrating the velocity over the section. Observing that $dA = 2\pi r dr$, eqn. (6) can be rewritten as

$$\frac{\partial u}{\partial t}+g\frac{\partial H}{\partial x}+\frac{2\pi}{\rho}\frac{\partial (r\tau)}{\partial A}=0 \tag{7}$$

that is more suitable for the numerical integration.

### 2.1.2 Stress models

A turbulent stress model should take into account the transport of stress and of some turbulent quantities (*e.g.* the turbulent kinetic energy and its dissipation rate). A poor sensitivity of head oscillations to turbulence model is found by Vardy & Hwang [1] who obtain that the extension to water hammer flow of a model valid for oscillating flow – in which the turbulent viscosity has algebraic expressions – gives good results and similar to those of simpler models. In particular low Reynolds number $k-\varepsilon$ turbulence models provide head oscillations very similar to those of mixing length models [5].

**2.1.2.1 Oscillating flow models** Vardy & Hwang [1] extend to water hammer flow the model of Kita *et al* [8] – valid for oscillating flow – in which the turbulent viscosity has the following algebraic expressions:

$$\textit{Region 1}: \quad 0<y^+<\frac{1}{a} \qquad \frac{\nu_t}{\nu}=1 \tag{8a}$$

$$\textit{Region 2}: \quad \frac{1}{c}<y^+<\frac{c}{C_b} \qquad \frac{\nu_t}{\nu}=ay^+ \tag{8b}$$

$$\textit{Region 3}: \quad \frac{c}{C_b}<y^+<\frac{\kappa}{C_b+\left(\dfrac{\kappa^2}{4C_m\,\mathrm{Re}^*}\right)} \qquad \frac{\nu_t}{\nu}=C_b y^{+2} \tag{8c}$$

$$\textit{Region 4}: \frac{\kappa}{C_b+\left(\dfrac{\kappa^2}{4C_m\,\mathrm{Re}^*}\right)}<y^+<\frac{2C_m\,\mathrm{Re}^*}{\kappa}\left(1+\sqrt{1-\frac{C_c}{C_m}}\right)$$

$$\frac{\nu_t}{\nu}=\kappa y^+\left(1-\frac{\kappa y^+}{4C_m\,\mathrm{Re}^+}\right) \tag{8d}$$

$$Region\ 5: \frac{2C_m \mathrm{Re}^*}{\kappa}\left(1+\sqrt{1-\frac{C_c}{C_m}}\right) < y^+ < \overset{*}{\mathrm{Re}} \qquad \frac{\nu_t}{\nu} = C_c \mathrm{Re} \quad (8e)$$

where $u^* = \sqrt{\frac{\tau_w}{\rho}}, y^+ = \frac{u^* y}{\nu}, \mathrm{Re}^* = \frac{u^* R}{\nu}, \tau_w$ wall shear stress, $y$ distance from the wall, $R$ pipe radius, $\nu$ and $\nu_t$ the kinematic viscosity and the eddy viscosity, respectively. The values of the numerical coefficients are those proposed by Kita *et al* [8]: $c = 0.19$, $C_b = 0.011$, $C_m = 0.077$ and $k = 0.41$. The parameter $C_c$ is a function of the Reynolds number and typically lies in the range $0.05 < C_c < 0.07$.

**2.1.2.2 Mixing length models** Models based on mixing length give good results as well [3, 4]. Silva-Araya & Chaudry [3] adopt the model of Granville [9] for the inner layer and an eddy viscosity for the outer region. The mixing length model was converted to an equivalent eddy viscosity to make models compatible [10]. Expressions for the dimensionless mixing length and eddy viscosity are:

$$l^+ = ky^+\sqrt{\tau^+}\left[1-\exp\left(-\frac{y^+}{\lambda_1}\right)\right] \tag{9}$$

$$\frac{\nu_t}{\nu} = \frac{1}{2}\left(\sqrt{1+2l^+\tau^+}-1\right) \tag{10}$$

where $l^+ = \frac{u_* l}{\nu}$, $\tau^+ = \frac{\tau}{\tau_w}, \lambda_1$ mixing length damping parameter, and $l$ mixing length. For the outer region the eddy viscosity is defined by eqn. (8e). The dependence of $C_c$ on the Reynolds number $Re$ is obtained via the following equation:

$$C_c = 0.4095 - 0.1390\log(\mathrm{Re}) + 0.137\left[\log(\mathrm{Re})\right]^2 \tag{11}$$

If $Re < 10^4$, $C_c = 0.07$, whereas if $Re < 2\cdot 10^6, C_c = 0.075$. The transition between the inner and the outer region is obtained by an exponential function.

Pezzinga [4] proposes a simple two-zone model based in the viscous sublayer on the Newton's viscosity and, in the turbulent region, on the Prandtl mixing length hypothesis, viz.

$$\tau = -\rho\nu\frac{\partial u}{\partial r} - \rho l^2\left|\frac{\partial u}{\partial r}\right|\frac{\partial u}{\partial r} \tag{12}$$

For the mixing length the following expression is adopted

$$\frac{l}{R} = k\frac{y}{R}e^{-\frac{y}{R}} \tag{13}$$

The parameter $k$ is determined by interpolation as a function of the initial Reynolds number $Re_0$, on the basis of the experimental data of Nikuradse valid for smooth pipes

$$k = 0.374 + 0.0132 \ln\left(1 + \frac{83100}{Re_0}\right) \tag{14}$$

The thickness of the viscous sublayer $\delta$ is obtained as the distance from the wall to the intersection between the velocity profiles in the viscous sublayer and in the turbulent zone, assuming that in the sublayer the velocity profile is linear and in the turbulent region the profile is locally logarithmic:

$$\frac{u^*\delta}{\nu} = 2.5\frac{\delta}{k_s} + B \tag{15}$$

with $k_s$ being the equivalent sand roughness. $B$ is a parameter considered dependent on $k_s^* = \frac{u^* k_s}{\nu}$ by the expression:

$$B = 8.5 - 2.5 \ln\left(1 + \frac{3.32}{k_s^*}\right) \tag{16}$$

Pezzinga [5] considers also the Van Driest model in which the equivalent eddy viscosity is:

$$\nu_t = f_\mu l^2 \left|\frac{\partial u}{\partial r}\right| \frac{\partial u}{\partial r} \tag{17}$$

with the mixing length $l$:

$$l = ky \tag{18}$$

being $k = 0.4$. The coefficient $f_\mu$, takes into account for the reduction of mixing length near the wall due to viscous stress effect; for smooth wall Van Driest [11] proposes the following expression:

$$f_\mu = \left[1 - \exp\left(-\frac{y^+}{26}\right)\right]^2 \tag{19}$$

Van Driest [11] proposes also a modification of function $f_\mu$ to take into account of wall roughness, that reduces the viscous sublayer and then increases turbulent fluctuations:

$$f_\mu = \left[1 - \exp\left(-\frac{y^+}{26}\right) + \exp\left(-2.3\frac{y^+}{k_s^*}\right)\right]^2 \tag{20}$$

Krogstaad [12] proposes a different expression of function $f_\mu$ to correctly reproduce the logaritmic law for high roughness:

$$f_\mu = \left\{1 - \exp\left(-\frac{y^+}{26}\right) + \exp\left[-\frac{y^+}{26}\left(\frac{70}{k_s^*}\right)^{\frac{3}{2}}\right]\sqrt{1 + \exp\left(-\frac{70}{k_s^*}\right)}\right\}^2 \tag{21}$$

Such models give results very similar to those of more complex $k - \varepsilon$ models. But a limitation of mixing length models is that the equivalent viscosity is zero on the pipe axis. On the other hand, both mixing length models and the five zone viscosity model do not give information on the turbulence quantities, as more complex models do.

**2.1.2.3 Low Reynolds number $k$–$\varepsilon$ models** Eichinger & Lein [2] and Pezzinga [13] use such an approach for water hammer flow obtaining numerical head oscillations that agree with sufficient accuracy with the experimental ones. The tested models are different versions of low Reynolds number $k-\varepsilon$ models. In $k-\varepsilon$ models the turbulent viscosity is defined by the expression:

$$\nu_t = f_\mu c_\mu \frac{k^2}{\varepsilon} \tag{22}$$

The turbulent kinetic energy $k$ and its dissipation rate $\varepsilon$ are obtained by the respective transport equations that can be written as:

$$\frac{\partial k}{\partial t} - \frac{1}{r}\frac{\partial}{\partial r}\left[r\left(\nu + \frac{\nu_t}{\sigma_k}\right)\frac{\partial k}{\partial r}\right] - \nu_t\left(\frac{\partial u}{\partial r}\right)^2 + \varepsilon - D = 0 \tag{23}$$

$$\frac{\partial \epsilon}{\partial t} - \frac{1}{r}\frac{\partial}{\partial r}\left[r\left(\nu + \frac{\nu_t}{\sigma_\epsilon}\right)\frac{\partial \epsilon}{\partial r}\right] - c_{\epsilon 1} f_1 \nu_t\left(\frac{\partial u}{\partial r}\right)^2 + c_{\epsilon 2} f_2 \frac{\epsilon^2}{k} - E = 0 \tag{24}$$

The function $f_\mu, f_1, f_2, D$ and $E$ are needed for the extension to low Reynolds number [14]. The constants and functions for the Launder-Sharma [15] and Nagano-Hishida [16] models are reported in Table 1a, whereas those for the Lam-Bremhorst [17] and Nagano-Tagawa [18] models are given in Table 1b, with $B_1 = \left[1 - \exp\left(-\frac{y^+}{6}\right)\right]^2$.

The models can be classified in two categories: those in which $\varepsilon$ is the isotropic part of the dissipation rate (if $D \neq 0$), that, for the authors that propose its use, has the advantage of being zero at the wall, and those in which $\varepsilon$ represents the dissipation rate (if $D = 0$). Then two wall conditions are possible for $\varepsilon$, $\varepsilon_w = 0$ (models of Launder-Sharma and of Nagano-Hishida) and $\varepsilon_w \neq 0$ (models of Lam-Bremhorst and of Nagano-Tagawa).

Extensions have been proposed to take into account the wall roughness. Zhang *et al* [19] propose a low Reynolds number $k-\varepsilon$ model for both smooth and rough walls. In particular the following function $f_\nu$ is proposed:

$$f_\mu = 1 - \exp\left[\left(-\frac{y^+}{42}\right)^2\right] + \sqrt{\frac{k_s^*}{200}} \exp\left(-25\frac{y^+}{k_s^*}\right) \tag{25}$$

Furthermore the following modification of the function $f_1$ is proposed:

$$f_\mu = 1 + \exp\left(-\frac{1}{0.1 + 1/k_s^*}\right)\left(\frac{9.2}{1 + y^+}\right)^6 \tag{26}$$

Table 1a: Constant and functions for $k-\varepsilon$ models (adapted from [5]).

| Model | Launder-Sharma | Nagano-Hishida |
|---|---|---|
| $c_\mu$ | 0.09 | 0.09 |
| $\sigma_k$ | 1.0 | 1.0 |
| $\sigma_\varepsilon$ | 1.3 | 1.3 |
| $c_{\varepsilon 1}$ | 1.45 | 1.45 |
| $c_{\varepsilon 2}$ | 1.92 | 1.90 |
| $f_\mu$ | $\exp\left[\frac{-3.4}{(1+R_t/50)^2}\right]$ | $\left[1-\exp\left(-\frac{y^+}{26.5}\right)\right]^2$ |
| $f_1$ | 1.0 | 1.0 |
| $f_2$ | $1-0.3\exp\left(-R_t^2\right)$ | $1-\exp\left(-R_t^2\right)$ |
| $D$ | $2\nu\left(\frac{\partial\sqrt{k}}{\partial r}\right)^2$ | $2\nu\left(\frac{\partial\sqrt{k}}{\partial r}\right)^2$ |
| $E$ | $2\nu\nu_t\left(\frac{\partial^2 U}{\partial r^2}\right)^2$ | $\nu\nu_t\left(1-f_\mu\right)\left(\frac{\partial^2 U}{\partial r^2}\right)^2$ |
| $\varepsilon_w$ | 0 | 0 |

Table 1b: Constant and functions for $k-\varepsilon$ models (adapted from [5]).

| Model | Lam-Bremhorst | Nagano-Tagawa |
|---|---|---|
| $c_\mu$ | 0.09 | 0.09 |
| $\sigma_k$ | 1.0 | 1.4 |
| $\sigma_\varepsilon$ | 1.3 | 1.3 |
| $c_{\varepsilon 1}$ | 1.44 | 1.45 |
| $c_{\varepsilon 2}$ | 1.92 | 1.90 |
| $f_\mu$ | $\left[1-\exp\left(-0.0165y^*\right)\right]^2\left(1+\frac{20.5}{R_t}\right)$ | $\left[1-\exp\left(-\frac{y^+}{26.5}\right)\right]^2\left(1+\frac{4.1}{R_t^{3/4}}\right)$ |
| $f_1$ | $1+\left(\frac{0.05}{f_\mu}\right)^3$ | 1.0 |
| $f_2$ | $1-\exp\left(-R_t^2\right)$ | $\left\{1-0.3\exp\left[-\left(\frac{R_t}{6.5}\right)^2\right]\right\}B_1$ |
| $D$ | 0 | 0 |
| $E$ | 0 | 0 |
| $\varepsilon_w$ | $\nu\frac{\partial^2 k}{\partial r^2}$ | $\nu\frac{\partial^2 k}{\partial r^2}$ |

## 3 Friction

A simple as well as reliable criterion for evaluating energy dissipation in unsteady-state pipe flow is important from both the technical and theoretical point of view.

As a consequence, since the mid 1950s, literature has been counting numerous contributions about the improvement of performance of one-dimensional (1-D) models, the most attractive when dealing with engineering problems. Inevitably, the approach to such a problem – *i.e.* the derivation of an unsteady-state friction formula – has been strongly influenced by existing results concerning uniform flow. In other words, the aim of most researchers was to extending Moody diagram – an essential tool for engineers – to transient flow conditions. In such a frame, research activity was addressed to evaluating the influence of the initial Reynolds number, $Re_0$, flow acceleration and in less degree pipe roughness. Due to the lack of high frequency response velocimeters, attention was focused on the pressure time-history as related to energy dissipation. Moreover, the prejudice about the shape of transient velocity profiles – *i.e.* to assume that during transients velocity profiles evolve as a series of uniform flows – led to a naive definition of the mean flow velocity, $V$. In fact, it is quite questionable to calculate $V$ as the ratio between flow discharge and cross-section area when, for example, local velocities near the pipe wall have an opposite sign with respect to those in the central part of the flow (the so-called *annular effect*). The limits of such an approach, that: *i*) postulates the existence of a link between the gradient in time of energy and pressure as it is in uniform flow; *ii*) neglets peculiarity of transient velocity profiles; and *iii*) improperly uses discharge measurements carried out by means of low frequency response probes, clearly reflect in the inconsistency of experimental results and lack of universality of unsteady friction formulas, as it will be pointed out in the review of the existing literature reported below.

Unsteady-state friction term, $J$, may be written in the form:

$$J = J_s + J_u = J_s + \frac{K_i}{g}\Im\left(\frac{dV}{dt}\right) \tag{27}$$

*i.e.* as the sum of a quasi-steady component, $J_s$, and an unsteady-state component, $J_u$, depending on weighted past velocity changes through a coefficient $K_i$. The quasi-steady component, $J_s$, is usually evaluated by considering the instantaneous value of the mean flow velocity, $V$, and the proper friction formula according to the value of $Re$, as in uniform flow. Both coefficient $K_i$ and the structure of the functional relationship $\Im$ are related to flow regime before the transient (*i.e.* whether it is laminar or turbulent). The following five distinct types of unsteady friction models have been selected for discussion: Daily *et al* [20], Carstens & Roller [21], Zielke [22], Brunone *et al* [23–25] and Shuy [26]. In the equation reported below, subscripts refer to the specific unsyeady-state model (*i.e.* $D$ = Daily *et al*; $CR$ = Carstens & Roller; $Z$ = Zielke; $BG$ = Brunone *et al*; and $S$ = Shuy).

### 3.1 The Daily *et al* model [20]

In their pioneering paper, the total piezometric drop along the laboratory pipe – a smooth brass conduit with an internal diameter $d = 1''$ and a length equal to $27d$ – is measured for different rates of change of acceleration and deceleration and then compared with the equivalent steady-state one (*i.e.* with the same instantaneous flow rate). Due to the characteristics of the experimental set-up along with the chosen measurement techniques, experimental values – particularly those concerning decelerated flow – are considerably spread. As a consequence, Authors only suggest the possibility of a link between unsteady resistance and local temporal acceleration. Precisely, they note that for accelerated flow the resistance is not appreciably greater whereas for decelerated flow it is appreciably less and "*data can be represented by a family of lines, essential parallel, one for each deceleration.*" In both cases, however, "*it appears that unsteadiness does not result in marked changes from the equivalent steady-state flow.*" Consequently, the following relationships could be tentatively assumed for coefficent $K_D$ and functional relationship $\Im$, respectively:

$$K_D = \begin{cases} 0.01 - 0.015 & \frac{dV}{dt} > 0 \\ 0.62 & \frac{dV}{dt} < 0 \end{cases} \tag{28a}$$

$$\Im_D \left( \frac{dV}{dt} \right) = \frac{dV}{dt} \tag{28b}$$

### 3.2 The Carstens & Roller model [21]

By assuming small magnitude of acceleration as well as high initial Reynolds number, Carstens and Roller postulate that velocity distribution is the same for the unsteady flow as for the equivalent steady flow. Precisely, they assume the validity of the power law velocity distribution for turbulent flow in a smooth pipe. Within a simplified linear-momentun equation – *i.e.* by considering the fluid as incompressible and the flow as a uniform unsteady one – they obtain:

$$K_{CR} = K_{CR}\,(\mathrm{Re}) \tag{29a}$$

$$\Im_{CR} \left( \frac{dV}{dt} \right) = \frac{dV}{dt} \tag{29b}$$

when $Re < 10^5$, it can be assumed $K_{CR} = 0.224$. Unfortunately, results of tests carried out by Cartsens and Roller in a smooth brass pipe with $d = 0.5''$ *are not sufficiently precise to either prove or disprove* the validity of eqns. (29a) and (29b). Also in this case, the poor reliability of data can be ascribed mainly to the technique used for measuring temporal acceleration.

### 3.3 The Zielke model [22]

By applying the Laplace transform to the axial component of the Navier-Stokes equation, Zielke obtains the following analytical expressions for laminar transients:

$$K_Z = \frac{16\nu}{d^2} \tag{30a}$$

$$\Im_Z\left(\frac{dV}{dt}\right) = \int_0^t \frac{\partial V}{\partial t}(x,t)\,W(x,t-\xi)\,d\xi \tag{30b}$$

where: $\xi =$ time (used in convolution integral); $W$ = weighting function for past velocity changes. Function $W$ can be evaluated by means of a series of values of the dimensionless time $\theta = 2\nu t/d$:

$$W(\theta) = e^{-26.3744\theta} + e^{-70.8493\theta} + e^{-135.0198\theta} + e^{-218.9216\theta} + e^{-322.5544\theta} \tag{30c}$$

for $\theta > 0.02$ and

$$W(\theta) = 0.282095\theta^{-0.5} - 1.25 + 1.057855\theta^{0.5} + 0.9375\theta + 0.396696\theta^{3/2} + 0.351563\theta^2 \tag{30d}$$

when $\theta < 0.02$.

Zielke's solution is successfully checked by laboratory test carried out by Holmboe & Rouleau [27]. Since numerical solution of eqn. (30b) requires the storage of all computed past velocities, several approximate expressions of the weighting function $W$ have been proposed. Trikha [28] approximates the function $W$ by the sum of three exponential terms whose coefficients are obtained by fitting the exact values in the original formulation of $W$ by Zielke. Such terms are introduced at each computational time, but only the change in velocity since the previous time step is needed. A larger computer storage is required by the approximate relationships obtained by Suzuki *et al* [29] and Schohl [30], where function $W$ is a sum of five exponential terms.

### 3.4 The Brunone *et al* model [23]

In order to take into account the difference with respect to uniform flow velocity profiles and the friction term, in the momentum equation two additional terms are introduced, both function of local acceleration, $\partial V/\partial t$:

$$\frac{\partial H}{\partial x} + \frac{V}{g}\frac{\partial V}{\partial x} + \frac{1}{g}\frac{\partial V}{\partial t} + \frac{\eta}{g}\frac{\partial V}{\partial t} + \frac{\phi}{g}\frac{\partial V}{\partial t} + J_s = 0 \tag{31}$$

In eqn. (31), $\eta = \frac{A\int \frac{u^2 dA}{A}}{V^2 A} - 1$ is the excess over unity of the Coriolis momentum flux correction coefficient, whereas $\frac{\phi}{g}\frac{\partial V}{\partial t} = J - J_s$ is the difference between the actual friction term and the one obtained within the quasi steady-state approach. Brunone and Golia [31] analyzing the experimental velocity profiles given by Hino *et al* [32], with regard to the lumped term $\frac{\eta+\phi}{g}\frac{\partial V}{\partial t}$, it is pointed out that:

$i$) a significant correction with respect to uniform flow conditions holds when $V\frac{\partial V}{\partial t} > 0$; and $ii$) a constitutive equation can be put in the form $\eta + \phi = 1 - a/\Omega_V$ where $K_{BG} =$ decay coefficient and $\Omega_V = \frac{\partial V/\partial t}{\partial V/\partial x}$ is the propagation speed of a given value of $V$. In the original formulation [23–25], the decay coefficient $K_{BG}$ is assumed as a constant for low initial Reynolds number transients and is estimated within a trial and error procedure by considering numerical and experimental pressure traces. Carravetta *et al* [33] point out that $K_{BG}$ is linked to the difference between two successive pressure peaks. Consequently, the Brunone *et al* model may be written as follows:

$$\Im_{BG} = 1 - \frac{a}{\Omega_V} \tag{31a}$$

with $K_{BG}$ ranging from 0.03 to 0.1.

With regard to the special case of uniform acceleration, Vardy & Brown [34] show that: *i*) coefficient $K_{BG}$ is equal to the limiting value of the unsteady friction coefficient, $f_{u,L}$:

$$f_{u,L} = \frac{4\tau_{w,u}}{\rho d \left(\frac{dV}{dt}\right)_c} \tag{32}$$

that can be considered when a uniform acceleration, $\left(\frac{dV}{dt}\right)_c$, has existed for a *sufficient* period of time; and *ii*) $f_{u,L}$ varies strongly with Reynolds number (precisely, $f_{u,L}$ decreases with increasing $Re$); in eqn. (32) $\tau_{w,u}$ is the additional wall shear stress due to unsteadiness. By comparing quasi 2-D and 1-D numerical model results, Pezzinga [13] develops diagram charts for $K_{BG}$ that present analogies with the Moody's diagram. In particular, for turbulent flow all curves tend to have the same behavior for low values of $Re_0$ and tend to values depending only on the relative roughness for high values of $Re_0$. For laminar flow, the value of $K_{BG}$ does not depend on $Re_0$. The graphs also show that $K_{BG}$ strongly depends, both for laminar and turbulent flow, on a dimensionless parameter of the installation.

Such a model has been extensively checked by other researchers with good results (*e.g.* Bergant & Simpson [35], Wylie [36], Bughazem & Anderson [37] [38], Vitkosky *et al* [39], Louriero & Ramos [40]). Moreover, it was extended by Pezzinga [13] and Bergant *et al* [41] also to "upstream transients". More precisely this extension generalizes the model for transients caused by valve closure at $x = 0$. Specifically, Pezzinga proposes:

$$\Im_{BG} = 1 + sign\left(V\frac{\partial V}{\partial x}\right)\frac{a}{\Omega_V} \tag{31b}$$

Adopting this form of the momentum equation the model provides always additional dissipations, with no dependence on the position of the valve in the coordinate system.

### 3.5 The Shuy model [26]

During a series of experiments carried out on a 16m long acrylic pipe with $d =$ 50.9mm, Shuy measures the unsteady wall shear stress during both accelerating and decelerating flow by means of two different techniques. Precisely, he uses both a shcar tube – which gives a direct measurement – as well as the pressure gradient along the pipe which is measured by means of a differential pressure transducer, as related to wall shear stress. Such different techniques were calibrated only by means of steady-state tests. In the Author's opinion, experimental data – that show a large amount of scatter – are reasonably well represented in the explored range of flow conditions ($-0.2 < \frac{2d}{f_s V^2}\frac{dV}{dt} < 0.3$, with $f_s$ being the steady-state friction factor) when assuming:

$$K_s = \begin{cases} -0.0165 & \frac{dV}{dt} > 0 \\ -0.26 & \frac{dV}{dt} < 0 \end{cases} \tag{33a}$$

with

$$\Im_S\left(\frac{dV}{dt}\right) = \frac{dV}{dt} \tag{33b}$$

Apparently, such a result contradicts the mentioned previous ones, specifically those by Daily *et al* and Carstens and Roller. According to eqn. (33a), wall shear stress decreases in accelerating flow and increases in decelerating flow, as compared to the steady-state values. Consequently, Shuy's paper generated a very interesting discussion even if no conclusive result was reached. In such a context, it should be pointed out the following weak points in Shuy's experiments that strongly limit the validity of his results: *(i)* the use of a differential pressure transducer for measuring unsteady pressure gradient; and *(ii)* the measurement of mean flow velocity – and then flow acceleration and deceleration – by differentiating the supply tank water level with respect to time. In fact, some doubts arise when considering the low frequency response of such techniques and the rapidity of flow changes in the carried out tests. Moreover, further attention should be paid to test pipe material: as a matter of fact, since it was an acryilic (*i.e.* plastic) pipe, the different behavior of viscoelastic materials should be taken into account.

## 4 Energy dissipation other than friction

### 4.1 Viscoelastic pipes

In some cases dissipations other than friction are important. In particular, in polymeric pipe transients the oscillations damping are to be attributed mainly to viscoelastic behavior of material. Studies on viscoelastic behavior of polymeric pipes were carried out to develop numerical or analytical methods suitable for the study of unsteady flow (*e.g.* Rieutord & Blanchard [42]; Franke & Seyler [43]; Guney [44]; Ghilardi & Paoletti [45]), or for the evaluation of viscoelastic parameters by means of unsteady flow runs [46]. Other studies have aimed at examing the reduction of unsteady flow pressure oscillation by short deformable pipes with viscoelastic behavior (Pezzinga & Scandura [47]; Pezzinga [48]).

## 4.2 Mathematical model

For viscoelastic material, it is possible to adopt a Kelvin-Voigt model, by which the mechanical behavior of the material is represented by a serie of elements, each constituted by a spring coupled with a viscous damper joined to a spring element representing the instantaneous deformation. The total deformation can be expressed as the sum of an instantaneous deformation and a delayed one:

$$\varepsilon_t = \varepsilon_i + \varepsilon_r \tag{34}$$

By using a Kelvin-Voigt model with $n$ elements one obtains:

$$\varepsilon_r = \sum_{k=1}^{n} \varepsilon_k \tag{35}$$

The instantaneous deformation can be expressed by:

$$\varepsilon_i = \frac{\sigma}{E_0} \tag{36}$$

The delayed deformation is obtained by the differential equation:

$$\sigma = E_k \varepsilon_k + \eta_k \frac{d\varepsilon_k}{dt} \tag{37}$$

being $\eta_k$ and $E_k$ the viscosity and the modulus of elasticity of the Kelvin-Voigt element, respectively. In the hypothesis of a single Kelvin-Voigt element and negligible convective term, one can write:

$$\frac{\partial \varepsilon_1}{\partial t} = \frac{1}{\theta_1}\left(\frac{d\,p}{2sE_1} - \varepsilon_1\right) \tag{38}$$

where $s$ is the wall thickness and $\theta_1 = \eta_1/E_1$ a relaxation time. Then the continuity equation assumes the following form:

$$\frac{\partial p}{\partial t} + \frac{\rho a^2}{A}\frac{\partial Q}{\partial x} + 2\rho a^2 \frac{\partial \varepsilon_1}{\partial t} = 0 \tag{39}$$

where the pressure wave speed $a$ has to be computed by the modulus of elasticity relevant to instantaneous deformation. Introducing a variable $\psi$ defined as:

$$\psi = \frac{p}{\gamma} + \frac{2a^2}{g}\varepsilon_1 \tag{40}$$

The continuity equation can be expressed in the following conservation form:

$$\frac{\partial \psi}{\partial t} + \frac{a^2}{gA}\frac{\partial Q}{\partial x} = 0 \tag{41}$$

# 5 Analysis of results

## 5.1 Considerations on turbulence models and velocity and turbulence quantities

Experimental runs were carried out on the laboratory installation at the Department of Civil and Environmental Engineering of the University of Catania. The installation is costituted by a zinc plated pipe DN 50 (internal diameter 53.9mm, thickness 3.2mm, modulus of elasticity 206MN/m$^2$, roughness 0.11mm, length 145.2m) fed by a centrifugal electropump and ends in a 1m$^3$ pressure tank. The manual closure of a ball valve allows to intercept the discharge in about 0.04s. Measurements of total discharge are made by an electromagnetic flowmeter. The line pressure is measured by strain gauge pressure transducers. The transducers have a range of 0–10 bar and a maximum error of 0.5% of full-scale pressure. The signal from the pressure transducer was sampled with a frequency of 100Hz.

Pezzinga [5] examines the low Reynolds number $k-\varepsilon$ models firstly with reference to steady state Darcy-Weisbach friction factor values. Firstly the models valid for smooth pipe are considered. The results of this analysis show that models behavior is highly dependent on the considered wall condition for $\varepsilon$. In particular, the models for which $\varepsilon_w = 0$ (Launder-Sharma and Nagano-Hishida) are too dependent on radial meshes number, with computed friction factor values in good agreement with the experimental ones only for very low meshes number. On the contrary, when the grid is dense, as it is needed to make thorough evalutations in unsteady flow conditions, friction factor values are less than the experimental ones, and numerical errors can verify, probably due to negative values of $\varepsilon$ near the wall. The models for which $\varepsilon_w \neq 0$ (Lam-Bremhorst and Nagano-Tagawa) have more suitable behavior, given that the friction factor values are less dependent on radial grid and more close to the experimental ones. Subsequently the formulations proposed to take in account wall roughness are examined. In particular, the modifications of $f_\mu$ for the Van Driest model proposed by Van Driest and by Krogstaad are compared, with reference to steady-state Darcy-Weisbach friction factor values again. The former gives better results. Then the model of Zhang *et al* associated to Lam-Bremhorst is considered; as a result, such a model provides friction factor values different from the experimental ones. Then the Van Driest $f_\mu$ function for rough wall associated to Lam-Bremhorst model is tested, giving very good results. The comparisons of computed head oscillations by Van Driest and Lam-Bremhorst models and measured ones show very small differences between numerical results, and higher differences between computed oscillations and measured ones, increasing with time (fig. 1).

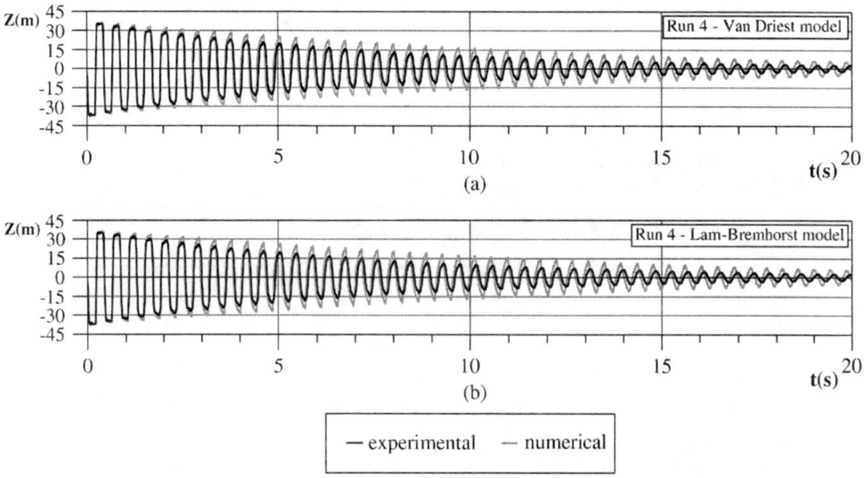


Figure 1: Comparison of measured heads and computed ones by two-dimensional models: (a) Van Driest; (b) Lam-Bremhorst (adapted from Pezzinga [5]).

This behavior verifies also considering, in both models, the Van Driest rough wall function $f_\mu$. Furthermore, the numerical results are very close to those of the two-zone mixing length model [4]. Then it seems that it is very difficult to test the performance of different turbulence models for water hammer flow on the basis of the comparison with experimental head oscillations. These considerations would induce to use simple models, to save computer time, but $k - \varepsilon$ models can give information on turbulence quantities useful for deeper theoretical evaluations. In fig. 2 the profiles of $u, k$, and $\varepsilon$ computed by the Lam-Bremhorst model are given at different dimensionless times $at/L$. The analysis of velocity profiles confirms that they differ from the steady-state velocity profiles, presenting recirculation zones. The $k$ and $\varepsilon$ profiles show that these quantities are little influenced by unsteady flow.

## 5.2 Comparisons with experimental results for water hammer flow

In the same experimental set-up described in Section 5.1, at a distance from the valve of about 81.4m, was recently inserted a short transparent polymethyl-methacrylate pipe (length 18cm, internal diameter 54mm, thickness 3mm, modulus of elasticity 2.94MN/m$^2$), to allow the velocity measurement by a laser Doppler velocimeter, with laser power between 10mW and 500mW, "frequency shifter" and Bragg cell. The lens has a 160mm focal distance and is mounted on an optic fiber probe, moving by a computer controlled system. The signal is received in "back scattering" and processed by a "Particle Dynamics Analyzer".

A comparison between numerical and experimental head (a) for $x = 0$ and numerical and experimental velocity (b) for $x = 81.5$m and a distance from the axis $r = 24.3$mm (2.7mm from the wall) is shown in fig. 3.

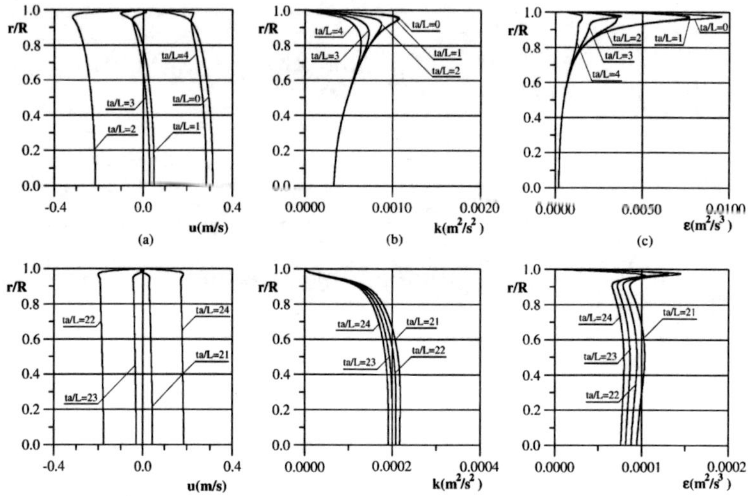


Figure 2: Profiles of velocity (a), turbulent energy (b), dissipation rate (c), computed by the Lam-Bremhorst model (adapted from Pezzinga [5]).

The mixing length model by Pezzinga [4] is used. The comparisons are very good. The first oscillations are almost perfectly coinciding. Subsequently, the experimental head oscillations are slightly more damped. Furthermore the comparison between head phases are very good, but the velocity phases are slightly shifted more and more with time. The data rate becomes very poor as time goes on, giving rise to unreliable results in the long term. A data rate reduction is observed also when the distance from the wall increases. For example, in fig. 4 the comparison is reported between velocity for $r = 10.8$mm (16.2mm from the wall). It seems that at this moment neither head measurements nor velocity measurements allow to compare the performance of different turbulence models for water hammer flow.

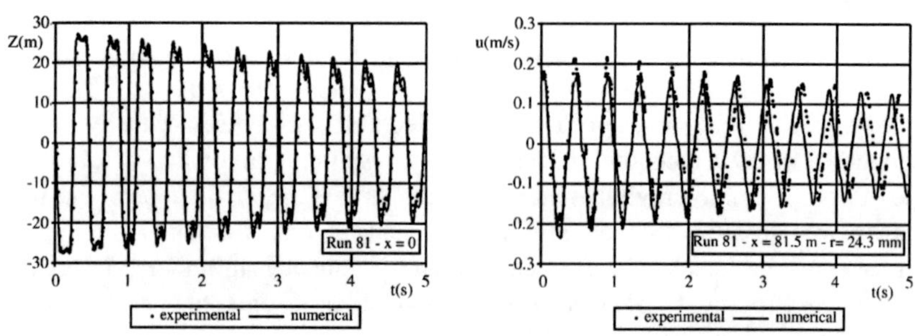


(a) (b)

Figure 3: Comparison between numerical and experimental values: (a) heads; (b) velocities (adapted from Nicosia & Pezzinga [49]).

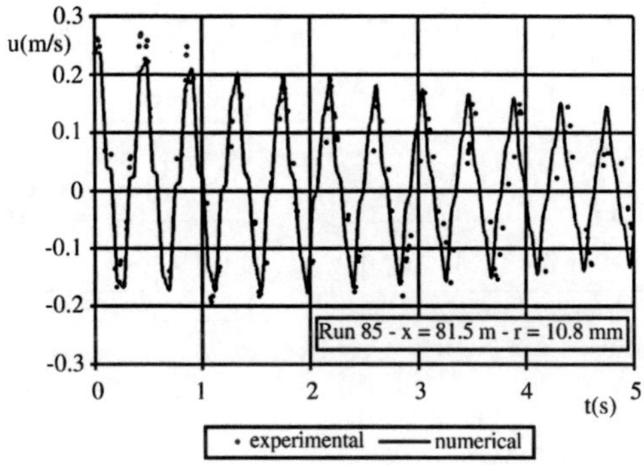


Figure 4: Decay of data rate for increasing distance from the wall (adapted from Nicosia & Pezzinga [49]).

## 5.3 Comparisons with experimental results for water hammer in viscoelastic pipes

Cannizzaro *et al* [50] analyze water hammer flow in a viscoelastic pipe. The comparisons between mathematical models, one-dimensional or two-dimensional, elastic or viscoelastic, and experimental data are made to evaluate the relative weight of unsteady friction and viscoelasticity on unsteady flow dissipation. The models results are compared with the results of experimental runs of water hammer on a high density polyetilene (HDPE) pipe carried out at the Water Engineering Laboratory of the University of Perugia. The runs were carried out on a pipe with length $L$ = 350.55m and internal diameter $d$ = 93.8mm. More details on the experimental installation are reported by Brunone *et al* [51]. The calibration of viscoelastic parameters $E_0$, $E_1$ and $\theta_1$ is carried out by trial and error, giving the best results for 1900N/mm$^2$, 3300N/mm$^2$ and 0.18s, respectively. In fig. 5a the comparison

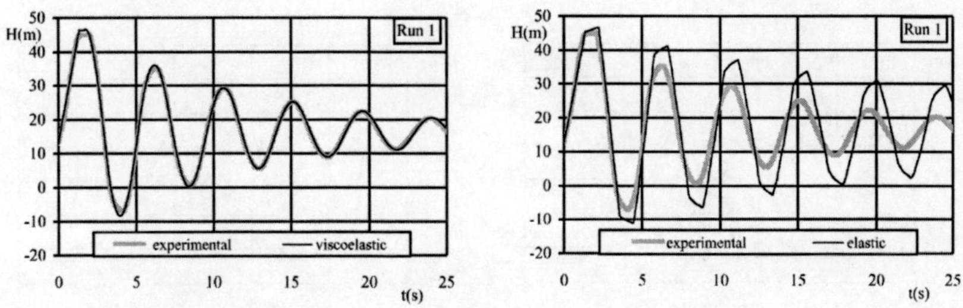


(a) (b)

Figure 5: Comparison between measured heads and computed ones by the two-dimensional model (mixing length): (a) viscoelastic; (b) elastic (adapted from Cannizzaro *et al* [50]).

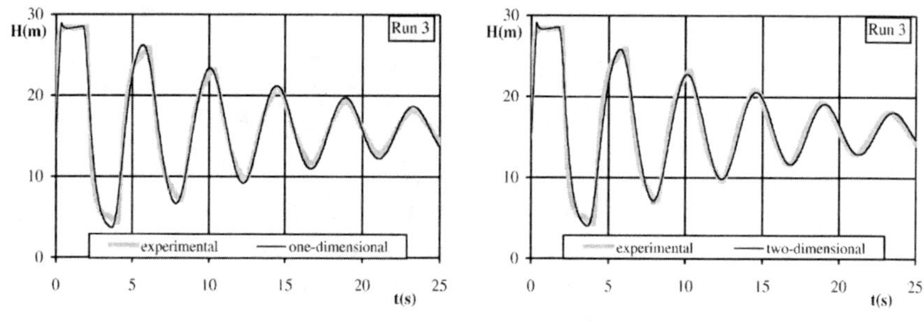


(a) (b)

Figure 6: Comparison between measured heads and computed ones by the viscoelastic model: (a) quasi-steady one-dimensional; (b) two-dimensional (adapted from Cannizzaro *et al* [50]).

is reported between measured heads and computed ones by the two-dimensional (mixing length) viscoelastic model. To put into evidence the influence of viscoelasticity of HDPE, the comparison between measured heads and computed ones by the two-dimensional elastic model is reported in fig. 5b. It can be observed that the viscoelastic model greatly improves the reproduction of experimental data. The calibration of modulus of elasticity $E$ (1200N/mm$^2$) to obtain the experimental period of oscillations, gives good prediction of first maximum head oscillation, but the observed damping is not simulated.

To show the influence of friction, the comparison between measured heads and computed ones by a quasi-steady resistance one-dimensional viscoelastic model (fig. 6a) and by the two-dimensional viscoelastic model (fig. 6b). The 1-D model gives results very close to those of the 2-D model (fig. 7), and then the influence of viscoelasticity on dissipation is higher than friction. To take into account the

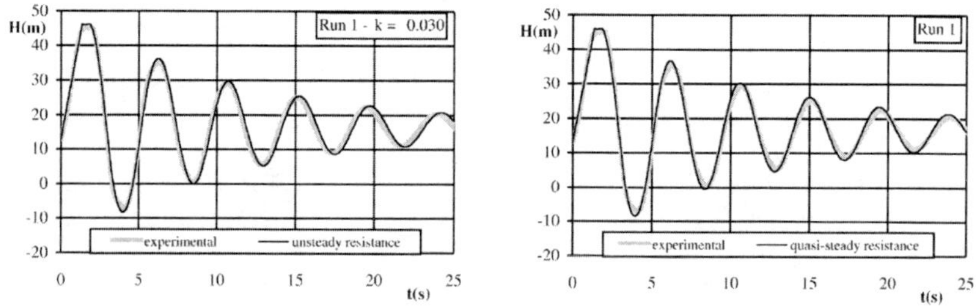


(a) (b)

Figure 7: Comparison between measured heads and computed ones by the one-dimensional viscoelastic model: (a) unsteady resistance; (b) quasi-steady resistance (adapted from Cannizzaro *et al* [50]).

1-D unsteady friction eqn. (31b) gives rise to results very similar to quasi-steady resistance. The calibration of unsteady friction resistance coefficient $K_{BG}$, gives values between 0.02 and 0.03, of the same order of magnitude of those of other studies [13], but its exact definition is difficult [52, 53].

## 6 Conclusions

In this paper the problem of energy dissipation in transient pipe flow is examined. For transients in elastic pipes, the energy dissipation is mainly due to friction. The correct evaluation of friction force can be made by two-dimensional models taking into consideration the behavior of velocity profiles. The comparison between different low Reynolds number $k - \varepsilon$ turbulence models shows that the Lam-Bremhorst model gives very good results. Simpler mixing length models as well give good results in the comparison of calculated and experimental velocity, and can be used if one is not interested in detailed turbulence characteristics.

Alternatively, unsteady friction one-dimensional models can correctly evaluate the friction force, if calibrated values of the model parameters are used. However, the generalization of these parameters is problematic.

For transients in polymeric pipes, energy dissipation is mainly due to the viscoelastic behavior of the material. As a matter of fact, an elastic model can predict only the maximum oscillations, whereas a Kelvin–Voigt viscoelastic model can reproduce the observed damping by means of calibrated values of the model parameters.

The problem of different velocity profiles in elastic or viscoelastic pipes is open and should be examined by means of experimental and theoretical future studies.

## 7 Acknowledgements

The writers gratefully acknowledge the support of the Italian Ministry of Education, University and Research (National Project on "Influence of vorticity and turbulence in interactions of water bodies with their boundary elements and effects on hydraulic design"). They also aknowledge the kind advice of anonymous reviewers.

## References

[1] Vardy, A.E. & Hwang, K., A characteristics model of transient friction in pipes. *J. Hydr. Res.*, IAHR, **29(5)**, pp. 669–684, 1991.

[2] Eichinger, P. & Lein, G., The influence of friction on unsteady pipe flow. *Unsteady Flow and Fluid Transients*, Balkema, Rotterdam (NL), pp. 41–50, 1992.

[3] Silva-Araya, W.F. & Chaudhry, M.H., Computation of energy dissipation in transient flow. *J. Hydr. Engrg.*, ASCE, **123(2)**, pp. 108–115, 1997.

[4] Pezzinga, G., Quasi-2D model for unsteady flow in pipe networks. *J. Hydr. Engrg.*, ASCE, **125**, pp. 676–685, 1999.

[5] Pezzinga, G., Affidabilità di modelli semi-empirici di turbolenza per la valutazione delle resistenze di attrito in condizioni di moto vario, *L'Acqua*, AII, **1**, pp. 29–38, 2000 (in Italian).

[6] Betamio de Almeida, A. & Koelle, E., *Fluid Transients in Pipe Networks*, Computational Mechanics Publications, Southampton, Elsevier Applied Science: London, 1992.

[7] Wylie, E.B., & Streeter, V.L., *Fluid Transients in Systems*, Prentice Hall, Englewood Cliffs: New Jersey, 1993.

[8] Kita, Y., Adachi, Y. & Hirose, K., Periodically oscillating turbulent flow in a pipe. *Bull. JSME*, **23**, pp. 656–664, 1980.

[9] Granville, P.S., A modified Van Driest for the mixing length of turbulent boundary layers in pressure gradient. *J. Fluids Engrg.*, ASME, **111**, pp. 94–97, 1989.

[10] Granville, P.S., A near-wall eddy viscosity formula for turbulent boundary layers in pressure gradients suitable for momentum, head or mass transfer. *J. Fluids Engrg.*, ASME, **112**, pp. 240–243, 1990.

[11] Van Driest, E.R., On turbulent flow near a wall. *J. Aero. Sci.*, **23**, pp. 1007–1011, 1956.

[12] Krogstad, P., Modification of the van Driest damping function to include the effects of surface roughness. *AIAA J.*, **29**, pp. 888–894, 1991.

[13] Pezzinga, G., Evaluation of unsteady flow resistances by quasi-2D or 1D models. *J. Hydr. Engrg.*, ASCE, **126**, pp. 778–785, 2000.

[14] Hallback, M., Johansson, A.V. & Burden, A.D., The basics of turbulence modelling, *Turbulence and Transition Modelling*, ERCOFTAC Series, Kluwer Academic Publishers: Dordrecht (NL), pp. 81–154, 1995.

[15] Launder, B.E. & Sharma, B.I., Application of the energy-dissipation model of turbulence to the calculation of flow near a spinning disk. *Lett. Heat Mass Transfer*, **1**, pp. 131–138, 1974.

[16] Nagano, Y. & Hishida, M., Improved form of the $k$ - $\varepsilon$ model for wall turbulent shear flow. *J. Fluids Engrg.*, ASME, **109**, pp. 156–160, 1987.

[17] Lam, C.K.G. & Bremhorst, K.A., Modified form of the $k$ - $\varepsilon$ model for predicting wall turbulence. *J. Fluids Engrg.*, ASME, **103**, pp. 456–460, 1981.

[18] Nagano, Y. & Tagawa, M., An improved $k$ - $\varepsilon$ model for boundary layer flows. *J. Fluids Engrg.*, ASME, **112**, pp. 33–39, 1990.

[19] Zhang, H., Faghri, M. & White, F.M., A new low-Reynolds-number $k$ - $\varepsilon$ model for turbulent flow over smooth and rough surfaces. *J. Fluids Engrg.*, ASME, **118**, pp. 255–259, 1996.

[20] Daily, J.W., Hankey, W.L., Olive, R.W. & Jordaan, J.M., Resistance coefficient for accelerated and decelerated flows through smooth tubes and orifices. *Trans.*, ASME, **78**, pp. 1071–1077, 1956.

[21] Carstens, M.R. & Roller, J.E., Boundary-shear stress in unsteady turbulent pipe flow. *J. Hydr. Div.*, ASCE, **85**, pp. 67–81, 1959.

[22] Zielke, W., Frequency-dependent friction in transient pipe flow. *J. Basic Engrg.*, ASME, **90**, pp. 109–115, 1968.

[23] Brunone, B., Golia, U.M. & Greco, M., Some remarks on the momentum equation for fast transients. *Proc. Int. Meeting on Hydraulic Transients and Water Column Separation*, Valencia (E), pp. 201–209, 1991.

[24] Brunone, B., Golia, U.M. & Greco, M., Modelling of fast transients by numerical methods. *Proc. Int. Meeting on Hydraulic Transients and Water Column Separation*, Valencia (E), pp. 273–282, 1991.

[25] Brunone, B., Golia, U.M. & Greco, M., Effects of two-dimensionality on pipe transients modeling. *J. Hydr. Engrg.*, ASCE, **121**, pp. 906–912, 1995.

[26] Shuy, E.B., Wall shear stress in accelerating and decelerating turbulent pipe flows. *J. Hydr. Res.*, IAHR, **34**, pp. 173–183, 1996.

[27] Holmboe, E. & Rouleau, W.T., The effect of viscous shear on transients in liquid lines. *J. Basic Engrg.*, ASME, **89**, pp. 174–180, 1967.

[28] Trikha, A.K., An efficient method for simulating frequency-dependent friction in liquid flow. *J. Fluids Engrg.*, ASME, **97**, pp. 97–105, 1975.

[29] Suzuki, K., Taketomi, T. & Sato, S., Improving Zielke's method of simulating frequency-dependent friction in laminar liquid pipe flow. *J. Fluids Engrg.*, ASME, **113(4)**, pp. 569–573, 1991.

[30] Schohl, G.A., Improved approximate method for simulating frequency-dependent friction in transient laminar flow. *J. Fluids Engrg.*, ASME, **115(3)**, pp. 420–424, 1993.

[31] Brunone, B. & Golia, U.M., Some considerations on velocity profiles in unsteady pipe flows. *Proc. Int. Conf. on Entropy and Energy Dissipation in Water Resources*, Maratea (I), pp. 481–487, 1991.

[32] Hino, M., Masaki, S. & Shuji, T., Experiments on transition to turbulence in an oscillatory pipe flow. *J. Fluid Mech.*, **75**, part 2, pp. 193–207, 1976.

[33] Carravetta, A., Golia, U.M. & Greco, M., Sull'attenuazione spontanea delle fluttuazioni di pressione durante i transitori di colpo d'ariete. *Atti XXIII Convegno di Idraulica e Costruzioni Idrauliche*, Florence (I), pp. E67–E79, 1992 (in Italian).

[34] Vardy, A. & Browne, J., On turbulent, unsteady, smooth-pipe friction. *Proc. 7th Int. Conf. on Pressure Surges and Fluid Transients in Pipelines and Open Channels,* BHR Group, pp. 289–311, 1996.

[35] Bergant, A. & Simpson, A.R., Estimating unsteady friction in transient cavitating pipe flow. *Proc. 2nd Int. Conf. on Water Pipeline Systems*, BHR Group, Edinburgh (UK), pp. 3–16, 1994.

[36] Wylie, E.B., Frictional effects in unsteady turbulent pipe flow. *Applied Mechanics in the Americas*, The University of Iowa, Iowa City, 5, pp. 29–34, 1996.

[37] Bughazem, M.B. & Anderson, A., Problems with simple models for damping in unsteady flow. *Proc. 7th Int. Conf. on Pressure Surges and Fluid Transients in Pipelines and Open Channels*, BHR Group, Harrogate (UK), pp. 537–548, 1996.

[38] Bughazem, M.B. & Anderson, A., Investigation of an unsteady friction model for waterhammer and column separation. *Proc. 8th Int. Conf. on Pressure Surges*, BHR Group, The Hague (NL), pp. 483–495, 2000.

[39] Vitkosky, J.P., Lambert, M.F., Simpson, A.R. & Bergant, A., Advances in unsteady friction modelling in transient pipe flow. *Proc. 8th Int. Conf. on Pressure Surges*, The Hague (NL), BHR Group, pp. 471–481, 2000.

[40] Louriero, D. & Ramos, H., A modified formulation for estimating the dissipative effect of 1-D transient pipe flow. *Proc. Int. Conf. on Pumps, Electromechanical Devices and Systems Applied to Urban Water Management PEDS 2003*, Valencia (E), II, pp. 755–763, 2003.

[41] Bergant, A., Simpson, A.R. & Vitkovsky, J., Developments in Unsteady Pipe Flow Friction Modelling. *J. Hydr. Res.*, IAHR, **39(3)**, pp. 249–257, 2001.

[42] Rieutord, E. & Blanchard, A., Ecoulement non permanent en conduite viscoelastique – coup de bélier. *J. Hydr. Res.*, IAHR, **17**, pp. 217–229, 1979.

[43] Franke, P.G. & Seyler, F., Computation of unsteady pipe flow with respect to viscoelastic material properties. *J. Hydr. Res.*, IAHR, **21**, pp. 345–353, 1983.

[44] Guney, M.S., Waterhammer in viscoelastic pipes where cross-section parameters are time dependent. *Proc. 4th Int. Conf. on Pressure Surges*, BHRA, Cranfield (UK), pp. 189–204, 1983.

[45] Ghilardi, P. & Paoletti, A., Problemi relativi al moto vario in condotti viscoelastici. *L'Energia Elettrica*, AEI, **64**, pp. 273–282, 1987 (in Italian).

[46] Ghilardi, P. & Paoletti, A., I parametri viscoelastici per lo studio dei transitori idraulici in condotte di materiale polimerico. *Atti XX Convegno di Idraulica e Costruzioni Idrauliche*, Padova (I), pp. 993–1006, 1986 (in Italian).

[47] Pezzinga, G. & Scandura, P., Unsteady flow in installation with polymeric additional pipe. *J. Hydr. Engrg.*, ASCE, **121**, pp. 802–811, 1995.

[48] Pezzinga, G., Unsteady flow in hydraulic networks with polymeric additional pipe. *J. Hydr. Engrg.*, ASCE, **128**, pp. 238–244, 2002.

[49] Nicosia, S. & Pezzinga, G., Analisi del campo di velocità di correnti in pressione in moto vario. *Atti XXIX Convegno di Idraulica e Costruzioni Idrauliche*, Trento (I), 1, pp. 229–235, 2004 (in Italian).

[50] Cannizzaro, D., Pezzinga, G., Brunone, B. & Ferrante, M., Dissipazioni nei transitori in condotte viscoelastiche. *Atti XXVIII Convegno di Idraulica e Costruzioni Idrauliche*, Potenza (I), **1**, pp. 93–100, 2002 (in Italian).

[51] Brunone, B., Karney, B.W., Mecarelli, M. & Ferrante, M., Velocity profiles and unsteady pipe friction in transient flow. *J. Water Resour. Plng. Mgmt.*, ASCE, **126**, pp. 236–244, 2000.

[52] Pezzinga, G., Discussion of "Velocity profiles and unsteady pipe friction in transient flow". *J. Water Resour. Plng. Mgmt.*, ASCE, **128**, p. 85, 2002.

[53] Brunone, B., Karney, B.W., Mecarelli, M. & Ferrante, M., Closure to "Velocity profiles and unsteady pipe friction in transient flow". *J. Water Resour. Plng. Mgmt.*, ASCE, **128**, p. 86, 2002.

## Nomenclature

The following symbols have been used in this paper:

| | |
|---|---|
| $a$ | pressure wave speed; |
| $A$ | cross-sectional area of the cylinder with radius R; |

| | |
|---|---|
| $A_0$ | total cross-sectional area of the pipeline; |
| $B$ | parameter of logarithmic law of the wall; |
| $d$ | pipe diameter; |
| $E_0$ | instantaneous modulus of elasticity of Kelvin–Voigt model; |
| $E_k$ | modulus of elasticity of k-th Kelvin–Voigt element; |
| $f_{u,L}$ | limiting value of unsteady friction factor; |
| $\Im$ | function of unsteady friction models; |
| $g$ | gravitational acceleration; |
| $H$ | piezometric head; |
| $J$ | friction force per unit weight; |
| $J_s$ | steady state friction force per unit weight; |
| $k$ | turbulent kinetic energy; |
| $K_i$ | coefficient of unsteady friction; |
| $k_s$ | equivalent sand roughness; |
| $k_s^* = \frac{u^* k_s}{\nu}$ | dimensionless sand roughness; |
| $l$ | mixing length; |
| $l^+ = \frac{u^* l}{\nu}$ | dimensionless mixing length; |
| $p$ | pressure; |
| $Q$ | discharge; |
| $r$ | distance from the axis; |
| $R$ | pipe radius; |
| $R_t = \frac{k^2}{\nu\varepsilon}$ | dimensionless parameter in low Reynolds number $k-\varepsilon$ models; |
| Re | Reynolds number; |
| $\mathrm{Re}_0$ | initial Reynolds number; |
| $\mathrm{Re}^* = \frac{u^* R}{\nu}$ | friction Reynolds numbers; |
| $s$ | pipe wall thickness; |
| $t$ | time; |
| $u$ | velocity component in the longitudinal direction; |
| $u^* = \sqrt{\frac{\tau_w}{\rho}}$ | friction velocity; |
| $V$ | mean velocity; |
| $v$ | velocity component in the radial direction; |
| $W$ | weighting function for past velocity changes; |
| $x$ | abscissa; |

| | |
|---|---|
| $y$ | distance from the wall; |
| $y^+ = \frac{u^* y}{\nu}, y^* = \frac{\sqrt{k}y}{\nu}$, | dimensionless distances from wall; |
| $\delta$ | thickness of the viscous sublayer; |
| $\eta$ | Coriolis momentum flux correction coefficient; |
| $\varepsilon$ | turbulent kinetic energy dissipation rate; |
| $\varepsilon_i$ | instantaneous strain; |
| $\varepsilon_k$ | strain of k-th Kelvin–Voigt element; |
| $\varepsilon_r$ | retarded strain; |
| $\varepsilon_t$ | unit strain; |
| $\eta_k$ | viscosity of k-th Kelvin–Voigt element; |
| $\lambda_1$ | mixing length damping parameter |
| $\nu$ | kinematic viscosity; |
| $\nu_t$ | eddy viscosity; |
| $\rho$ | density; |
| $\psi$ | auxiliary variable; |
| $\sigma$ | normal stress in pipe material; |
| $\sigma_r, \sigma_x, \sigma_\theta$ | normal stresses in $r, x$, and $\theta$ direction; |
| $\tau$ | shear stress; |
| $\theta = \frac{2\nu t}{d}$ | dimensionless time; |
| $\tau_w$ | wall shear stress; |
| $\tau^+ = \frac{\tau}{\tau w}$ | dimensionless shear stress; |
| $\Omega_V$ | propagation speed of $V$; |
| $c$, $C_b$, $C_G$, $C_m$, $c_{1\varepsilon}$, $c_{2\varepsilon}$, $c_\mu$, $k$, $\sigma_k$, $\sigma_\varepsilon$ | empirical constants of turbulence models; |
| $D$, $E$, $f_1$, $f_2$, $f_\mu$ | transition functions of low-Reynolds number $k-\varepsilon$ models. |

# CHAPTER 10

## Scalar dispersion within canopies: new challenges and frontiers

D. Poggi[1,2,3], A. Porporato[3], L. Ridolfi[1] & G.G. Katul[2,3]
[1]*Dipartimento di Idraulica, Trasporti ed Infrastrutture Civili Politecnico di Torino, Torino, Italy.*
[2]*Nicholas School of the Environment and Earth Sciences, Duke University, Durham, U.S.A.*
[3] *Department of Civil and Environmental Engineering, Pratt School of Engineering, Duke University, Durham, U.S.A.*

### Abstract

Scalar dispersion inside canopies has moved from the margins of micro-meteorology to a rich scientific discipline that integrates fundamental fluid mechanics principles with hydrologic, radiative, soil, chemical, and eco-physiological processes. An inclusive review of all these topics is well beyond the scope of a single chapter. The compass of this work is on approaches developed to infer biological sources and sinks within canopies without resorting to gradient-diffusion formulations (or k-theory). This chapter reviews recent developments in multi-layer methods that compute distribution and strength of scalar biological sources and sinks within the canopy volume using both Lagrangian and Eulerian framework. Planar homogeneous and stationary turbulent flows in the absence of buoyant forces are considered. Two types of model formulations are reviewed: 1) forward methods that require vertical foliage distribution along with canopy radiative, physiological, biochemical, and drag properties and 2) inverse methods that require measured mean scalar concentration distribution. Based on numerous field and laboratory studies, these approaches appear to reproduce measured turbulent fluxes within and above the canopy reasonably well without relying on empirical relationship between turbulent scalar fluxes and mean concentration gradients. Both approaches share the need for detailed description of the second moments of the velocity statistics inside the canopy. Methods for inferring these velocity statistics from leaf area density measurements are also discussed.

## 1 Introduction

The emergence of new societal problems pertaining to the goods and services of the biosphere are motivating national and international science agendas (*e.g.* U.S. Global Climate Change Program) that call for long-term continuous monitoring initiatives of scalar concentration and/or fluxes at a spatial and temporal scale never attempted before. This demand for continuous monitoring is moving the study of scalar dispersion within canopies from the margins of micro-meteorology to a major research thrust in Earth System Sciences. Scalar dispersion within canopies remains among the most vexing and complex problems in hydrological, atmospheric, and ecological sciences. Progress in this field requires a comprehensive approach that involves a number of disciplines including fluid mechanics, surface hydrology, radiative transfer, soil physics, atmospheric chemistry, and physiological ecology. The outcome from scalar dispersion studies now have direct bearing on a number of disciplines such as climate change, air and water quality, agricultural management, landscape ecology, and decision making for environmental compliances and policy formulation.

From the climate change perspective, concerns about increased anthropogenic $CO_2$ emissions and the potential role of the biosphere as a carbon sink resulted in a proliferation of long-term eddy-covariance flux measurements of carbon dioxide, heat and water vapor across different biomes and climate (*e.g.* FluxNet, see [1]). Within the context of this initiative, how to connect biological sources and sinks to the monitored turbulent fluxes in the atmosphere remains a problem that can only be approached through fundamental understanding of scalar transport within canopies and is the focus of this review.

However, other equally pressing problems also benefit from studies of scalar dispersion. Air quality and linkages between atmospheric chemistry and turbulent transport is another major research question in biosphere-atmosphere exchange with major implications to food production and forest health. It is now estimated that about 10 to 35 percent of the world's grain production may occur in regions where ozone pollution can potentially reduce crop yields [2]. Ozone deposition, which can be harmful to plant stomata, or the production and transport of $O_3$ precursors (*e.g.* isoprene) remains a computationally high-dimensional research problem in which hundreds of chemical reactions governing $O_3$ production and destruction occur at time scales comparable to turbulent transport.

An analogous problem is predicting volatilization of several substances commonly used or produced in agriculture. For example: 1) the application of pesticides to crops and soils is a major source of persistent organic pollutants in the environment, 2) atmospheric ammonia (NH3) is recognized as a major pollutant for semi-natural ecosystems, as its deposition leads to soil acidification and ecosystem eutrophication [3]. Both measurement programs and modelling studies are now proposed to track their transport, transformation and deposition to water bodies and other terrestrial surfaces [4].

Long-distance dispersal (LDD) of seeds and pollen by wind is yet another important topic that involves scalar dispersion in ecology. LDD has many implications for gene flow, pest control, species expansion, recolonization of disturbed areas, and population dynamics. Previous modelling approaches that did not consider the role of turbulence within canopies failed to simulate LDD. There is now a clear recognition that seed and pollen escape from the canopy is a necessary condition for LDD and hence progress in this area must explicitly deal with scalar dispersion within canopies [5].

When all these example problems are taken together, it is clear that a complete theory for scalar transport within canopies must address questions at spatial and temporal scales ranging from centimeters to tens of kilometers and from fractions of seconds to several decades, respectively [6]. Given such a wide space-time domain, a "modular" research approach must be adopted in which the spatial structure of scalar dispersion can be studied over short periods of time (hereafter referred to as spatial studies), and conversely, the long-term structure of scalar dispersion is studied over limited spatial domains (hereafter referred to as temporal studies). Below, we summarize on-going work in each of these two categories:

1. Spatial studies:
   (a) COMPUTATIONAL EXPERIMENTS: The computational experiments often utilize high-resolution Large Eddy Simulations (LES) or Boltzman type equations to investigate how spatial heterogeneity in boundary conditions (*e.g.* canopy non-uniformity) express themselves in the scalar flow statistics, usually on short time scales (hourly time scale). The use of LES in canopy flows, in which energetic eddies along with many attributes of the energy cascade are explicitly resolved, has progressed significantly over the last decade to permit exploration of such problems though some thorny issues about subgrid models and computational grid size remain [7].
   (b) FIELD EXPERIMENTS: The main advantage of field experiments is that the velocity and scalar flow field can be collected and analyzed *in situ*. However, costs and logistics require that data be primarily collected at a single tower thereby only sampling the vertical structure of the flow field reasonably well. Field experiments often provide little horizontal information about the flow with few notable exceptions [8].
   (c) LABORATORY EXPERIMENTS: The flume and wind tunnel experiments have some indisputable advantages over field experiments, for example they can be carried out under controlled conditions. Also, the time and spatial resolution (in 2- and 3-D) of laboratory experiments can be very high. Traditionally, flume and wind tunnel experiments focused on the velocity field and comparatively less attention was paid to scalar transport. This deficit in flume and wind tunnel data, at least when compared to field experiments, is often attributed to the challenges in simulating realistic scalar source distribution within canopies and being able to measure concentrations at the necessary high frequency.

2. Temporal Studies:

   (a) MODELS: The use of simplified models, such as $K - \epsilon$ or Reynolds-averaged equations, are now one possible method to integrate scalar transport equations from time scales of seconds to hours and longer. Lagrangian Dispersion Models (LDM) is another computationally viable alternative for such temporal aggregation. How well these approaches reproduce measured turbulent fluxes and mean concentration across a broad range of canopy morphology remains an on-going research topic.

   (b) LABORATORY EXPERIMENTS: These experiments permit investigating how variations in particular boundary conditions (*e.g.* leaf area density) express themselves in time and spatially averaged flow statistics within and above the vegetation. Such experiments are often used to produce bench-mark data sets for investigating or testing closure models or LDM (*e.g.* [9, 10]).

   (c) FIELD EXPERIMENTS: Here, the examples are numerous given the monitoring initiatives dictated by many science research agendas previously discussed [1]. Perhaps among the most comprehensive initiative in scalar transport is FLUXNET, now providing scalar fluxes at time scales ranging from fractions of seconds to tens of years and over a wide range of biomes. However, these experiments yield measurements at one level, often in the canopy sublayer or the atmospheric surface layer (ASL) thereby providing a limited spatial perspective about vertical or planar source-sink variations.

It is clear that a review of all these topics is well beyond the scope of a single chapter; hence, the compass of this work is restricted to "temporal studies". Even within this restrictive scope, we only consider high Reynolds and Peclet numbers for stationary, and planar homogeneous flows within extensive and rigid canopies. We focus on these idealized conditions because they form the building blocks of a more comprehensive treatment of scalar dispersion in heterogeneous canopies and on complex terrain.

## 2 Theory

Before proceeding to the theory of scalar dispersion, the main attributes of turbulent flows within the canopy sublayer (CSL) are presented.

### 2.1 Turbulent flow field

Historically, CSL flows were thought to be analogous to rough-wall boundary layers [11]; however, in the last two decades, substantial theoretical and field research suggested that CSL flows possess many unique attributes. These attributes can be summarized as follows:

1) Turbulent flows near the canopy top are not subject to the no-slip boundary constraint as rough-wall boundary layers due to the finite porosity of the vegetation.

An immediate consequence of this attribute is that the dominant organized coherent structures in the CSL are due to the strong inflection point in the mean velocity profile leading to Kelvin-Helmholtz vortices rather than attached eddies dominating rough-wall boundary layers.

2) The short-circuiting of the energy cascade inside canopies due to wakes by canopy elements plays an important role in the time-space averaged momentum equations and the spectral properties of the flow. This short-circuiting leads to direct interaction between large and small scales thereby disrupting the classical statistical treatment of small-scale eddies using locally homogeneous and isotropic theories [12, 13].

Incorporating these two features remains a basic challenge for canopy momentum transport models. Hence, theoretical developments and experiments on canopy flows with elementary morphology must serve as a necessary first step towards solving the broader problem of scalar dispersion in natural systems [12].

## 2.2 Scalar transport

The time and spatially averaged scalar budget equation for high Peclet number is given by

$$\frac{D\overline{c}}{Dt} + \frac{\partial \overline{F_j}}{\partial x_j} = \overline{S}, \tag{1}$$

where $\overline{c}$ is the mean concentration of a scalar entity (*e.g.* $H_2O$, $CO_2$, $O_3$, or air temperature $T_a$), $D\overline{c}/Dt$ is the material derivative $= \partial\overline{c}/\partial t + \overline{u_i}\partial\overline{c}/\partial x_i$, and $u_i$ are the three velocity components along $x_i$ with $x_1 = x, x_2 = y$, and $x_3 = z$ being the longitudinal, lateral, and vertical directions, respectively, $\overline{S}$ is the mean biological source (or sink) strength, and $\overline{F_j}$ is the mean turbulent flux in direction $x_j$. All mean quantities are subject to both time and horizontal averaging as described by Raupach and Shaw [14]. For simplicity, we use an overbar to indicate both time and horizontal averages. Throughout, the velocity coordinate system is rotated such that $\overline{u}_2 = 0$. Also, both index and meteorological notation are used interchangeably in keeping with the relevant literature (*i.e.* $u_1 = u$, $u_2 = v$, and $u_3 = w$).

For planar homogeneous flows and in the absence of subsidence (*i.e.* $\overline{u}_3 = 0$), this equation reduces to

$$\frac{\partial\overline{c}}{\partial t} + \frac{\partial\overline{F}_c}{\partial z} = \overline{S}_c, \tag{2}$$

where $\overline{F}_c = \overline{u_3'c'}$ is the mean vertical flux of the scalar entity (*e.g.* $\overline{F}_{CO_2}$, $\overline{F}_{H_2O}$, and $\overline{F}_T$ are the $CO_2$, $H_2O$, and sensible heat turbulent fluxes at height $z$, respectively), and $\overline{S}_c$ is the mean vegetation source strength (*i.e.* sink implies $\overline{S}_c \leq 0$) at time $t$ and height $z$ above the soil surface. The balance between all these terms is shown in fig. 1 for a uniform and dense canopy.

The scalar continuity eqn. (2) is one equation in three unknowns ($\overline{c}$, $\overline{F}_c$, $\overline{S}_c$). Hence, to predicted sources, sinks, fluxes, or concentration, at least 2 additional

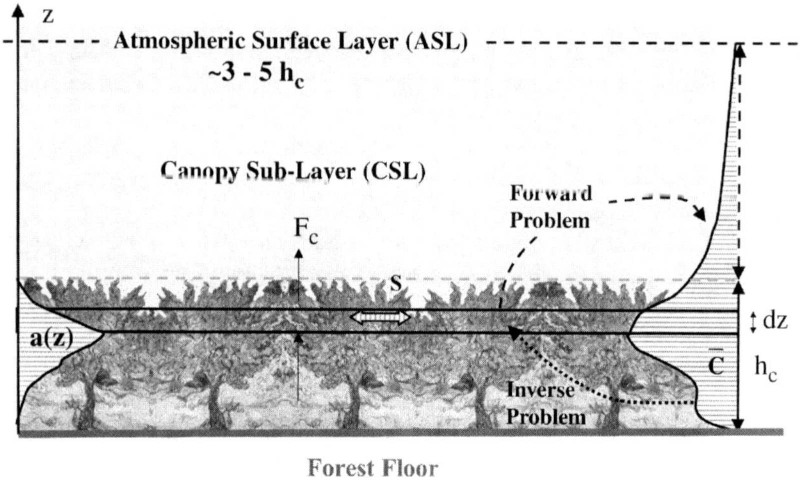


Figure 1: A schematic representation of the scalar mass balance at a given layer $(dz)$ within an extensive canopy. The forward problem seeks to address the estimation of the mean concentration profile knowing the source distribution, and the inverse problem seeks to address the estimation of the source profile from a measured concentration distribution.

equations must be available. The formulation of these two equations in the context of the so-called "forward" and "inverse" problems is presented next [15–17]. Henceforth, we refer to the "forward" problem as the problem in which the source strength is known (or measured) and the mean concentration distribution is desired, and the "inverse" problem as the problem in which the source strength (or the flux) is desired but the mean concentration profile is known (or measured) as shown in fig. 1.

### 2.2.1 k-Theory: the genesis of canopy turbulence models

Early attempts to establish an additional equation that links the flux with the mean concentration made use of the so-called "k-theory", which relates $\overline{F}_c$ to $d\overline{c}/dz$ via an eddy diffusivity $(K_m(z))$. With this approximation,

$$\overline{F}_c = -K_m(z)\frac{\partial \overline{c}}{\partial z}, \tag{3}$$

When k-theory is used in eqn. (2) for a known $K_m(z)$, both "forward" and "inverse" problems can be explicitly solved for stationary conditions using:

$$\overline{S}_c = K_m(z)\frac{\partial^2 \overline{c}}{\partial z^2} + \frac{\partial K_m(z)}{\partial z}\frac{\partial \overline{c}}{\partial z} \tag{4}$$

Hence, it is of no surprise that much of the early efforts in CSL research was to determine $K_m(z)$ from the velocity statistics and to establish the necessary conditions for the application of such theory. Based on numerous theoretical and field research, limitations of k-theory are now well recognized [15, 18] and have been documented in many field experiments [19] and laboratory studies [20]. These measurements revealed that co-gradient flow of heat, water vapor, and $CO_2$ do exist near the canopy top but counter- or zero-gradient flow may exist in the mid to lower canopy levels thereby negating the application of k-theory. Later studies [18, 21, 22] also revealed that counter- and zero-gradient flows primarily occur because 1) the variable $\overline{S}_c$ within the canopy strongly impacts the apparent diffusivity (*i.e.* the ratio of the local flux to the local gradient), as discussed in Raupach [16, 17, 23, 24] and Wilson [18], 2) much of the vertical transport occurs by eddy motion whose size is comparable to the canopy height ($h_c$) rather than height from the ground surface ($z$) [18, 25, 26], and 3) canopy turbulence lacks any local equilibrium (*i.e.* a region in which local production of turbulent kinetic energy is balanced by local viscous dissipation) as discussed in Shaw [27] and later by Maitani & Seo [28].

Because "k-theory" may not even correctly predict the direction of the flux, two other approaches aimed at addressing the limitations of k-theory emerged over the last 15–20 years: higher-order Eulerian closure models and Lagrangian dispersion models (LDM), considered next.

### 2.2.2 Beyond k-theory: second-order Eulerian closure models

The time and horizontally averaged steady-state scalar flux budget equation for planar homogeneous – high Reynolds and Peclet number flows (neglecting molecular diffusion) can be written as [15, 29]:

$$\begin{aligned} \frac{\partial \overline{w'c'}}{\partial t} &= -\overline{w'^2}\frac{\partial \overline{c}}{\partial z} - \frac{\partial \overline{w'w'c'}}{\partial z} - \overline{c'\frac{\partial p'}{\partial z}} \\ 0 &= \quad P_1 \quad + \quad P_2 \quad + \quad P_3, \end{aligned} \tag{5}$$

where $p'$ is the static pressure turbulent perturbation. In eqn. (5), the buoyancy term was neglected. While the first term, $P_1$, on the right side of eqn. (5) is the production term and is expressed in quantities that are known or being modelled, the second and third terms, $P_2$ and $P_3$, respectively, require closure approximations. The flux transport term, $P_2$, is usually modeled using [30]

$$\overline{w'w'c'} = \frac{\tau}{C_S}\left[-\overline{w'w'w'}\left(\frac{\partial \overline{c}}{\partial z}\right) - \overline{w'c'}\left(\frac{\partial \overline{w'w'}}{\partial z}\right) - 2\overline{w'w'}\left(\frac{\partial \overline{w'c'}}{\partial z}\right)\right], \tag{6}$$

and the dissipation term, $P_3$, is modelled according to Finnigan [29] as

$$\overline{c'\frac{\partial p'}{\partial z}} = C_4\frac{\overline{w'c'}}{\tau}, \tag{7}$$

where $C_4$ and $C_S$ are closure constants and $\tau$ is a Eulerian time scale given by $\tau = K/\epsilon$ [31], where $K = \frac{1}{2}\overline{u_i'u_i'}$ is the turbulent kinetic energy, and $\epsilon$ is the mean turbulent kinetic energy dissipation rate. When $P_2 \ll P_1$, the flux-transport term can be neglected and the classical k-theory is recovered by setting $P_1 = -P_3$

$$\overline{w'c'} = -\frac{\tau\overline{w'w'}}{C_4}\frac{\partial\bar{c}}{\partial z} \tag{8}$$

and gives a formal estimate of $K_m(z)$

$$K_m = \frac{\tau\overline{w'w'}}{C_4}. \tag{9}$$

Hence, the failure of k-theory is commonly attributed to a large flux transport term within the canopy volume. In fact, for some canopy layers, $P_2$ is the only term balancing the dissipation (or term $P_3$) and can be much larger than $P_1$. Naturally, the term $P_2$ is often linked with the ejection-sweep cycle, and hence, the asymmetry in the relative contributions of ejections and sweeps to the overall mass (or heat) transport is the reason why $P_2$ is large. This point has been explored by Cava *et al* [32] who combined cumulant expansion methods and higher order closure models and derived an explicit linkage between the sensible heat flux, mean air temperature profiles, and the ejection-sweep statistics to demonstrate this point.

Upon combining the standard closure approximations with the scalar budget in eqn. (1), a second order ordinary differential equation (ODE) can be derived to link the variations in the scalar turbulent fluxes with height:

$$A_1(z)\frac{\partial^2\overline{w'c'}}{\partial z^2} + A_2(z)\frac{\partial\overline{w'c'}}{\partial z} + A_3(z)\overline{w'c'} = A_4(z) \tag{10}$$

where:

$$\begin{aligned}
A_1(z) &= \frac{2\tau}{C_8}\overline{w'w'} \\
A_2(z) &= \frac{\tau}{C_8}\frac{\partial\overline{w'w'}}{\partial z} + 2\frac{\partial}{\partial z}\left(\frac{\tau}{C_8}\overline{w'w'}\right) \\
A_3(z) &= \frac{\partial}{\partial z}\left(\frac{\tau}{C_8}\frac{\partial\overline{w'w'}}{\partial z}\right) - \frac{C_4}{\tau} \\
A_4(z) &= \overline{w'w'}\frac{\partial\bar{c}}{\partial z} - \frac{\partial}{\partial z}\left(\frac{\tau}{C_8}\overline{w'w'w'}\frac{\partial\bar{c}}{\partial z}\right) \\
&\quad - \frac{\tau}{C_8}\overline{w'w'w'}\frac{\partial^2\bar{c}}{\partial z^2}
\end{aligned} \tag{11}$$

Hence, if the velocity statistics are known, the above ODE can be readily used to solve both the forward and the inverse problem. This approach was first tested for $CO_2$ by Katul & Albertson [33] and later revised by Siqueira & Katul [34] to account for the buoyancy term. More recently, Siqueira *et al* [35] and Katul *et al* [36] tested this approach for a number of scalars and for a wide range of canopy morphologies. One major theoretical criticism to Eulerian closure models is that they still use a flux-gradient closure approximation for the triple correlations; hence, if k-theory is not valid for "closing" $\overline{w'c'}$, why should it be valid for closing $\overline{w'w'c'}$? This question motivated the development of alternative models, considered next.

### 2.2.3 Beyond k-theory: Lagrangian models

In describing scalar transport within the CSL, it was suggested by Raupach [16, 17, 37] that Lagrangian transport approaches are better suited than their Eulerian counterpart given their ability to overcome flux-gradient closure model limitations [25, 38]. Among the various three-dimensional LDM approaches proposed in the past two decades, the models by Thomson [39] and Pope [40, 41] are now widely used because they satisfy the so-called well mixed condition *wmc* for strongly inhomogeneous flows. The well-mixed condition states that if the concentration of a material is uniform at some time (*t*) it will remain so at all later times if there are no sources or sinks. The Thomson model [39] is particularly popular in ASL and mixed layer (ML) turbulence applications because it only requires first and second moments of the Eulerian velocity field, which can be measured or readily estimated from similarity theory [42], and because it is simple to implement [43, 44]. While much of the theory is already presented in Rodean [42] and Reynolds [45], a review of the Thomson model is provided for completeness.

The formulation of LDM for the trajectories of tracer-particles in turbulent flow is now, in terms of *wmc*, well established. This condition is currently the most rigorous theoretical framework for the formulation of LDM and ensures consistency with prescribed Eulerian velocity statistics. With this framework, Thomson [39] showed that the Lagrangian velocity of a tracer-particle can be described by a generalized Langevin equation

$$du_i = a_i(x_k, u_k, t)dt + b_{ij}(x_k, u_k, t)d\Omega_i \tag{12}$$

where $x$ and $u$ are the position and velocity vectors of the tracer-particle at time $t$ (rather than defining them at a stationary point as was done with the Eulerian approach). The terms $a_i(x_k, u_k, t)$ and $b_{ij}(x_k, u_k, t)$ are the drift and diffusion coefficient, respectively. The quantities $d\Omega_i$ are increments of a vector-valued Wiener process (Brownian walk) with independent components, mean zero, and variance $dt$.

The specification of the drift and diffusion terms is sufficient to determine how a fluid particle moves. While $b_{ij}(x_k, u_k, t)$ can be determined by requiring that the Lagrangian velocity structure function in the inertial sub-range satisfies Kolmogorov's similarity theory, the determination of $a_i(x_k, u_k, t)$ is more difficult. It requires the use of the well-mixed condition in three-dimensions. From the *wmc*, if the distribution of tracer particles in position-velocity space is proportional to the Eulerian probability distribution function ($P(u, x, t)$) of the fluid velocity, it must remain so for all later times. This condition requires that $P(u, x, t)$ be a solution of the Fokker-Planck equation,

$$\frac{\partial P}{\partial t} + \frac{\partial(u_i P)}{\partial x_i} = -\frac{\partial(a_i P)}{\partial u_i} + \frac{1}{2}\frac{\partial^2(b_{ij}b_{jk}P)}{\partial u_i \partial u_j} \tag{13}$$

The solution of this Fokker-Planck equation for Gaussian turbulence consists of a set of three equations (see also [42]) and constitute the well-known Thomson model [39]. For two- and three-dimensional turbulence, Thomson [39] showed that the drift term $a_i(x, u, t)$, can be constrained but not completely determined by requiring consistency with prescribed Eulerian velocity statistics.

We note that a variant on the Thomson model for homogeneous and isotropic turbulence was also derived from the so-called Eulerian PDF methods. The latter method utilizes the Navier-Stokes equations to arrive at a transport equation for the entire probability distribution of the velocity components [41].

As was done for the Eulerian flux budget equation, the LDM can be derived to take into account the vertical variability of the flow statistics. To illustrate, we consider the simplest case of a planar homogeneous source emitting substances in a stationary and planar homogeneous turbulent flow having a very low turbulent intensity (*i.e.* $\sigma_u/\overline{u} < 0.1$ where $\sigma_u$ is the horizontal velocity standard deviation). For these idealized conditions, the Langevian and the Fokker-Planck equations reduce to:

$$dw = a_i(z, w, t)dt + b(z, w, t)d\Omega_i, \tag{14}$$

and

$$\frac{\partial P}{\partial t} + \frac{\partial (wP)}{\partial z} = -\frac{\partial (aP)}{\partial w} + \frac{1}{2}\frac{\partial^2 (b^2 P)}{\partial w^2}, \tag{15}$$

where the drift coefficient now reduces $b = \sqrt{C_o\,\epsilon}$ and $C_o = 5.5$ is a similarity constant (related to the Kolmogorov constant). Moreover, assuming that the probability distribution function $P(z, w, t)$ is Gaussian,

$$P(z, w, t) = \left(\frac{2}{\pi\overline{w'w'}}\right)^{-1/2} \exp\left(-\frac{1}{2}\frac{w^2}{\overline{w'w'}}\right) \tag{16}$$

it is a simple matter to obtain the drift coefficient

$$a(z, w, t) = -\frac{w}{T_L} + \frac{1}{2}\left[1 + \frac{w^2}{\overline{w'w'}}\right]\frac{\partial \overline{w'w'}}{\partial z} \tag{17}$$

where $T_L$ is the Lagrangian integral time scale. Hence, for the "forward" problem, once the source and the Lagrangian velocity are statistically known, the evolution of the concentration can be computed by intergrading horizontally the vertical trajectory of every single fluid element, given by

$$z(t + \Delta t) = z(t) + \int_t^{t+\Delta t} w(p)dp. \tag{18}$$

The Eulerian model earlier described can incorporate non-Gaussian statistics (*e.g.* nonzero $(\overline{w'w'w'})$ and advective scalar transport, two mechanisms clearly neglected in the Lagrangian approach. However, the Eulerian model calculations of $\overline{S}_c$ and $\overline{F}_c$ for the "inverse" problem were shown to be sensitive to measurement errors in $\bar{c}$ [33, 46]. The development of "inverse" approaches sufficiently robust to such unavoidable measurement errors became a research priority for practical applications of these models.

In the late 1980, Raupach and coworker [16, 17, 37] introduced a new approach, called $LNF$ or localized near field theory, to explicitly evaluate the interdependency between $\bar{c}$ and $\overline{S}_c$ via Lagrangian dispersion theory in a robust way (*i.e.* not overly sensitive to measurement errors). Over the last decade, the LNF became the first popular alternative to *k-theory* and has been successfully applied over a wide range of vegetation types [47–52].

In this theory, Raupach [15–17] introduced a dispersion matrix ($D_{ij}$) that relates mean concentration difference between a given level inside the canopy and a reference level above the canopy ($z_R \geq h$) to the scalar source strength by

$$\bar{c}(z_i) - \bar{c}(z_R) = \sum_{j=1}^{N} S_{cj} D_{ij} \Delta z_j$$

$$D_{ij} = \frac{\widehat{c_i} - \widehat{c_r}}{\widehat{s}\Delta z_j} \tag{19}$$

where $i$ and $j$ are the indices for concentration and source locations, respectively, $\Delta z$ is the discrete layer thickness within the canopy, $\widehat{c_i}$ is the resulting concentration at a layer $i$ above a reference concentration $\widehat{c_r}$ resulting from a unit source $\widehat{s}$ placed at layer $j(=1, 2, \cdots N)$ inside the canopy, and, as before, $N$ is the number of layers within the canopy volume. The $\widehat{c_i} - \widehat{c_r}$ is calculated from the velocity statistics inside the canopy by following the trajectory of an ensemble of fluid parcels released uniformly from $s$ placed at the $j$th layer. The random walk algorithm of Thomson can be used to compute the elements of $D_{ij}$. To use the Lagrangian dispersion algorithm of Thomson, $\overline{w'w'}$ and $T_L$ must be described. Raupach [15] also proposed idealized analytical profiles for these two quantities thereby further popularizing the application of LNF. We describe the application of LDM and higher order Eulerian closure models to solving the forward and inverse problems.

# 3 Practical considerations for solving the forward and inverse problems

While the mathematical formulation for solving the forward and inverse problems have been established in the previous sections, we focus here on practical aspects and case studies starting with the forward problem.

## 3.1 Forward problem: merging physiology and turbulence

In biosphere-atmosphere exchange, the source strength is rarely measured but is often calculated from the physiological and properties of the foliage, which must be assumed or, preferably, measured. Hence, the definition of the forward problem here is modified from the classical version so that physiological, biophysical, and radiative properties of the canopy are specified rather than the source strength distribution itself. As we show next, this definition implies that $\overline{S}_c$ is not independent of $\bar{c}$.

One common approach to modeling $\overline{S}_c$ assumes that the transfer of mass and heat from the leaf surface to the atmosphere at a given $z$ is governed by molecular (or Fickian) diffusion

$$\overline{S}_c = -\rho_a a(z)\frac{\bar{c}(z) - \bar{c}_i(z)}{r_b + r_s} \tag{20}$$

where $\rho_a$ is the mean air density, $a(z)$ is the leaf area density, $\bar{c}_i$ is the mean intercellular scalar concentration at height $z$, $r_b$ is the boundary layer resistance, and $r_s$ is the stomatal resistance. Equations (19) and (20) permit a complete mathematical description of $\bar{c}$, $\overline{S}_c$ and $\overline{F}_c$ if $\bar{c}_i$, $r_s$, and $r_b$ are known or parameterized. Much of the present research in forward methods (*e.g.* the *CANVEG* model by Baldocchi and Meyers [53] or Lai and coworkers [54, 55]) is aimed at predicting $\bar{c}$, $\overline{S}_c$ and $\overline{F}_c$ through coupling the radiative, physiological, and biochemical properties of the foliage with eqn. (20) to arrive at a complete description of $\bar{c}$, $\overline{S}_c$, $\overline{F}_c$, $\bar{c}_i$, $r_s$, and $r_b$. We briefly review the canonical form of the parameterizations for the unknowns $r_s$, $r_b$, and $\bar{c}_i$ without resorting to any explicit formulations.

The leaf boundary layer resistance is often computed from flat plate theory using the modelled $\overline{u_1}$. Monteith & Unsworth [56] and Campbell & Norman [57] derived explicit equations for heat and mass boundary layer transfer subject to forced and free convection. After an appropriate characteristic leaf length scale ($l_d$) is specified for a particular vegetation type (*e.g.* needle diameter for a pine forest), the leaf boundary layer resistance is given by

$$r_b = \frac{l_d}{d_m S_h} \tag{21}$$

where $d_m$ is the molecular diffusivity of the scalar entity, and $S_h$ is the Sherwood number, which requires the mean longitudinal velocity inside the canopy. Models and formulations for $\bar{c}_i$ and $r_s$ are much more complex and require knowledge of the enzymatic biochemistry of carbon assimilation in leaves. The stomatal conductance $r_s^{-1}$ is often expressed as a function of the leaf photosynthesis $A$, relative humidity $(rh)$ or vapor pressure deficit and mean $CO_2$ concentration at the leaf surface ($\bar{c}_s$), and is given

$$g_s(z) = m\frac{A(z) \times rh(z)}{\bar{c}_s(z)} + B \tag{22}$$

where $m$ and $B$ are physiological parameters that vary with vegetation type. Biochemical models can be used to couple $A$ with internal $CO_2$ concentration (*i.e.* the net mathematical outcome is a relationship between $A$ and $\bar{c}_i$ for $CO_2$). For water vapor and heat, the leaf energy balance provides the necessary equation to couple leaf surface temperature ($T_{sl}$), surface water vapor concentration, and the sensible and latent heat fluxes. Also, because the model requires photosynthetically active radiation to estimate the biochemical kinetic constants at all levels within the canopy, a radiation attenuation model is needed. Gas exchange experiments, commonly performed using porometry techniques, can be employed to determine independently the parameters of the so-called Collatz model for $g_s$ (*e.g.* $m$ and $b$) and the Farquhar photosynthesis model. Note that $\bar{c}_i$, $rh$ and $T_{sl}$ are all required to calculate $g_s$; therefore, all three scalars ($H_2O$, $CO_2$ and $T_a$) must be simultaneously considered at each canopy layer. Also, the modeled $T_{sl}$ at different levels within the canopy are needed to estimate plant respiration which adjusts the modeled $A$. For many vegetation types, these parameters may not be statistic but vary with leaf nitrogen and age. The relative importance of the non-stationarity in these parameters *vis-a-vis* the leaf area density remains a subject of research in forward problems.

## 3.2 Inverse problem: inferring biological activity from turbulence theory

The interest in inverse models is primarily driven by the fact that mean concentration profiles inside the canopy can be readily measured (or monitored) and, in practice, it is $\overline{S}_c$ – the biological source or sink that is desired. Few examples of the application of the scalar transport methods introduced above are shown in the next sections.

### 3.2.1 Lagrangian models: LNF

Based on the LNF theory, the difference in the mean scalar concentration $\bar{c}(z)$ at any height $z$ and a reference value $\bar{c}_R$ measured above the canopy at ($z_R > h$) is calculated by super-imposing near field ($C_n$) and far field ($C_f$ ) contributions:

$$\bar{c}(z) - \bar{c}_R = C_n + C_f \tag{23}$$

The near field contribution is computed via a kernel function:

$$\begin{aligned} C_n &= \int_0^\infty \frac{\overline{S}_c(z_o)}{\sigma_w(z_o)} \left(f_1(z, z_o) + f_2(z, z_o)\right) dz_o \\ f_1(z, z_o) &= k_n \left[\frac{z - z_o}{\sigma_w(z_o) T_L(z_o)}\right] \\ f_2(z, z_o) &= k_n \left[\frac{z + z_o}{\sigma_w(z_o) T_L(z_o)}\right] \end{aligned} \tag{24}$$

where $z_o$ is a dummy variable. An analytical approximation for the kernel function $k_n$ is given by:

$$k_n(\xi) = -0.39894 \log\left(1 - e^{-|\xi|}\right) - 0.15623\, e^{-|\xi|} \tag{25}$$

The far field contribution is calculated using results from the near field and a gradient-diffusion relationship given by:

$$C_f(z) = \bar{c}_R(z_R) - C_n(z) + \int_z^{z_R} \frac{\overline{F}_c(z)}{\sigma_w(z) T_L(z)} dz \tag{26}$$

The Lagrangian time scale can be estimated from:

$$\frac{T_L \times u_*}{h} = \beta \tag{27}$$

where $u_*$ is the friction velocity at the canopy top and $\beta$ is a constant ($\sim 0.1 - 0.3$). With these formulations for the near and far field concentrations, and with direct concentration measurements within the canopy, it is possible to estimate the sources and sinks via eqn. (19). However, to avoid numerical instability in the source profile calculations because of measurement errors, redundant concentration measurements are necessary (*i.e.* the number of concentration measurements must exceed the number of source layers). As shown by Raupach [16], such redundancy reduces the source inference problem to a regression problem with the source strengths calculated by a least-squares approach given by

$$\sum_{k=1}^{m} A_{jk} S_k = B_j$$

$$A_{ij} = \sum_{i=1}^{n} D_{ij} \Delta z_j D_{ik} \Delta z_k$$

$$B_j = \sum_{l=1}^{n} \left( \bar{C}_l - \bar{C}_R \right) D_{lj} \Delta z_j \tag{28}$$

Once $A_{jk}$ and $B_k$ are determined from the measured concentration and modelled dispersion matrix, the estimation of the source strength can be readily achieved. The LNF approach does not allow for non-zero vertical velocity skewness, strong inhomogeneity in vertical source strength variation and mean horizontal velocity variation within the canopy. These limitations can be relaxed in practice if one invokes a second-order Eulerian closure "inverse" model (EUL) as earlier described at the expense of sensitivity to measurement errors. Hence, one logical question to pursue is whether a model that retains the advantages of both EUL and LNF can be derived for such an inverse problem.

### 3.2.2 Hybrid methods: merging Lagrangian and Eulerian methods

Recently, a Hybrid Eulerian Lagrangian method that combines many of the advantages of EUL and LNF was developed and tested by Siqueira *et al* [58]. The method adopts a second-order closure model, similar to the one described above, to estimate the elements of $D_{ij}$ but computes $\overline{S}_c$ from the "robust" regression algorithm of Raupach [16, 17]. The second order ODE describing the concentration profile from a prescribed unit source (and hence the flux) is given by:

$$B_1(z)\frac{\partial^2 \bar{c}}{\partial z^2} + B_2(z)\frac{\partial \bar{c}}{\partial z} = B_3(z) \tag{29}$$

where:

$$\begin{aligned} B_1(z) &= \tfrac{\tau}{C_8}\overline{w'w'w'} \\ B_2(z) &= -\overline{w'w'} + \tfrac{\partial}{\partial z}\left(\tfrac{\tau}{C_8}\overline{w'w'w'}\right) \\ B_3(z) &= -\tfrac{\partial}{\partial z}\left(\tfrac{\tau}{C_8}\left[\overline{w'c'}\frac{\partial\overline{w'w'}}{\partial z} + \overline{w'w'}\frac{\partial\overline{w'c'}}{\partial z}\right]\right) \\ &\quad + C_4\frac{\overline{w'c'}}{\tau} \end{aligned} \tag{30}$$

In eqn. (29), $\overline{w'c'}$ is calculated by simply integrating the known unit source placed at one layer. The scheme embeds the robustness of LNF in a more physically sound forward calculation of $D_{ij}$, which includes the effects of $\overline{w'w'w'}$ on $D_{ij}$ through the coefficients $B_1$ and $B_2$. We note that the Thomson algorithm does not uniquely define $D_{ij}$ in forward methods (described earlier) and the above approaches can also be used to establish a relationship between $\overline{S}_c$ and $\bar{c}$.

### 3.3 How to evaluate the flow statistics needed for scalar transport calculations?

To implement all the forward and inverse models discussed above, the velocity statistics within the canopy, particularly the vertical velocity standard deviation and the Lagrangian integral time scale, must be assumed or specified *a priori*. Field experiments demonstrated that the flow statistics vary across canopy types with leaf area distribution being a dominant factor governing this variability [16, 17, 30]. Detailed velocity statistics profiles such as $\sigma_w$ are rarely measured within the canopy and cannot be readily specified for an arbitrary leaf area density distribution. Several methods are now available to evaluate the necessary velocity statistics from leaf area density. Some of the most promising methodologies that link foliage distribution with such flow statistics are:

1. simplified analytical models to classical second order closure equations [51],
2. higher-order closure models [8, 30, 59, 60, 61],
3. large Eddy Simulations (LES) techniques [62].

While the details of these methods are beyond the scope of this review, we briefly highlight the main advantages of each.

One of the most detailed and physically sound approach to compute the flow statistics within canopies is LES [7]. However, the time scales required for modelling landscape dynamics and biogeochemical cycles far exceeds the computational ability of LES, thus necessitating the use of computationally efficient but simpler transport models such as Reynolds averaged closure approaches. The symbiotic use of LES and closure models may be a desirable strategy in the future for which LES output across different classes of canopy heterogeneities are used to "train" or evaluate closure models (or similarity constants). The most popular closure model that generates desired velocity statistics for scalar transport are second-order models. While these models were originally developed and have been used for more than 30 years [63], their application to canopy turbulence continues to be a major research area. After Wilson & Shaw's seminal paper [60], numerous second-order closure models for the CSL were proposed [8, 18, 30, 51, 59, 64] but the reported successes have been mixed [8, 59]. Some of the weaknesses of second-order closure models is attributed to the inability of eddy-viscosity models to simulate flows subject to body forces or flows with significant non-local momentum exchange, as is the case within the CSL [12, 15, 18, 27, 63, 64]. Attempts to rectify these limitations by increasing the closure order have not translated into significantly improved predictive skill [59, 63].

An underlying reason for such inaccuracies is the dependence on a single (and typically local) length scale to characterize the entire effect of turbulence on the statistical moments being modelled. Moreover, the second order models can become computationally expensive and require complex numerical algorithms in 3-dimensional transport problems especially if multiple scalar species must be treated. Therefore, it is natural to explore a practical question: what is the minimum turbulence closure model necessary to simulate the mean and measures of second-order flow statistics yet is sufficiently general and computationally efficient to be incorporated in complex models of the environment [65]? A logical choice is a 1.5 closure model in which a budget equation for $K$ must be explicitly considered. Such models, known as 2-equation models or $K-\epsilon$ models are among the most popular computational models in engineering applications but have received rather limited attention in canopy turbulence [66]. Katul *et al* [67], explored different classes of $K-\epsilon$ models for a broad range of canopy morphologies, and showed that, a simple one-equation model ($K-U$) performs no worse than the classical $K-\epsilon$ models and, for that matter, second-order closure models within the canopy. A computationally efficient algorithm may be analytical solutions to the second-order model subject to few idealizations. For example, Massman & Weil [51] proposed such a simple analytical model describing the within-canopy velocity variances, turbulent intensities and dissipation rates for dense canopies in which the mixing length is constant and the mean velocity decays exponentially. By extending the work of Massman [68] for momentum transfer and using the second-order closure model of Wilson & Shaw [60], Massman & Weil were able to show that this analytical model captures most of the observed features of few data sets and agreed reasonably well with other independent modelling approaches. As a further test, we conducted here an analysis on the analytical solution for 6 canopies including rice, corn, an aspen forest, a loblolly pine forest, a Scots pine forest, and a mature oak-hickory hardwood forest, and we found good agreement between measured and modelled statistics for $\overline{u}$, and $\overline{u_i'u_j'}$ as shown in Section 2.

While this review primarily focused on "Temporal studies", numerous issues must also be addressed for the "Spatial studies". We conclude this chapter by describing some of these issues and needed progress for the spatial problem.

## 4 Spatial heterogeneity: the new frontier

As earlier stated, one of the most detailed approaches to investigate spatial heterogeneity (*e.g.* canopy non-uniformity) is LES [7]. However, there are numerous thorny issues about their routine usage: 1) sub-grid modelling: Given the short-

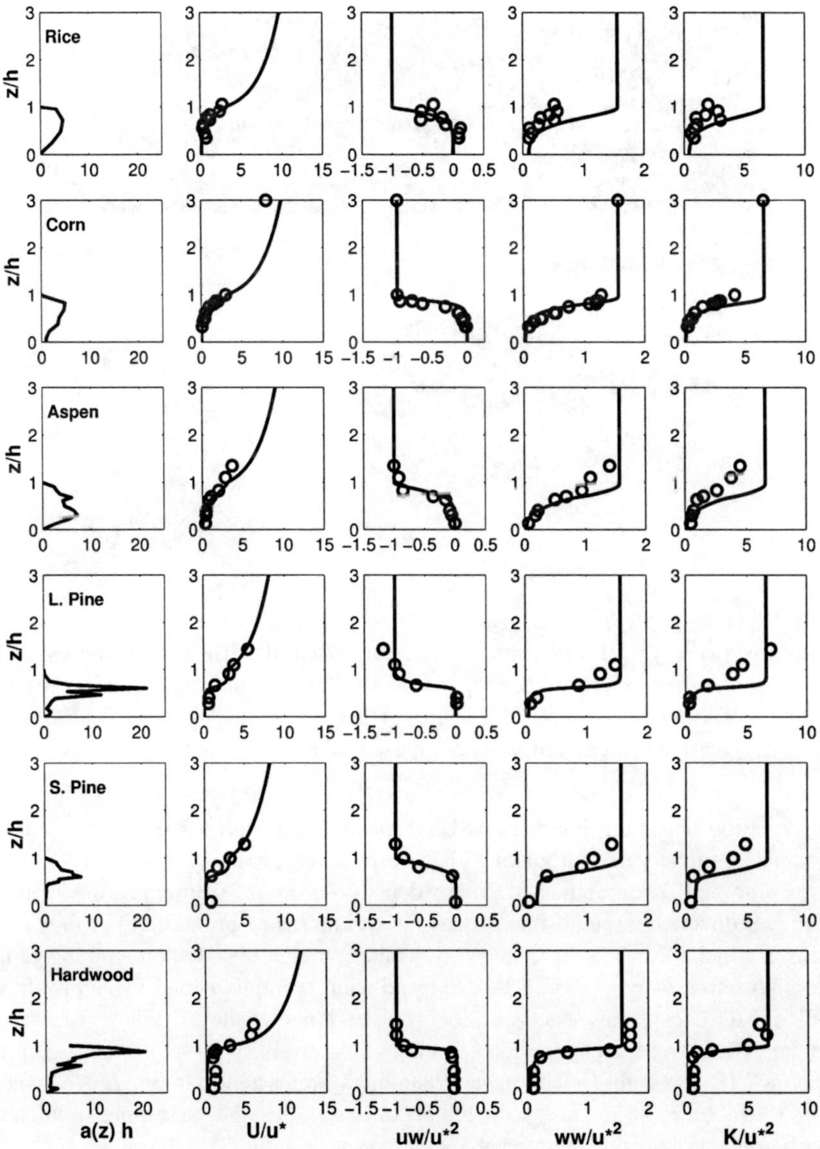


Figure 2: Comparison between measured (open circle) and predicted (solid line) mean velocity and Reynolds stresses for 6 canopy types with LAI ranging from 2 to 5.5 and $h_c$ ranging from 0.75m to 30m.

circuiting of the energy cascade, standard eddy-viscosity sub-grid models may not realistically reflect local wake production or dissipation. 2) Grid resolution: For large domains with complex boundary conditions (*e.g.* flow inside a tall canopy on complex terrain), the problem of how to optimize the grid size to retain as much

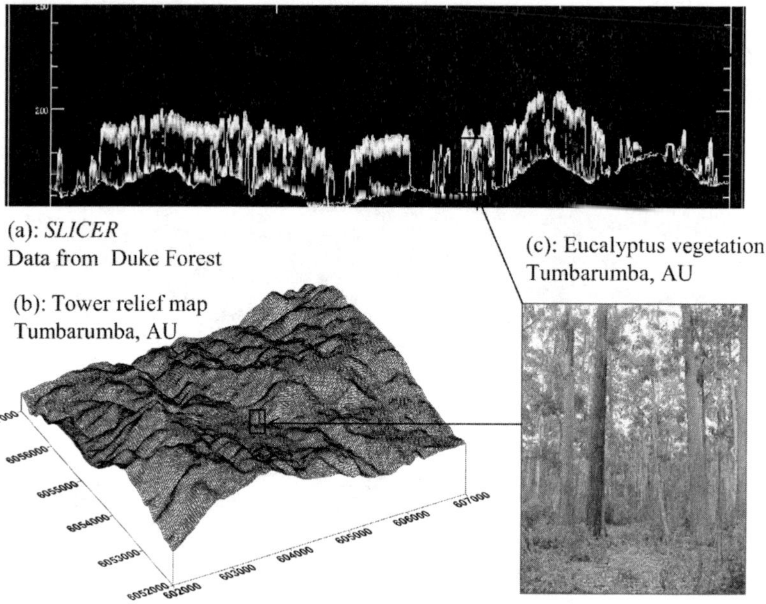


Figure 3: (a) A typical waveform power return from SLICER at Duke Forest, near Durham, North Carolina, U.S.A; (b) Tower relief topographic variation for one of the FLUXNET sites at Tumbarumba, Australia; (c) the Tumbarumba site is primarily composed of 30m tall Eucalyptus trees.

information about the interactions between low and high wave-number (due to complex boundary and turbulent eddies, respectively) is likely to remain.

Combined Topographic and canopy density variation: Historically, much attention was devoted to scalar transport on flat terrain (as reviewed here) or flow over hills in which the canopy height ($h_c$) was much smaller than the topographic variation. With the emergence of FLUXNET and with its emphasis on $CO_2$ uptake from tall forested ecosystems, a new class or problems emerged: how to link $\overline{F}_c$ measured in the CSL above the canopy to $\overline{S}_c$ on complex terrain. With the advancement of canopy LIDAR (*Li*ght *D*etection and *R*anging) systems and *S*canning *Li*dar *I*mager of *C*anopy by *E*cho *R*ecovery (SLICER), it is now possible to sample simultaneously topographic and canopy density variations (*e.g.* fig. 3a; see [69]).

However, incorporating such invaluable measurements in forward or inverse problems have not yet been attempted. The solution to this problem is crucial for interpreting biological controls on net ecosystem carbon exchange measured above the canopy on complex terrain. Several long-term sites have already positioned meteorological towers on tall vegetation in complex terrain (see *e.g.* fig. 3b,c, for the Tumbarumba Eucalyptus site in Australia) to quantify long-term carbon sequestration by such managed forests. The solution to this problem must be addressed by a sequence of carefully planned laboratory experiments, LES runs, and field experiments.

Currently, our group is taking the first steps to address this problem by initiating a long-term experimental program to measure and model the flow in sparse and dense canopies situated on wavy terrain at the OMTIT recirculating channel, a large flume at the Giorgio Bidone hydraulics Laboratory, DITIC Politecnico di Torino (Italy). The wavy terrain serves as a logical starting point because spectrally it constitutes the most elementary topographic variability.

## 5 Acknowledgements

Funding was provided by the Italian MURST Program through the PRIN 2002 project "Influence of vorticity and turbulence in interactions of water bodies with their boundary elements and effect on hydraulic design". Katul and Poggi acknowledge the support of the U.S. National Science Foundation (*NSF-EAR*-02 – 08258), the Biological and Environmental Research (*BER*) Program, U.S. Department of Energy, through the Southeast Regional Center (*SERC*) of the National Institute for Global Environmental Change (*NIGEC*), and through the Terrestrial Carbon Processes Program (*TCP*) and the *FACE* project. We thank the anonymous referees of this paper for their useful comments.

## References

[1] Baldocchi, D., Falge, E., Gu, L., Olson, R., Hollinger, D., Running, S., Anthoni, P., Bernhofer, C., Davis, K., Evans, R., Fuentes, J., Goldstein, A., Katul, G., Law, B., Lee, X., Malhi, Y., Meyers, T., Munger, W., Oechel, W., Paw U., K., Pilegaard, K., Schmid, H., Valentini, R., Verma, S., Vesala, T., Wilson, K. & Wofsy, S., Fluxnet: A new tool to study the temporal and spatial variability of ecosystem-scale carbon dioxide, water vapor, and energy flux densities. *Bull. Amer. Meteorol. Soc.*, **82**, pp. 2415–2434, 2001.

[2] Chameides, W., Kasibhatla, P., Yienger, J. & Levy, H., Growth of continental-scale metro-agro-plexes, regional ozone pollution, and world food-production. *Science*, **264**, pp. 74–77, 1994.

[3] Sutton, M., Fowler, D. & Moncrieff, J., The exchange of atmospheric ammonia with vegetated surfaces .1. unfertilized vegetation. *Q. J. R. Meteorol. Soc.*, **119**, pp. 1023–1045, 1993.

[4] Bidleman, T., Harner, T., Wiberg, K., Wideman, J., Brice, K., Su, K., Falconer, R., Aigner, E., Leone, A., Ridal, J., Kerman, B., Finizio, A., Alegria, H., Parkhurst, W. & Szeto, S., Chiral pesticides as tracers of air-surface exchange. *Environ. Pollut.*, **102**, pp. 43–49, 1998.

[5] Nathan, R., Katul, G., Horn, H., Thomas, S., Oren, R., Avissar, R., Pacala, S. & Levin, S., Mechanisms of long-distance dispersal of seeds by wind. *Nature*, **418**, pp. 409–413, 2002.

[6] Katul, G., Lai, C., Schafer, K., Vidakovic, B., Albertson, J., Ellsworth, D. & Oren, R., Multiscale analysis of vegetation surface fluxes: from seconds to years. *Adv. Water Resour.*, **24**, pp. 1119–1132, 2001.

[7] Albertson, J., Kustas, W. & Scanlon, T., Large-eddy simulation over heterogeneous terrain with remotely sensed land surface conditions. *Water Resources Res.*, **37**, pp. 1939–1953, 2001.

[8] Katul, G. & Chang, W., Principal length scales in second-order closure models for canopy turbulence. *J. Appl. Meteorol.*, **38**, pp. 1631–1643, 1999.

[9] Poggi, D., Porporato, A., Ridolfi, L., Katul, G. & Albertson, J., The effect of vegetation density on canopy sublayer turbulence. *Boundary-Layer Meteorol.*, p. to appear, 2004.

[10] Poggi, D., Katul, G. & Albertson, J., Moment transfer and turbulent kinetic energy budgets within a dense model canopy. *Boundary-Layer Meteorol.*, p. to appear, 2004.

[11] Raupach, M. & Thom, A., Turbulence in and above plant canopies. *Annu. Rev. Fluid Mech.*, **13**, pp. 97–129, 1981.

[12] Finnigan, J., Turbulence in plant canopies. *Ann. Rev. Fluid Mech.*, **32**, pp. 519–571, 2000.

[13] Poggi, D., Porporato, A., Ridolfi, L., Katul, G. & Albertson, J., Interaction between large and small scales in the canopy sublayer. *Geophys. Rs. Lett.*, **31–5**, p. L05102, 2004.

[14] Raupach, M. & Shaw, R., Averaging procedures for flow within vegetation canopies. *Boundary-Layer Meteorol.*, **22**, pp. 79–90, 1982.

[15] Raupach, M., *Canopy transport processes, in Flow and Transport in the Natural Environment*, Vol. I. Springer-Verlag: New York, pp. 95–127, 1988.

[16] Raupach, M., Applying lagrangian fluid-mechanics to infer scalar source distributions from concentration profiles in plant canopies. *Agric. For. Meteor.*, **47**, pp. 85–108, 1989.

[17] Raupach, M., A practical lagrangian method for relating scalar concentrations to source distributions in vegetation canopies. *Quart. J. Roy. Meteorol. Soc.*, **115**, pp. 609–632, 1989.

[18] Wilson, J., *Turbulent transport within the plant canopy, in Estimation of Areal Evapotranspiration*, Vol. 177. IAHS, pp. 43–80, 1972.

[19] Deanmead, O. & Bradley, E., *Flux gradient relationships in The Forest-Atmosphere Interaction*, volume I. D. Reidel, pp. 95–127, 1985.

[20] Coppin, P., Raupach, M. & Legg, B., Experiments on scalar dispersion within a model-plant canopy .2. an elevated plane source. *Boundary-Layer Meteorol.*, **35**, pp. 167–191, 1986.

[21] Thurtell, G., *Comment on using K-theory within and above the plant canopy to model diffusion processes*, volume 177. IAHS, pp. 43–80, 1989.

[22] Kaimal, J. & Finnigan, J., *Atmospheric Boundary Layer Flows: their Structure and Measurements*. Oxford Univ. Press.: New York, p. 289, 1994.

[23] Raupach, M., Coppin, P. & Legg, B., Experiments on scalar dispersion within a model-plant canopy .1. the turbulence structure. *Boundary-Layer Meteorol.*, **35**, pp. 21–52, 1986.

[24] Raupach, M., A lagrangian analysis of scalar transfer in vegetation canopies. *Quart. J. Roy. Meteorol. Soc.*, **113**, pp. 107–120, 1987.

[25] Corrsin, S., Limitations of gradient transport models inrandom walks and in turbulence. *Adv. Geophys.*, **18a**, pp. 25–60, 1974.

[26] Finnigan, J., Turbulence in waving wheat .1. mean statistics and honami. *Boundary-Layer Meteorol.*, **16**, pp. 181–211, 1979.

[27] Shaw, R., Secondary wind speed maxima inside plant canopies. *J. Appl. Meteorol.*, **16**, pp. 514–521, 1983.

[28] Maitani, T. & Seo, T., Turbulent transport of scalar quantities within and above a paddy field. *Bound-Layer Meteor.*, **33**, pp. 197–208, 1985.

[29] Finnigan, J., *Turbulent Transport in Plant Canopies*. B.A. Hutchinson and B.B. Hocks: Norwell, MA, pp. 443–480, 1985.

[30] Meyers, T. & Paw, K., Testing of a higher-order closure model for modeling airflow within and above plant canopies. *Boundary-Layer Meteorol.*, **37**, pp. 297–311, 1986.

[31] Poggi, D., Katul, G. & Albertson, J., Scalar dispersion within a model canopy: Measurements and three dimensional lagrangian models. *Adv. Water Res.*, p. submitted, 2005.

[32] Cava, D. & Katul, G., Buoyancy and the sensible heat flux budget within dense canopies. *Bound-Layer Meteor.*, submitted, 2005.

[33] Katul, G. & Albertson, J., Modeling $CO_2$ sources, sinks, and fluxes within a forest canopy. *J. Geophys. Res.*, **104**, pp. 6081–6091, 1999.

[34] Siqueira, M. & Katul, G., Estimating heat sources and fluxes in thermally stratified canopy flows using higher-order closure models. *Boundary-Layer Meteorol.*, **103**, pp. 125–142, 2002.

[35] Siqueira, M., Kolle, R.L.O., Kelliher, F. & Katul, G., Modeling sources and sinks of $CO_2$, $H_2O$ and heat within a siberian pine forest using three inverse methods. *Quart. J. Roy. Meteorol. Soc.*, **129**, pp. 1373–1393, 2003.

[36] Katul, G., Leuning, R., Kim, J., Denmead, O., Miyata, A. & Harazono, Y., Estimating $CO_2$ source/sink distributions within a rice canopy using higher-order closure models. *Boundary-Layer Meteorol.*, **98**, pp. 103–125, 1998.

[37] Raupach, M. & Finnigan, J., Single-layer models of evaporation from plant canopies are incorrect but useful, whereas multilayer models are correct but useless - discuss. *Australian J. Plant Physiology*, **15**, pp. 705–716, 1988.

[38] Wilson, J., A 2nd-order closure-model for flow through vegetation. *Bound-Layer Meteor.*, **42**, pp. 371–392, 1988.

[39] Thomson, D., Criteria for the selection of stochastic-models of particle trajectories in turbulent flows. *J. Fluid Mech.*, **180**, pp. 529–556, 1987.

[40] Pope, S., Lagrangian pdf methods for turbulent flows. *Ann. Rev. Fluid Mech.*, **26**, pp. 23–63, 1994.

[41] Pope, S., Stochastic Lagrangian models of velocity in homogeneous turbulent shear flow. *Phys. Fluids*, **14**, pp. 1696–1702, 2002.

[42] Roden, H., *Stochastic Lagrangian Models of Turbulent Diffusion, Meteorological Monographs*, volume 26, No.48. American Meteorological Society: Boston, MA, pp. 1–100, 1996.

[43] Reynolds, A., On trajectory curvature as a selection criterion for valid lagrangian stochastic dispersion models. *Boundary-Layer Meteorol.*, **88**, pp. 77–86, 1998.

[44] Sawford, B., Rotation of trajectories in lagrangian stochastic models of turbulent dispersion. *Boundary-Layer Meteorol.*, **93**, pp. 411–424, 1999.

[45] Reynolds, A., A Lagrangian stochastic model for particle trajectories in non-Gaussian turbulent flows. *Fluid Dynamics Research*, **19**, pp. 277–288, 1997.

[46] Siqueira, M., Lai, C. & Katul, G., Estimating scalar sources, sinks, and fluxes in a forest canopy using lagrangian, eulerian, and hybrid inverse models. *J. Geophys. Res.*, **105**, pp. 29475–29488, 2000.

[47] Raupach, M. & Dunin, O.D.F., Challenges in linking atmospheric $CO_2$ concentrations to fluxes at local and regional scales. *Boundary-Layer Meteorol.*, **40**, pp. 697–716, 1992.

[48] Denmead, O. & Raupach, M., *Methods for measuring atmospheric gas transport in agricultural and forest system*, volume I. American Society of Agronomy: Madison, pp. 49–80, 1992.

[49] Denmeadand, O., Novel meteorological methods for measuring trace gas fluxes. *Philosophical Transactions of the Royal Society of London A*, **351**, pp. 383–396, 1995.

[50] Katul, G., Oren, R., Ellsworth, D., Hsieh, C., Phillips, N. & Lewin, K., A lagrangian dispersion model for predicting $CO_2$ sources, sinks, and fluxes in a uniform loblolly pine. *J. Geophys. Res.*, **102**, pp. 9309–9321, 1997.

[51] Massman, W. & Weil, J., An analytical one-dimensional second-order closure model of turbulence statistics and the Lagrangian time scale within and above plant canopies of arbitrary structure. *Bound-Layer Meteor.*, **91**, pp. 81–107, 1999.

[52] Leuning, R., Denmead, O., Miyata, A. & Kim, J., Source/sink distributions of heat, water vapour, carbon dioxide and methane in a rice canopy estimated using lagrangian dispersion analysis. *Agric. For. Meteorol.*, **104**, pp. 233–249, 2000.

[53] Baldocchi, D. & Meyers, T., On using eco-physiological, micrometeorological and biogeochemical theory to evaluate carbon dioxide, water vapor and trace gas fluxes over vegetation: a perspective. *Agric. For. Meteor.*, **90**, pp. 1–25, 1998.

[54] Lai, C., Schauer, A., Owensby, C., Ham, J. & Ehleringer, J., Isotopic air sampling in a tallgrass prairie to partition net ecosystem $CO_2$ exchange. *J. Geophys. Res. Atmos.*, **108**, 2003.

[55] Lai, C., Ellsworth, G.K.D. & Oren, R., Modelling vegetation-atmosphere $CO_2$ exchange by a coupled Eulerian-Lagrangian approach. *Boundary Layer Meteorol.*, **95**, pp. 91–122, 2003.

[56] Monteith, J. & Unsworth, M., *Principles of Environmental Physics*. Edward Arnold: London, 1990.

[57] Campbell, G. & Norman, J., *An Introduction to Environmental Biophysics*. Springer-Verlag: N.Y., pp. 148–259, 1998.

[58] Siqueira, M., Lai, C. & Katul, G., Estimating scalar sources, sinks, and fluxes in a forest canopy using Lagrangian, Eulerian, and hybrid inverse models. *J. Geophys. Res. Atmos.*, **105**, pp. 29475–29488, 2000.

[59] Katul, G. & Albertson, J., An investigation of higher-order closure models for a forested canopy. *Boundary-Layer Meteorol.*, **89**, pp. 47–74, 1998.

[60] Wilson, N. & Shaw, R., A higher order closure model for canopy flow. *J. Appl. Meteorol.*, **16**, pp. 1198–1205, 1977.

[61] Wilson, J., Lozowski, E. & Zhuang, Y., Comments on a relationship between fluid and immersed-particle velocity fluctuations proposed by walklate (1987). *Boundary-Layer Meteor.*, **43**, pp. 93–98, 1988.

[62] Albertson, J. & Parlange, M., Surface length scales and shear stress: Implications for land-atmosphere interaction over complex terrain. *Water Resources Res.*, **35**, pp. 2121–2132, 1999.

[63] Hanjalic, K., One-point closure models for buoyancy-driven turbulent flows. *Annual Review of Fluid Mechanics*, **34**, pp. 321–347, 2002.

[64] Ayotte, K., Finnigan, J. & Raupach, M., A second-order closure for neutrally stratified vegetative canopy flows. *Boundary-Layer Meteor.*, **90**, pp. 189–216, 1999.

[65] Wilson, J., Finnigan, J. & Raupach, M., A first-order closure for disturbed plant-canopy flows, and its application to winds in a canopy on a ridge. *Quart. J. Roy. Meteorol. Soc.*, **124**, pp. 705–732, 1998.

[66] Sanz, C., A note on k-epsilon modelling of vegetation canopy air-flows. *Boundary-Layer Meteor.*, **108**, pp. 191–197, 2003.

[67] Katul, G., Mahrt, L., Poggi, D. & Sanz, C., One and two equation models for canopy turbulence. *Boundary-Layer Meteorol.*, To appear, 2004.

[68] Massman, W., An analytical one-dimensional model of momentum transfer by vegetation of arbitrary structure. *Boundary-Layer Meteor.*, **83**, pp. 407–421, 1997.

[69] Lefsky, M., Cohen, W., Parker, G. & Harding, D., Lidar remote sensing for ecosystem studies. *Bioscience*, **52-1**, pp. 19–30, 2002.

# CHAPTER 11

# Flow solvers for liquid–liquid impacts

F. Trivellato[1], E. Bertolazzi[2] & A. Colagrossi[3]
[1] *Department of Civil and Environmental Engineering, and*
[2]*Department of Mechanical and Structures Engineering, University of Trento, Italy*
[3]*INSEAN, Roma, Italy*

## Abstract

The performance of two flow solvers is compared on liquid-liquid impacts. The two solvers are of completely different nature: one solver is a Mixed Finite Element scheme, based on the potential flow theory; the second solver is a fully Lagrangian scheme, based on a Smoothed Particle Hydrodynamics formulation. For the tested flow impacts, they compare well and adequately model free surface deformations after the impact. The analysis reveals what are the effects of computation conditions and clarifies the range of validity provided by the numerical models. In the case of a 2-D liquid column released onto a liquid pool at rest, it turned out a remarkable similarity between the two solutions. Air entrainment and rebound of the free surface along with the Lagrangian mixing may be vividly simulated.

## 1 Introduction

A cavity forms when a drop of water strikes a body of water at rest. As the interface recedes to its former location, the lower part of the cavity separates from the interface to evolve to a bubble.

Mass and heat transfers, phase changes, pollutant accumulation, capillary effects and other phenomena may take place on the interface between a liquid and a gas. This study is devoted to the most typical interface, namely the air-water interface. In some instances the interface may undergo spectacular deformations as a result of violent impacts, such as water drops falling on a liquid pool. Violent motion is referred to the fact that the local (or temporal) acceleration of the fluid particle is by far greater than the convective part of the acceleration, at least during the time period of the impact.

Liquid–liquid impacts, along with the associated air entrainment and drop formations, are of interest in several industrial and environmental instances, such as chemical and metallurgical processes, cavitation of marine propellers, oxygen exchanges on free surfaces and so forth.

A number of physical quantities can affect the splashing phenomenon; among them, the most important are: (i) the diameter and the velocity of the impinging drop; (ii) the dynamic viscosity, the mass density and the surface tension of the liquid.

It has been observed that some jets can be ejected from the crown along with secondary droplets. According to [1] four stages of the impact phenomenon can be recognized: the lamella formation after the impact and possibly a prompt splash; the formation of a crown and a cavity; the formation and break-up of a jet; and the collapse of the crown, followed by a possible rebound.

Subsurface vortices are often generated when a falling drop strikes a flat water surface. There is no optimal drop shape providing the best vortex penetration [2]: rather, experimental evidences suggest that the optimal shape depends on the surface tension of the falling drop. The shape of the cavity is basically controlled by the impact velocity.

The interface dynamics following the liquid impact was calculated by means of the boundary integral method by [3] who suggested that the bubble is entrapped when the downward momentum is such that the buoyancy force cannot reverse the motion before the lateral walls meet. When the drop meets the liquid surface and during the evolution of the cavity, the vorticity production is not supposed to affect the flow, which was then assumed to be basically irrotational. As a result, the flow field at any time step may be solved in the framework of the potential flow analysis.

High-speed motion pictures of bubble entrainment due to drop impacts were presented by [4]; by means of ultra-high-speed video camera [5], observed that the volume of entrapped air depends strongly on the bottom curvature of the drop. They concluded that the air volume which is entrained can be roughly estimated as 2% of the original raindrop volume and this is useful to assess gas transfer from the atmosphere to ocean.

The aim of this work is to simulate numerically the flow motion generated when a 2-D column-shaped liquid jet meets a liquid surface at rest. The interest in this study is to understand how much the phenomenon can be schematized without losing its physical essentials.

## 2 The Mixed Finite Element (MFE) solver

A synthetic description is given in this section on the background of the MFE formulation which deals with the most typical interface, namely the air–water interface; the problem under investigation has been already tackled by a variety of methods, ranging from Lagrangian formulations [6, 7], to Free Particle methods [8]. However, the present MFE formulation proposed hereafter is basically different and simpler, even though the computational domain undergoes a full re-meshing at any time step. Indeed the comparison with the popular Boundary Element Method –

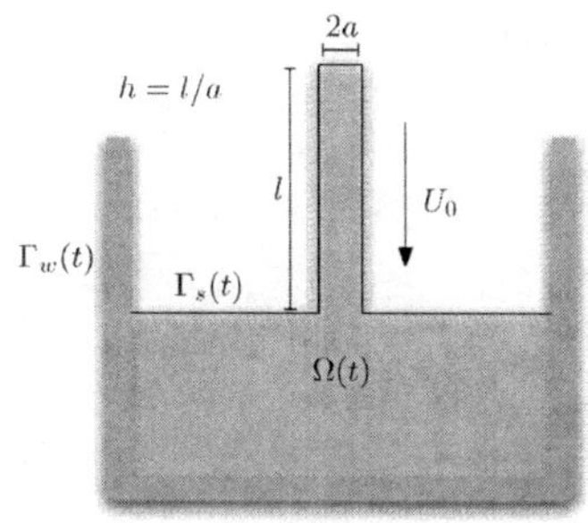


Figure 1: Two-dimensional liquid column impacting a liquid pool.

which does not need to perform the above mentioned re-gridding – is not the goal of this work: nevertheless it turned out that the present re-meshing routine does not cost more than 2% of the total CPU time.

The present MFE model is based on the assumptions of incompressible, inviscid liquid and irrotational flow. The time scale of impacts dominated by inertial effects is

$$t_0 = \frac{a}{U_0}$$

where $a$ is the length scale (namely, the liquid column half-width) and $U_0$ the impact velocity. Due to the highly transient nature of this kind of impact and to the large Reynolds number involved, viscosity is not expected to be important. As a first approximation, Weber number will also be neglected since surface tension is of interest for small drops.

Two parameters are considered in this study: the *aspect ratio*

$$h = \frac{l}{a}$$

and the *Froude number*

$$\mathsf{Fr} = \frac{U_0^2}{ga} \,.$$

It is assumed that the domain boundary $\partial\Omega(t) = \Gamma_w(t) \cup \Gamma_s(t)$ is suitably parameterized.

Figure 1 depicts a sketch of the model problem, which is formulated on the time-dependent computational domain $\Omega(t)$, a closed two-dimensional region bounded by the curves $\Gamma_w(t)$, a rigid wall, and $\Gamma_s(t)$, the moving interface between a gas and a liquid. Both the curves $\Gamma_w(t)$ and $\Gamma_s(t)$ are time dependent. The rigid wall $\Gamma_w(t)$ depends on the time variable $t$ because the intersection with the moving interface $\Gamma_s(t)$ may change in time.

The numerical simulations are started from a configuration where a 2-D column-shaped jet has a non-vanishing area of contact with the receiving liquid, *i.e.* at $t = 0$ a liquid column of half-width $a$ and length $l$ is impacting with velocity $\mathbf{v} = (0, U_0)$ a liquid surface at rest.

For simplicity's sake, the dependence on space variables will not be indicated explicitly, using instead a "dot" as a dummy argument; by contrast, the dependence on the time variable $t$ will always be indicated.

The first model equation is the Bernoulli's law:

$$\begin{aligned}\frac{\partial\phi}{\partial t}(t,\cdot) - \nabla\Phi(t,\cdot)\cdot\mathbf{v}^s(t,\cdot)\\ -\tfrac{1}{2}|\nabla\Phi(t,\cdot)|^2 - g\mathbf{x}\cdot\hat{\mathbf{z}}\,, \quad \text{on} \quad \Gamma_s(t),\end{aligned} \tag{1}$$

and relates the potential field $\phi(t,\cdot)$ defined along $\Gamma_s(t)$ to the potential field $\Phi(t,\cdot)$ defined within $\Omega(t)$ and to the velocity $\mathbf{v}^s(t,\cdot)$ on $\Gamma_s(t)$.

The last term in eqn. (1) takes into account the gravity contribution: $g$ is the scalar gravity constant and $\hat{\mathbf{z}}$ the vertical direction versor, positive upward.

The second model equation updates the position vector $\mathbf{x}^s(t,\cdot)$ of the points lying on $\Gamma_s(t)$ according to the velocity field $\mathbf{v}^s(t,\cdot)$:

$$\frac{\partial\mathbf{x}^s}{\partial t}(t,\cdot) = \mathbf{v}^s(t,\cdot), \quad \text{on } \Gamma_s(t). \tag{2}$$

Finally, a relation is needed between the fields $\mathbf{v}^s(t,\cdot)$ and $\Phi(t,\cdot)$ to close the model. Assuming that $\Gamma_s(t)$ is a material line, the closure is given by taking

$$\mathbf{v}^s(t,\cdot) = \nabla\Phi(t,\cdot) + \{\text{param. term}\}, \quad \text{on } \Gamma_s(t). \tag{3}$$

The above second term on the right-hand-side follows from the parameterization of $\Gamma_s(t)$. The domain potential $\Phi(t,\cdot)$ in eqn. (3) is the solution of the Laplace's problem:

$$\Delta\Phi(t,\cdot) = 0 \qquad \text{in } \Omega(t), \tag{4a}$$

$$\Phi(t,\cdot) = \phi(t,\cdot) \quad \text{on } \Gamma_s(t), \tag{4b}$$

$$\nabla\Phi(t,\cdot)\cdot\mathbf{n} = 0 \qquad \text{on } \Gamma_w(t), \tag{4c}$$

formulated using a Dirichlet boundary condition on $\Gamma_s(t)$, eqn. (4b), and an homogeneous Neumann boundary condition on $\Gamma_w(t)$, eqn. (4c). The time variable $t$ is discretized into a sequence of equi-spaced time steps of size $\Delta\, t$. The mathematical model is then numerically re-formulated in two nested loops.

The operations of the outer loop are illustrated in fig. 2, the inner loop in fig. 3.

Let $\Omega_h(t)$ indicate the polygonal approximation of the domain $\Omega(t)$. The domain $\Omega_h(t)$ is bounded by the rigid wall $\Gamma_w(t)$ and the poly-line approximating $\Gamma_s(t)$. This poly-line connects pair of consecutive interface nodes by using straight segments. A conformal triangulation covering $\Omega_h(t)$ is then introduced following the definition of [9].

In the outer loop the Laplace's problem (4a–4c) is first solved on $\Omega_h(t)$ using the potential field $\phi(t,\cdot)$ as the boundary condition along the moving interface

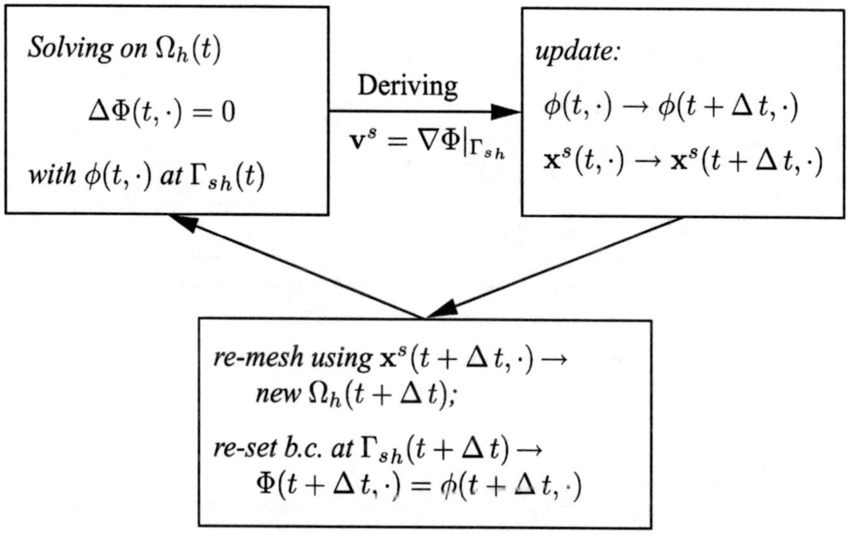


Figure 2: The outer approximation loop.

$\Gamma_s(t)$. Then, the moving interface velocity $\mathbf{v}^s(t,\cdot)$ is calculated and used to update the potential field $\phi(t,\cdot)$ and the position vector $\mathbf{x}^s(t,\cdot)$ by solving, respectively, eqns. (1) and (2).

The changes in $\mathbf{x}^s(t,\cdot)$ approximate the unsteady shape evolution of $\Omega(t)$, thus requiring the update of the computational domain $\Omega_h(t)$ along with its triangulation. A full re-meshing is performed at the end of each outer loop step. Finally, a new Laplace's problem is set up on $\Omega(t+\Delta t)$ by re-setting the Dirichlet boundary condition at $\Gamma_s(t+\Delta t)$ with the updated potential field $\phi(t+\Delta t,\cdot)$. This allows to re-start the next outer loop step at $t+\Delta t$.

The inner approximation loop is originated by the method of [10], which is used to compute the fields $\mathbf{x}^s$ and $\phi$ at $t+\Delta t$. The Crank-Nicholson method is an implicit time-stepping scheme, and the implicit dependence on the unknown fields introduces a non-linearity in each outer loop step. This non-linearity is tackled by fixed point iterations of the form depicted in fig. 3.

The inner steps are labeled using the superscript $l$. The $l$-th inner iteration calculates the approximate fields at $t+\Delta t$ by updating their $\ell$-th guess value to the $(\ell+1)$-th one. The inner iteration performs the same operations as the outer one; the most significant difference is that at the end of the $\ell$-th inner iteration the mesh $\Omega_h^{(\ell+1)}(t+\Delta t)$ is not re-calculated afresh, but modified by simply stretching the cells of $\Omega_h^{(\ell+1)}(t+\Delta t)$.

## 2.1 The Mixed Finite Element framework

The key point of mixed FEMs consists of splitting the second-order differential eqn. (4a) in two first-order equations:

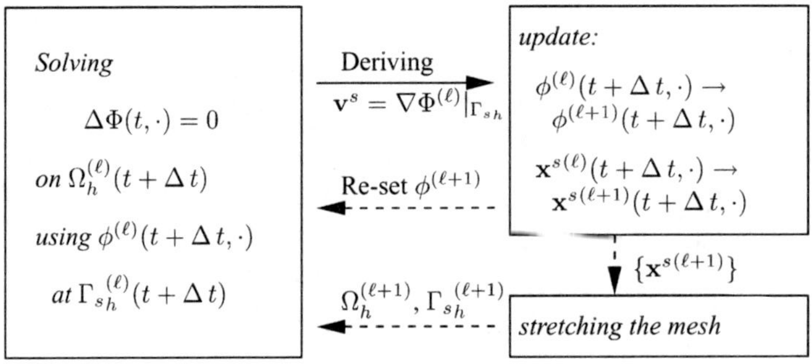


Figure 3: The inner approximation loop.

$$\nabla\Phi(t,\cdot)=\mathbf{v}(t,\cdot), \quad \text{in } \Omega(t), \tag{5a}$$

$$\nabla\cdot\mathbf{v}(t,\cdot)=0, \quad \text{in } \Omega(t). \tag{5b}$$

Equations (5a–5b) are solved simultaneously for $\Phi(t,\cdot)$ and $\mathbf{v}(t,\cdot)$ using the boundary conditions, eqns. (4b) and (4c).

As it is usual in Finite Element Method, eqns. (5a–5b) are first re-written in weak form by introducing suitable functional spaces of test functions. Following [11], the weak formulation of problem eqns. (5a–5b) for fixed $t$ in the domain $\Omega\equiv\Omega(t)$ has been stated in the present work as follows:
Find $\mathbf{v}\in\mathcal{V}$ and $\Phi\in L^2(\Omega)$ such that

$$\int_\Omega \mathbf{v}\cdot\mathbf{w}\,d\mathbf{x}+\int_\Omega \Phi\,\nabla\cdot\mathbf{w}\,d\mathbf{x}=\int_{\Gamma_s}\phi\mathbf{w}\cdot\mathbf{n}\,ds,$$
$$\int_\Omega(\nabla\cdot\mathbf{v})\psi\,d\mathbf{x}=0,$$

$\forall\mathbf{w}\in\mathcal{V}$ and $\forall\psi\in L^2(\Omega)$ where

$$\mathcal{V}=\left\{\mathbf{q}\in L^2(\Omega)^2,\nabla\cdot\mathbf{q}\in L^2(\Omega),\ \text{and}\ \mathbf{q}\cdot\mathbf{n}|_{\Gamma_w(t)}=0\right\}$$

and $L^2(\Omega)$ is the space of square integrable functions.

The discretization follows by considering discrete sub-spaces of the functional spaces used in the weak formulation.

The potential field $\Phi(t,\cdot)$ is discretized by the *piecewise-constant functional space* $Q_h=\{\psi(\mathbf{x}):\Omega\mapsto\mathbb{R},\ \psi(\mathbf{x})|_T=\text{const},\ \forall T\in\Omega_h\}$ and the velocity field $\mathbf{v}^s(t,\cdot)$ by the *lowest-order Raviart-Thomas space*

$$V_h=\Big\{\mathbf{w}\in\left(L^2(\Omega)\right)^2,\ \mathbf{w}(\mathbf{x})|_T=\gamma\mathbf{x}+\delta,$$
$$\forall T\in\Omega_h,\forall\gamma\in\mathbb{R},\forall\delta\in\mathbb{R}^2,\int_{e_{ij}\cap\Gamma_s(t)}\mathbf{w}\cdot\mathbf{n}=0\Big\}.$$

The corresponding unknowns describe respectively the piecewise constant potential field $\{\zeta_k\}$ in any triangular cell, and the normal component of the velocity field $\{u_j\}$ at any mesh edge. The discrete potential field $\Phi_h(t,\cdot)$ and the normal velocity field $\mathbf{v}_h(t,\cdot)$ may then be written as linear combinations of the corresponding sets of basis functions of $Q_h$ and $V_h$:

$$\Phi_h(t,\cdot) = \sum_{k=1}^{N_T} \zeta_k(t)\psi_k(\cdot), \tag{6a}$$

$$\mathbf{v}_h(t,\cdot) = \sum_{j=1}^{N_e} u_j(t)\mathbf{w}_j(\cdot) \tag{6b}$$

where $\{\psi_k\}$ = pressure basis functions, $\{\mathbf{w}_j\}$ = velocity basis functions, $N_T$ and $N_e$ are respectively the number of triangular cells and edges. In view of eqns. (6a–6b) the final algebraic system to be solved is

$$\begin{bmatrix} \mathbf{M} & \mathbf{A} \\ \mathbf{A}^T & \mathbf{0} \end{bmatrix} \begin{bmatrix} \{u_j\} \\ \{\zeta_k\} \end{bmatrix} = \begin{bmatrix} \{q_j\} \\ \mathbf{0} \end{bmatrix}, \tag{7}$$

where $\mathbf{M}$, $\mathbf{A}$, and $\mathbf{q}$ are defined as:

$$M_{ij} = \int_{\Omega(t)} \mathbf{w}_i \cdot \mathbf{w}_j \, d\mathbf{x}, \; A_{ij} = \int_{\Omega(t)} (\nabla \cdot \mathbf{w}_i)\psi_j \, d\mathbf{x}, \; q_j = \int_{\Gamma_s(t)} \phi \mathbf{w}_j \cdot \mathbf{n} \, ds.$$

The above integrals are numerically evaluated by the mid-point formula.

The solution of the linear system eqn. (7) may be difficult due to the unstable nature of the corresponding weak form; the system eqn. (7) is solved by applying the routine *MA47* of the *HSL-2000* library [12], which allows several strategies that are quite effective in numerically coping with this problem.

The advantage of using mixed FEM is twofold.

Firstly, this approach yields a method that is second order accurate, at least in principle. This is due to a super-convergence effect, which provides an approximation that is more than first order at the edge mid-points, namely where the mid-point integration formula takes its argument values.

The second important consequence of solving the Laplace's problem by the mixed FEM is that the no-flux condition, eqn. (4c), is exactly satisfied. This occurs because the normal component of the velocity field $\mathbf{v}(t,\cdot) \cdot \mathbf{n}$ is one of the two degrees of freedom of the discrete problem, and can thus be directly set to zero at the wall boundary.

The present numerical method differs remarkably from Prosperetti & Oguz [3], even though the mathematical formulation might appear quite similar. The mathematical model is re-formulated by [3] in order to use the boundary integral method; the problem then arises as how to deal with the removal of the logarithmic

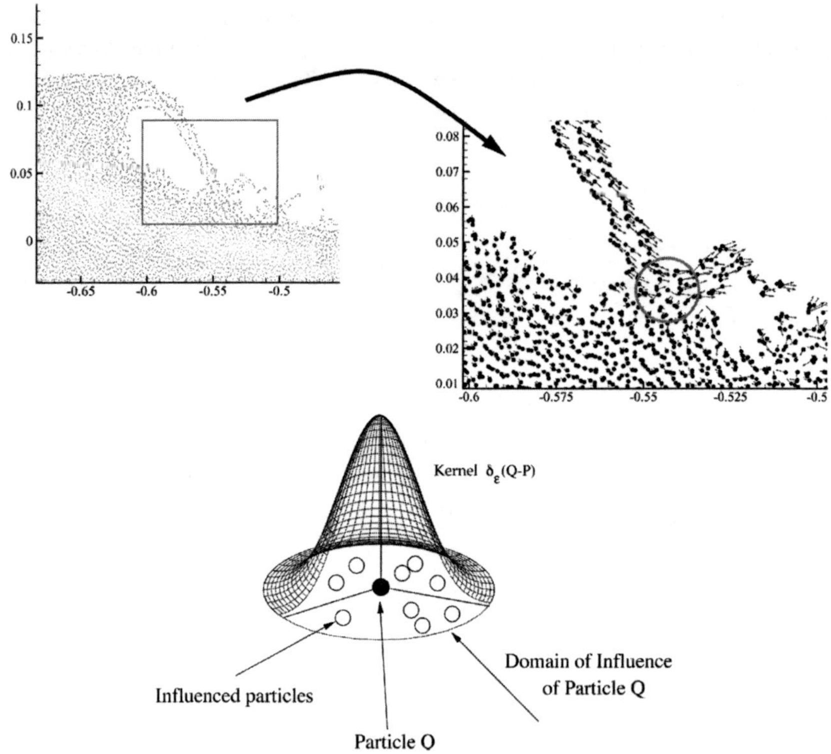


Figure 4: Top: typical distribution of particles, •, in a free-surface flow. In the enlarged detail, the circle represents the radius of interaction of a generic particle. Bottom: Schematic view of the kernel $\delta_\varepsilon$ centered on the particle $\boldsymbol{Q}$.

singularities, which is a rather standard practice in evaluating elliptic integrals, and how to introduce some arbitrary smoothing to manage the moving free surface.

None of the above problems is present in the MFE formulation. The boundary-integral formulation is actually replaced by the re-solution of a potential problem on a domain which is entirely re-meshed at any new time iteration. No arbitrary smoothing is introduced in the model formulation: as a matter of fact, the role of the curvature damping on short wavelengths is played by the parameterization term that naturally appears in the formulation of eqn. (3).

## 3 The Smoothed Particle Hydrodynamics (SPH) solver

The implementation of a Lagrangian flow solver based on the Smoothed Particle Hydrodynamics (SPH) method is described in this section.

The SPH method was originally introduced by Lucy [13]; since then it has been greatly enhanced in the last two decades.

The basic idea is to consider a set of $N$ particles with mass $m_j$ distributed over the bulk of fluid $\Omega$, as in the example of fig. 4. Each particle is characterized by a kernel $\delta_\epsilon$: a shape function which is symmetric, regular, non-negative, and centered on the particle position. In a naive interpretation of the method, mass and other physical information associated to each particle are "smeared out" through the kernel over its spatial support. Each particle moves in the force field generated by the whole particles system.

The essential feature of the resulting algorithm is the absence of a computational grid and the fully Lagrangian character.

### 3.1 Basic SPH equations

Considering the field $f(\boldsymbol{P})$ of a physical quantity, a regularized (smoothed) representation of $f(\boldsymbol{P})$ is given by the interpolation integral

$$\langle f(\boldsymbol{Q})\rangle = \int_\Omega f(\boldsymbol{P})\,\delta_\epsilon(\boldsymbol{Q}-\boldsymbol{P})\,\mathrm{d}\Omega_P\ . \tag{8}$$

The parameter $\epsilon$ represents a measure of the spatial spreading of the kernel and it is assumed that $\delta_\epsilon(\boldsymbol{Q}-\boldsymbol{P})$ converges to a Dirac delta-function as $\epsilon \to 0$. Even though computationally effective, the kernel does not need a compact support (*i.e.* it is identically zero beyond some distance from the particle). The choice of the kernel does affect remarkably the stability properties of the discrete method [14]. In the present computations, third- and fifth-order spline kernels and a Gaussian kernel have been tested.

Upon inserting eqn. (8) into the Euler's equations, the following evolution equations for the density $\rho_i$ and the velocity $\boldsymbol{v}_i$ of the $i$-th particle located at point $\boldsymbol{P}_i$ can be obtained:

$$\begin{aligned} \frac{\mathrm{d}\rho_i}{\mathrm{d}t} &= -\rho_i \textstyle\sum_{j=1}^N (\boldsymbol{v}_j-\boldsymbol{v}_i)\cdot\nabla_i\delta_\epsilon(\boldsymbol{P}_i-\boldsymbol{P}_j)\,dV_j \\ \frac{\mathrm{d}v_i}{\mathrm{d}t} &= -\frac{1}{\rho_i}\textstyle\sum_{j=1}^N (p_i+p_j+\Pi_{ij})\,\nabla_i\,\delta_\epsilon(\boldsymbol{P}_i-\boldsymbol{P}_j)\,dV_j+\boldsymbol{g} \end{aligned}\ , \tag{9}$$

where $p_i$ is the pressure and $\Pi_{ij}$ is the artificial viscosity term, which will be discussed below.

During its evolution, each particle carries a constant mass $m_j$ and therefore the particle volume is $dV_j = m_j/\rho_j$. The structure of the discretized momentum equation is symmetric and ensures linear and angular momentum conservation (see also eqn. (10) below).

In the present approach, the fluid is modeled as a weakly compressible fluid described by the state equation of the form

$$p = P_0\left[\left(\frac{\rho}{\rho_0}\right)^\gamma - 1\right],$$

with $\gamma = 7$ and $P_0$ tuned as to have a negligible compressibility [15]. In practical computations, the largest density fluctuations are of order $10^{-2}\rho_0$.

The use of an explicit formula for the pressure avoids to solve for the Poisson's equation, and therefore increases the efficiency and reduces the memory requirements of the method.

Consistently with the free surface dynamic boundary condition, the smoothed pressure field, eqn. (8), falls to zero approaching the free surface, which is spread over the support of the kernel, while the use of an evolution equation for the density ensures a correct evaluation of the smoothed density field [16].

The artificial viscosity term $\Pi_{ij}$ in the discretized momentum equation, the second of eqn. (9), has the purpose to increase the stability of the numerical algorithm. The following form, discussed in [17], has been adopted:

$$\mu_{ij} := \epsilon \frac{\boldsymbol{v}_{ij} \cdot \boldsymbol{P}_{ij}}{|\boldsymbol{P}_{ij}|^2 + 0.01\epsilon^2},$$

$$\Pi_{ij} := \begin{cases} \dfrac{-\alpha(c_i + c_j)\mu_{ij}}{1/\rho_i + 1/\rho_j} & \boldsymbol{v}_{ij} \cdot \boldsymbol{P}_{ij} < 0, \\ 0 & \boldsymbol{v}_{ij} \cdot \boldsymbol{P}_{ij} \geq 0. \end{cases} \tag{10}$$

In eqn. (10) the compact notation $f_{ij} := f_i - f_j$ is used, and $c$ is the speed of sound, which follows from the state equation.

In the present work, the influence of $\alpha$ has been checked in the range 0.005–0.03.

Finally, the motion of the particles is described by

$$\frac{\mathrm{d}\boldsymbol{P}_i}{\mathrm{d}t} = \boldsymbol{v}_i + \sum_{j=1}^{N} \frac{\boldsymbol{v}_{ij}}{(\rho_i + \rho_j)} \delta_\epsilon(\boldsymbol{P}_{ij})\, m_j \tag{11}$$

where the latter term weakly averages the velocity field around the particle. It prevents the particles from penetrating each other, and smoothes the field oscillations due to the weak compressibility of the liquid [16].

The evolution eqns. (9) and (11) can be stepped forward in time by any ODE integrator. In the present implementation, a second-order predictor-corrector scheme is adopted with a dynamic choice of the time step according to stability constraints related both to the local speed of sound and to the local value of the particle acceleration. The resulting stability requirements are rather demanding (a weak point of the pseudo-compressible SPH formulation) and the time step *locally* requested can be extremely small. To alleviate the problem, an individual time-stepping algorithm have been developed [18] to let the particles evolve hierarchically according to their own time step.

Although the particle equations are coupled, the right-hand side of each of them can be evaluated independently of the others, and without the solution of an algebraic system (as in most of the discretization methods for PDEs). Therefore, the memory requirement is simply proportional to the number $N$ of the particles, and the algorithm is well suited for use on parallel computers.

More technical details, such as the searching algorithm for the efficient particle interaction (which presently requires $\mathcal{O}(N)$ operations), enforcement of the no-penetration boundary condition on solid boundaries by ghost-particles and re-initialization, are not herein discussed, even though such details have a decisive role in the algorithm (see [8]).

In the SPH method the characteristic discretization parameters are:

(i) the ratio $\epsilon/\mathcal{L}$, where $\mathcal{L}$ is a typical length scale of the problem, and
(ii) the number $\mathcal{N}$ of particles within the interaction radius (*i.e.* the circle in fig. 4).

Roughly speaking, the ratio $\Delta x = 5\epsilon/\sqrt{\mathcal{N}}$ is equivalent to the grid spacing in mesh-based methods.

In the present study, the half-width of the liquid column $a$ has been taken as the length scale, $\epsilon/a = 0.133$–$0.267$ and $\mathcal{N}$ in the range 30 to 50. The resulting number of particles may turn out as great as $10^5$.

The above SPH method proved successful in a variety of violent free-surface flows such as dam-break flows [8, 19], sloshing flows with fluid-solid impacts [20], ship flows [21].

## 4. Numerical Experiments

### 4.1 Numerical experiments by MFE

All of the impact instances presented by [3] have been simulated using the above MFE scheme; however, only the simulations $\mathsf{Fr} = 8$ and 32 and aspect ratio $h = 10$ and 20 are presented herein. The cases $\mathsf{Fr} = 2$ and aspect ratio $h = 10$ and 20 have been already illustrated in [22].

To start the simulations with the very same initial conditions as [3], an half-circle has been considered on the top of the liquid column. This geometrical feature is in excess of the stated aspect ratios and has only a tiny influence on the subsequent development of the impact. However, for completeness' sake, the SPH simulation (see subsection 4.2 below) was performed without the half-circle. The symmetry of the problem *has not* been imposed in the present MFE simulations. The mesh is generated at each time step by the public domain mesh generator TRIANGLE [23]. All the geometrical and topological data related to the mesh are managed by the software P2MESH [24], a freeware library for Finite Volume and Finite Element development.

Figures 5 through 9 show the free surface time evolutions for $\mathsf{Fr} = 8$ and 32 and $h = 10$ and 20. In all of these pictures, times and lengths are in dimensionless units, the time scale being $a/U_o$ and the length scale is the column half-width $a$.

When the aspect ratio is increased to $h = 20$, the cavity lateral walls collapse against each other roughly at the time the impacting 2-D column has fully penetrated the liquid pool and a bubble is entrapped. For both of the aspect ratios, the bottom cavity has downward momentum at the time the closure of the cavity occurs.

The case depicted in figs. 7 and 8 shows no bubble entrapment. Rather a rising water column is ejected upward at the end of the impact, to fall down again.

The same is not true for $\mathsf{Fr} = 32$ and $h = 20$, where an air bubble is entrained instead.

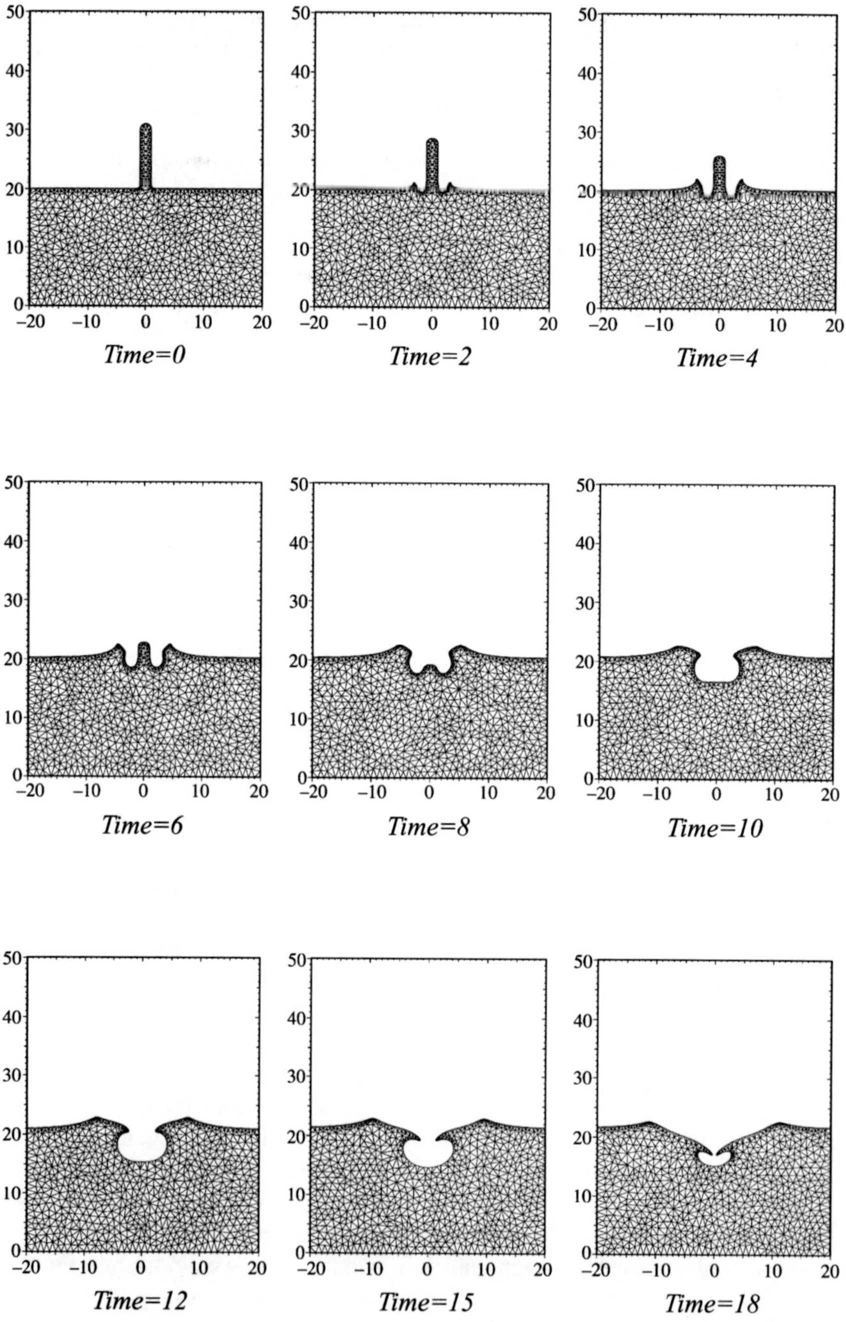

Figure 5: Free surface evolution (MFE, $\mathsf{Fr} = 8,\ h = 10$).

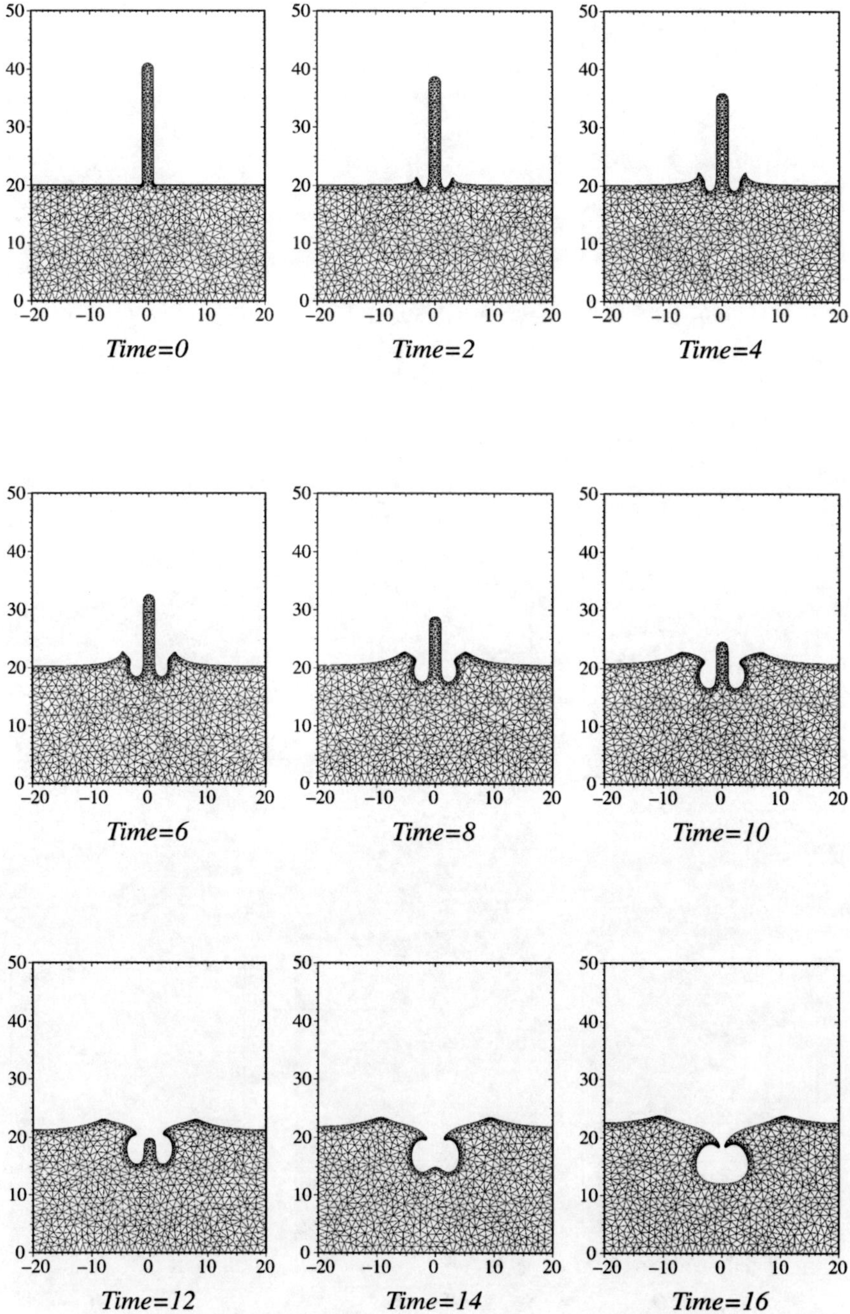


Figure 6: Free surface evolution (MFE, Fr $= 8,\ h = 20$).

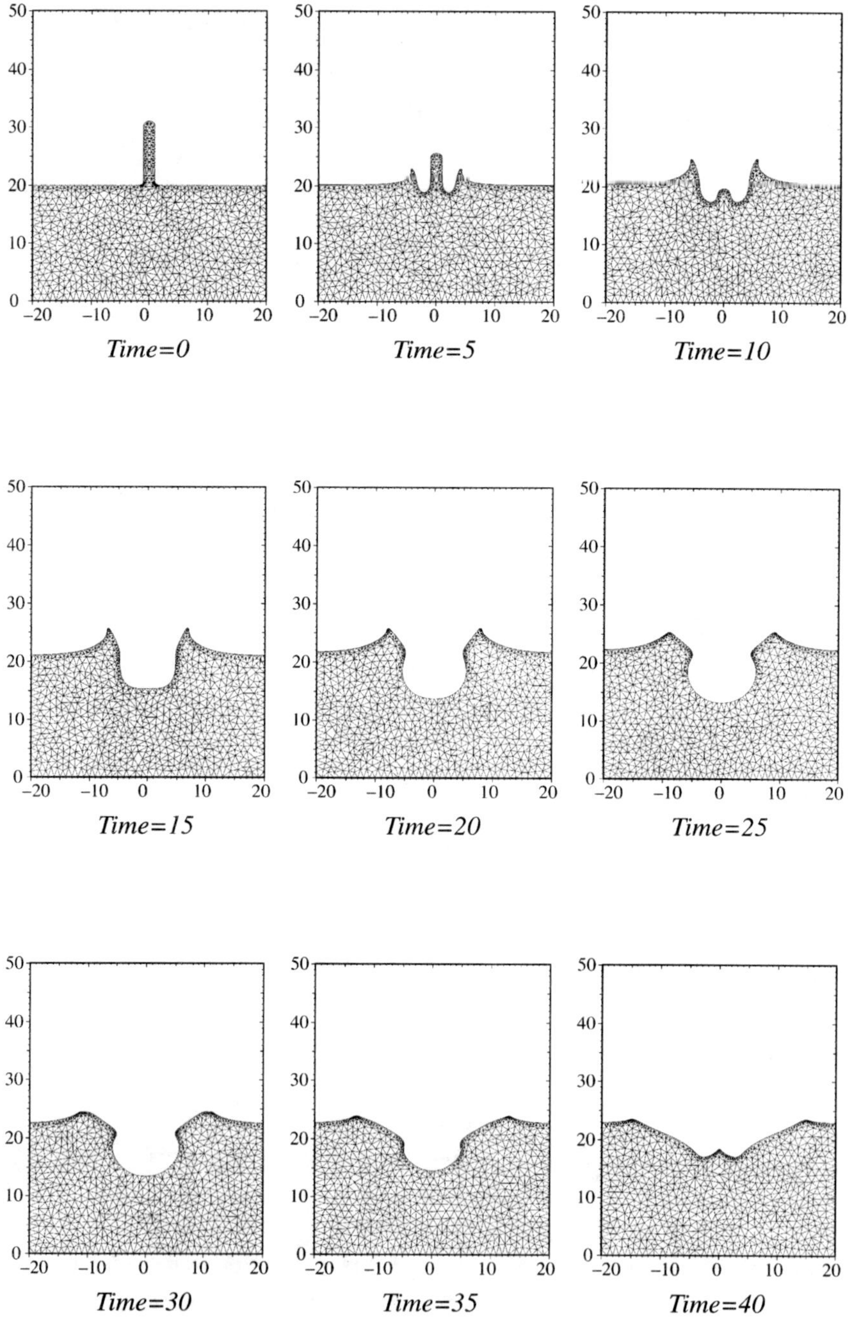


Figure 7: Free surface evolution (MFE, $\mathsf{Fr} = 32,\ h = 10$).

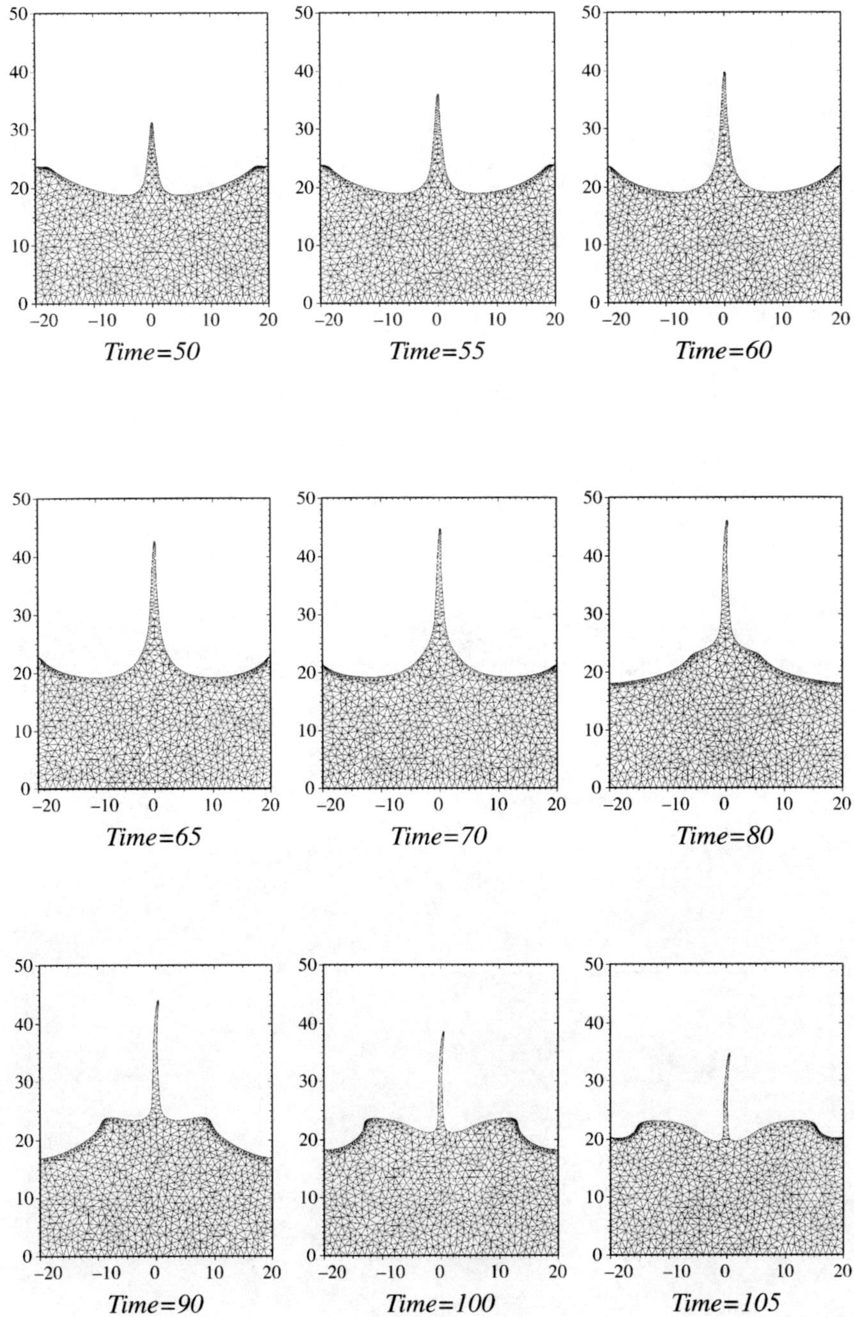


Figure 8: Free surface evolution (MFE, $\mathsf{Fr} = 32,\ h = 10$).

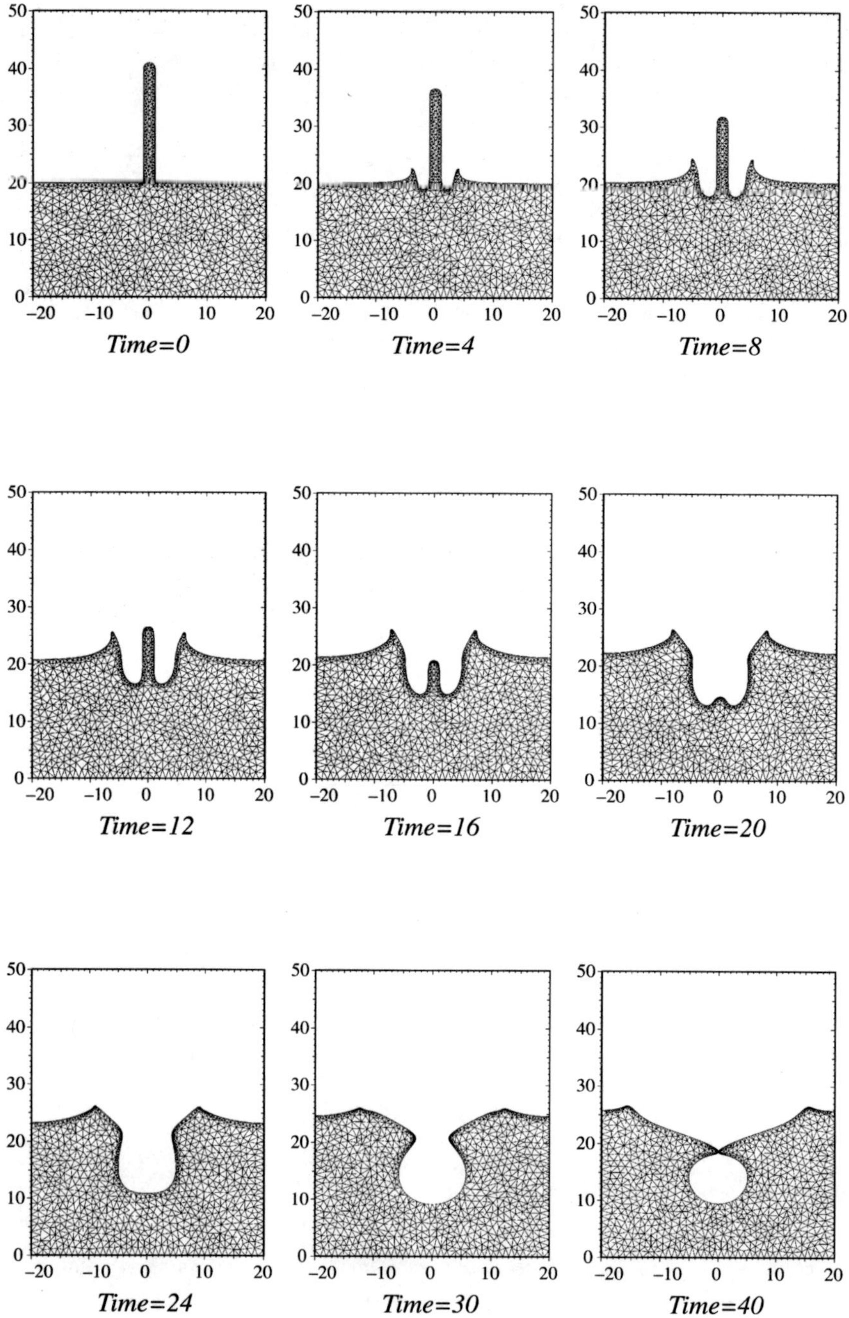


Figure 9: Free surface evolution (MFE, $\mathsf{Fr} = 32,\ h = 20$).

Table 1 displays the minimum, the maximum and the average number of equations forming the linear system eqn. (7) as well as the inner iterations of the coupled solver. The number of equations does not increase dramatically, while the number of iterations is roughly 10.

It is clear from table 2 that mesh sizes are stable as regards the number of triangular cells, vertices and edges.

Table 3 reports the total CPU time elapsed on a 1 GHz processor required to solve eqn. (7) in each simulation (second column) and the CPU time required to generate the mesh (third column). The last column indicates the percentage of the third column with respect to the second one. Even though the MFE method calls for a full re-meshing at each time step, it seems clear that re-meshing costs are not reflected, as a matter of fact, on the total CPU cost.

Energy conservation was satisfied by the MFE method within 2% in the worst case.

### 3.2 Numerical Experiments by SPH

The symmetry of the problem *has not* been imposed in the present SPH simulations. Also the half-circle has not been included on the top of the liquid column. The particles of the 2–D impacting liquid column have been marked in black, so as

Table 1: Computational cost for the Laplace's problem (MFE, Fr = 32).

| $h$ | # Equations | | | # Iterations | | |
|---|---|---|---|---|---|---|
| [/] | Min | Max | Avg | Min | Max | Avg |
| 10 | 4145 | 4710 | 4414 | 5 | 20 | 6.3 |
| 20 | 4808 | 5226 | 5016 | 5 | 10 | 6.3 |

Table 2: Computational cost for re-meshing (MFE, Fr = 32).

| $h$ | # Triangles | | | # Vertices | | | # Edges | | |
|---|---|---|---|---|---|---|---|---|---|
| [/] | Min | Max | Avg | Min | Max | Avg | Min | Max | Avg |
| 10 | 1594 | 1820 | 1702 | 958 | 1071 | 1010 | 2551 | 2890 | 2712 |
| 20 | 1844 | 2010 | 1926 | 1121 | 1207 | 1164 | 2964 | 3216 | 3089 |

Table 3: CPU cost (MFE, Fr = 32).

| $h$ [/] | CPU Time (s) | Re-Meshing (s) | % |
|---|---|---|---|
| 10 | 1222.14 | 19.63 | 1.6 |
| 20 | 554.01 | 8.33 | 1.5 |

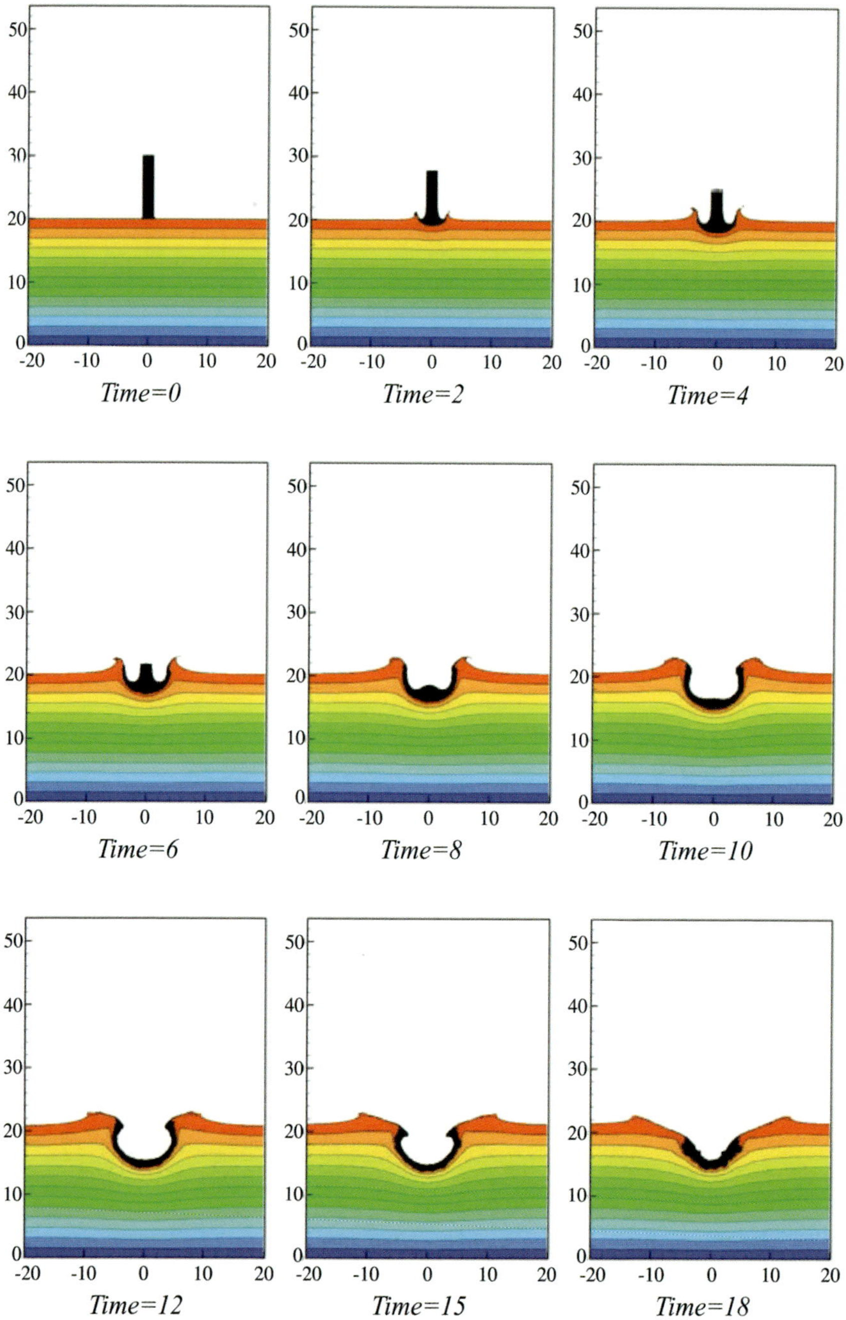


Figure 10: Free surface evolution (SPH, Fr $= 8,\ h = 10$).

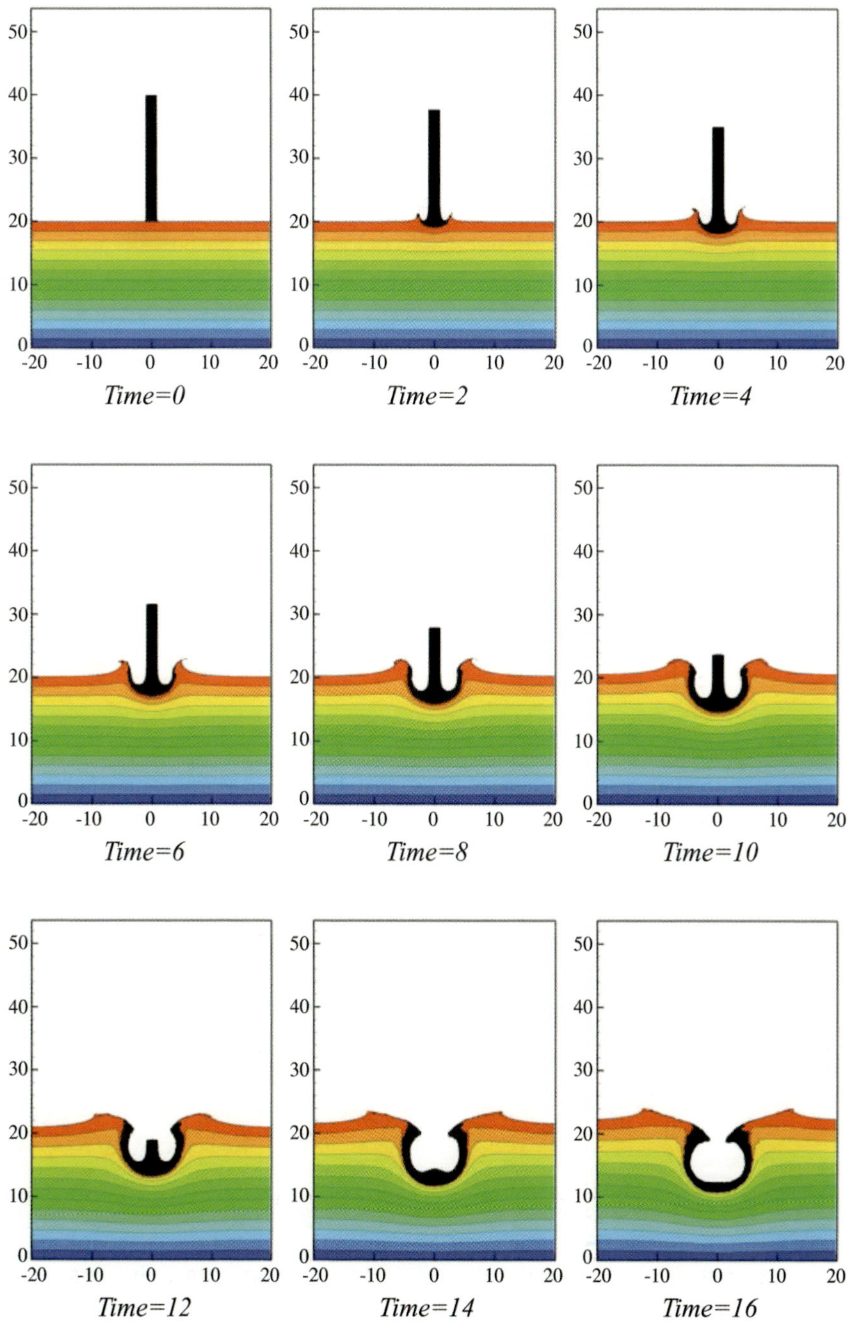


Figure 11: Free surface evolution (SPH, Fr $= 8$, $h = 20$).

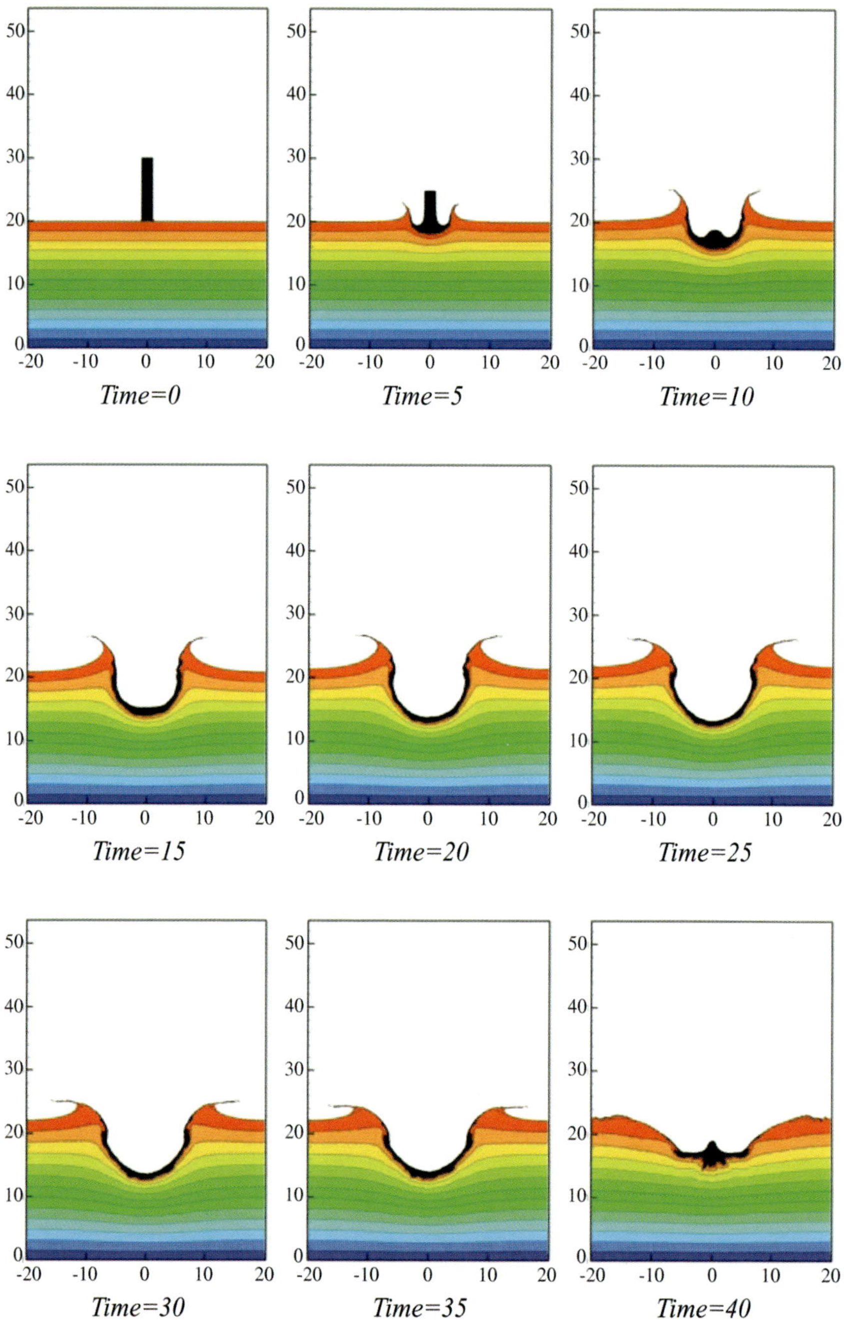


Figure 12: Free surface evolution (SPH, $\mathsf{Fr} = 32,\ h = 10$).

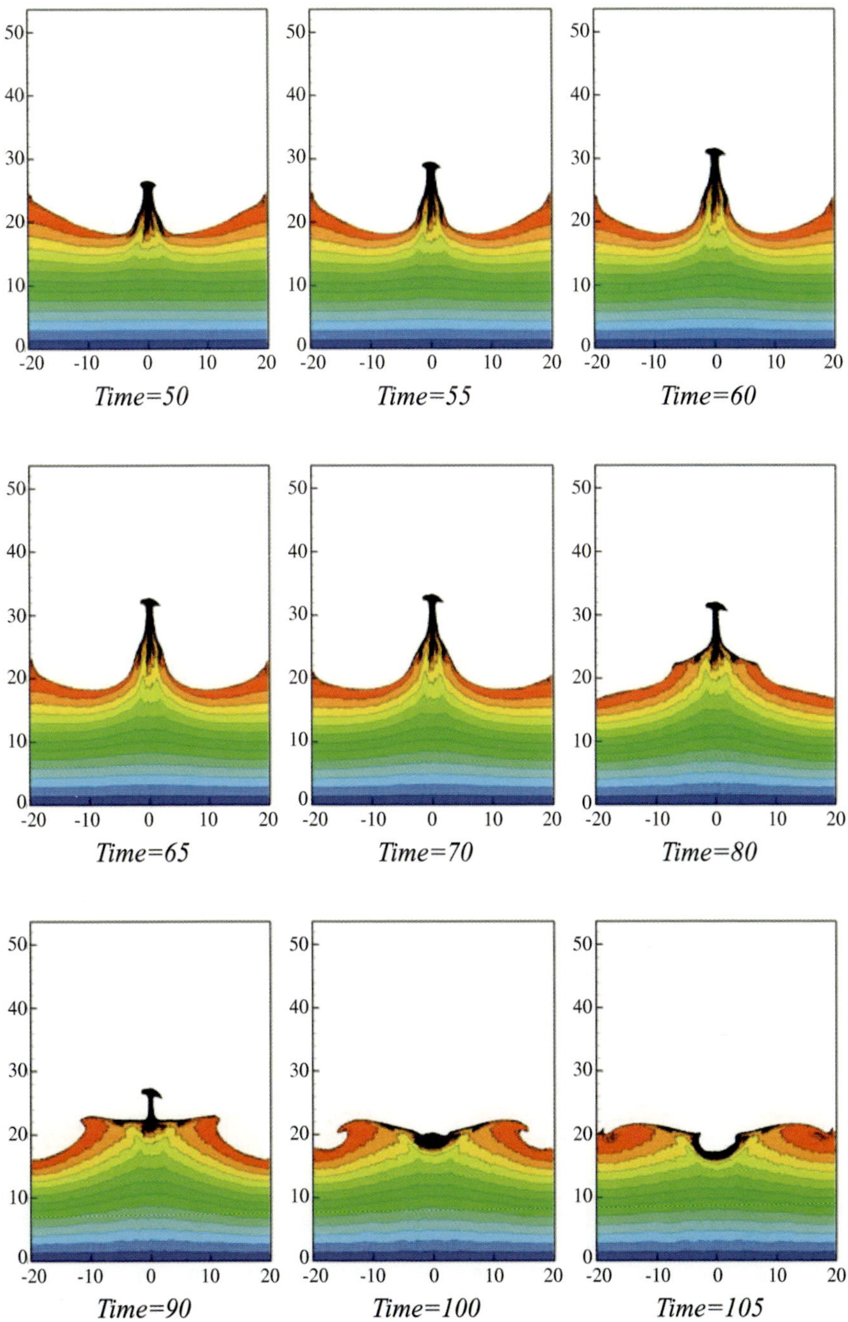


Figure 13: Free surface evolution (SPH, $\mathsf{Fr} = 32,\ h = 10$).

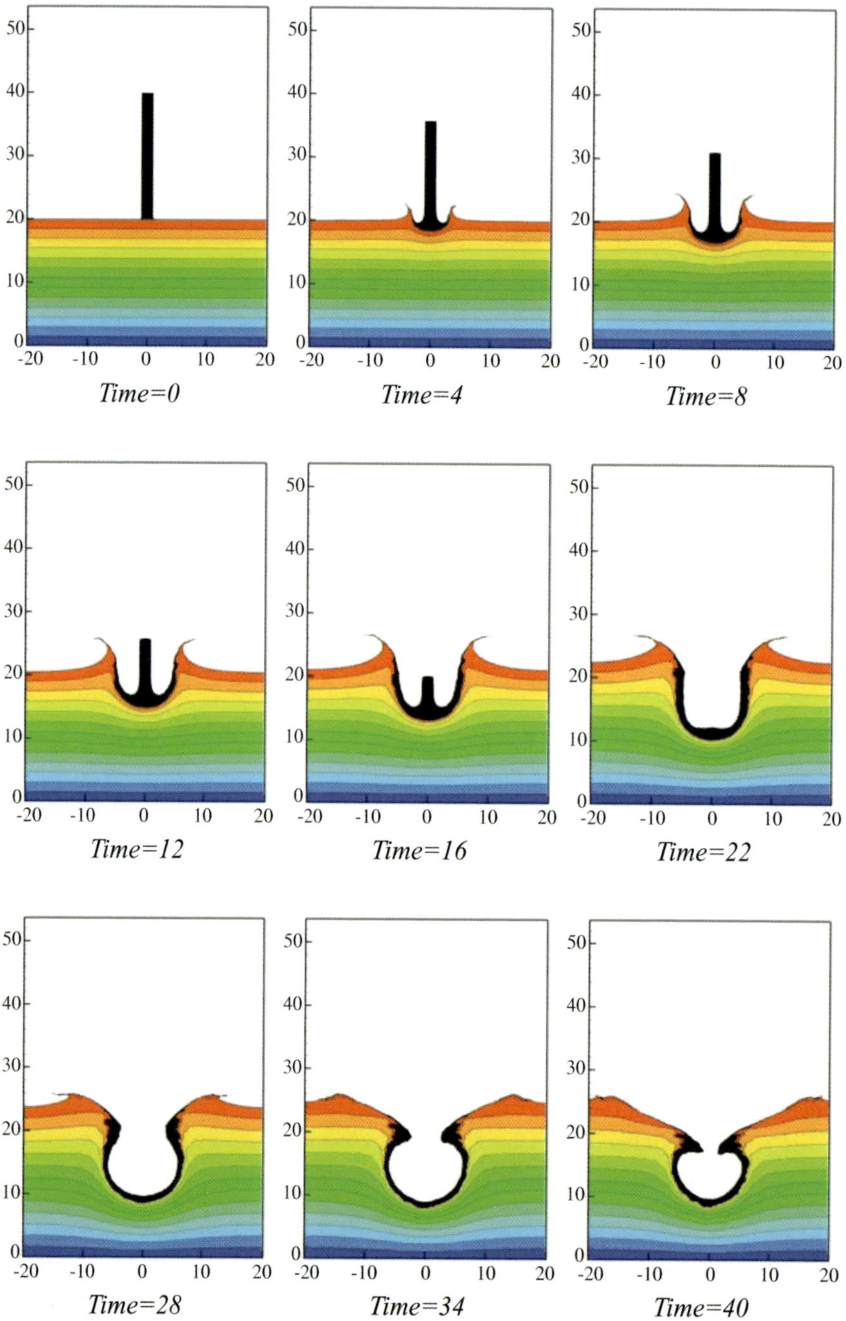


Figure 14: Free surface evolution (SPH, $\mathsf{Fr} = 32,\ h = 20$).

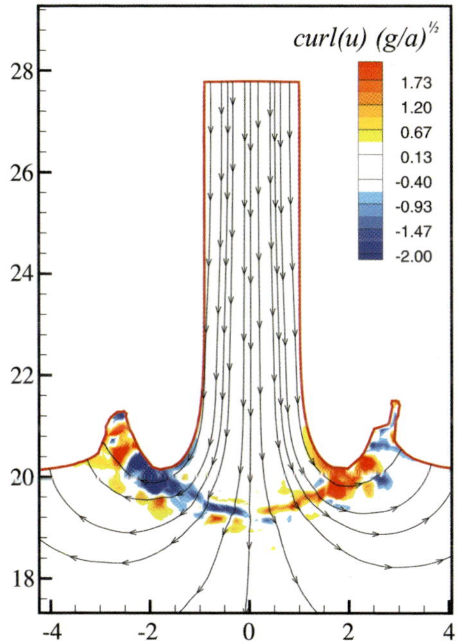


Figure 15: Vorticity field (SPH, $\mathsf{Fr} = 8$, $h = 10$, time$= 2$).

the Lagrangian mixing is easier to trace. The dimensionless time step used was $\delta t\sqrt{g/a} = 8 \times 10^{-4}$, the total number of particles 82000, the CPU time for the longest simulation ($T = 105$) is some 9 h on a 3.3GHz processor.

The SPH simulations for $\mathsf{Fr} = 8$ and 32 and $h = 10$ and 20 are shown in figs. 10 through 14. It seems clear that the overall pattern of the free surface deformation is basically identical to the MFE's solution. It is worth noting that the run-up elevation of the rebound and the pattern of the radiated waves are also similar to the MFE simulations.

The injection of the liquid column would generate vorticity. This fact cannot be accounted for by the Laplace's equation and the near field in the two simulations may be very different. Figure 15 illustrates for $\mathsf{Fr} = 8$, $h = 10$ at time $= 2$ that the vorticity field is confined at the base of the penetrating column.

The distribution of the particles for $\mathsf{Fr} = 8, h = 10$ at time$= 2$ is displayed in fig. 16. After the impact, the black particles located at the front edge of the penetrating column are no more lined up to the streamlines. This uneven arrangement of the particles produces a numerical noise, which is, however, confined to the narrow transition layer between the entering column and the liquid at rest. Since the above numerical vorticity is admittedly low, the global flow field is basically not affected and this is the main reason why SPH simulations are similar to MFE simulations.

As for the energy conservation, the SPH method demonstrated a relatively high percentage of energy numerical losses, ranging from 8% ($\mathsf{Fr} = 8$, $h = 20$) to

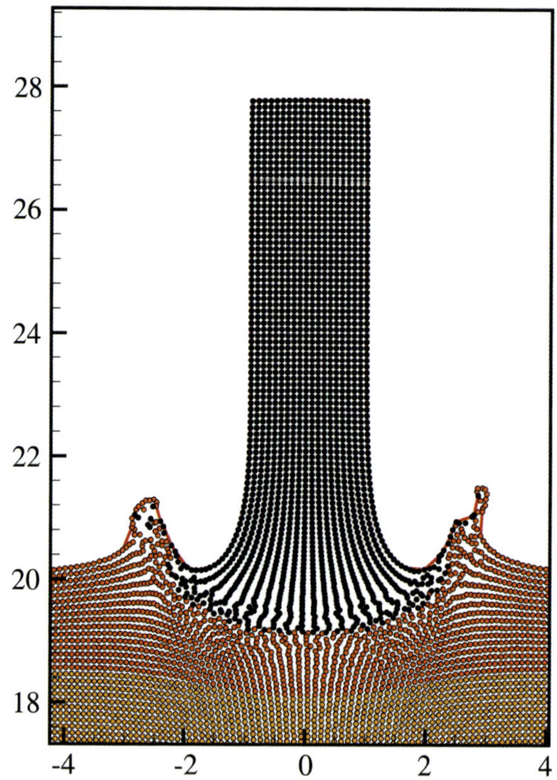


Figure 16: Particles arrangement (SPH, $\mathsf{Fr} = 8$, $h = 10$, time$= 2$).

$18\%$ ($\mathsf{Fr} = 32$, $h = 10$). However, a meaningful and above all rigorous comparison with the MFE method as far as the numerical dissipation is concerned is not straightforward and it has not been investigated in detail.

### 3.3 Comparison with the boundary integral method

The presentation of the comprehensive comparison with the boundary integral method [3] is beyond the scope of the present work. For brevity's sake, only the impact $\mathsf{Fr} = 8$, $h = 10$ is reported in fig.17; however, the other impact instances are available by the Authors. It is recalled that the present SPH simulation does not cap the column top with a half-circle, while both MFE and boundary integral method do. It seems clear there are no major discrepancies among the three different methods, apart from a well-defined attempt for the MFE solver to close the cavity earlier than the other two methods.

It seems useful to observe that a thorough comparison can only be performed by knowing in more detail how the numerical smoothing has been dealt with by any solver.

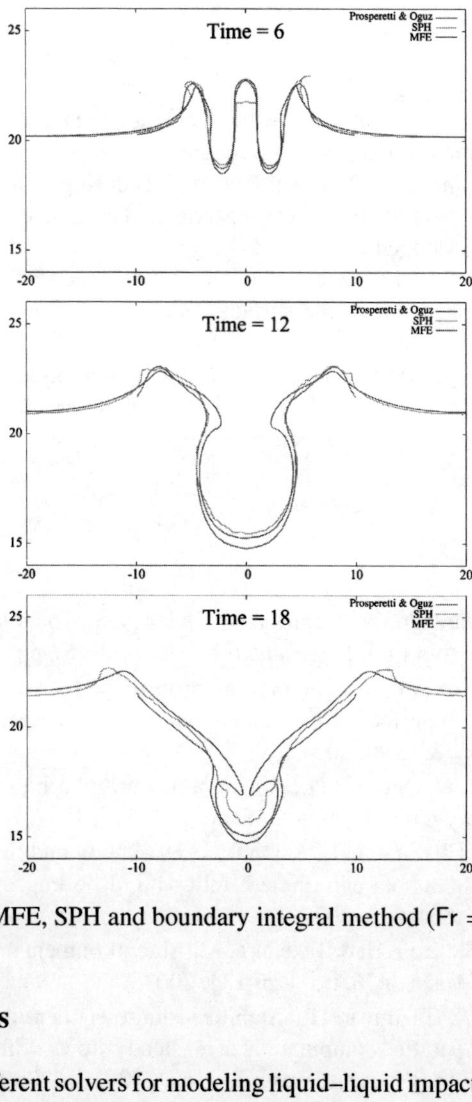


Figure 17: MFE, SPH and boundary integral method (Fr $= 8$, $h = 10$).

# 4 Conclusions

Two radically different solvers for modeling liquid–liquid impacts have been implemented. One solver is simply based on the potential flow theory: it is the simpler between the two solvers, requiring CPU minutes to provide a solution which is, as a matter of fact, reasonably similar to the one calculated by the SPH approach. The cost of this latter approach is by far greater than the MFE potential model in terms of memory requirements and CPU times.

It must be observed that the impact instances which have been chosen from [3] are relatively simple to model and in fact there are no complicated free surface deformations, such as breaking waves. Obviously the potential model could not describe those complicated motions, which by contrast the SPH solver can. One further remarkable feature of the SPH method which is worth mentioning is the capability of vividly tracing the Lagrangian mixing of the particles.

## Acknowledgments

The Authors are deeply grateful to professor Maurizio Brocchini (University of Genova, Italy) who has prompted this work, has promoted the cooperation between two distant Institutions and has offered thoughtful advices.

The helpful comments of Professor Robert A. Dalrymple (Johns Hopkins University, Baltimore, USA) and of Professor Mario Gallati (University of Pavia, Italy) are gratefully acknowledged.

The important scientific contribution of Doctor Maurizio Landrini (Insean, Rome, Italy) in the non-linear free surface hydrodynamics is acknowledged: his premature end did not stop his work.

This work has received financial support by the Italian Ministry of Scientific and Technology Research, project PRIN 2002 "Influence of vorticity and turbulence in interactions of water bodies with their boundary elements and effect on hydraulic design".

## References

[1] Cossali, G., Marengo, M., Coghe, A. & Zhdanov, S., The role of time in single drop splash on thin film. *Experiments in Fluids*, **36(6)**, pp. 888–900, 2004.

[2] Saylor, J. & Grizzard, N., The optimal drop shape for vortices generated by drop impacts: the effect of surfactants on the drop surface. *Experiments in Fluids*, **36**, pp. 783–790, 2004.

[3] Prosperetti, A. & Oguz, H.N., Air entrainment upon liquid impact. *Philos. Trans. Roy. Soc. London Ser. A*, **355**, pp. 491–506, 1997.

[4] Elmore, P.A., Chahine, G.L. & Oguz, H.N., Cavity and flow measurements of reproducible bubble entrainment following drop impacts. *Experiment in Fluids*, **31**, pp. 664–673, 2001.

[5] Thoroddsen, S., Etoh, G. & Takehara, K., Air entrainment under an impacting drop. *J. Fluid Mech.*, **478**, pp. 125–134, 2003.

[6] Aliabadi, S. & Tezduyar, T., Stabilized-finite-element/interface-capturing technique for parallel computation of unsteady flows with interface. *Computs. Methods Appl. Mech. Engrg.*, **190**, pp. 243–261, 2000.

[7] Cruchaga, M., Celentano, D. & Tezduyar, T., A moving Lagrangian interface for flow computations over fixed meshes. *Computs. Methods Appl. Mech. Engrg.*, **191**, pp. 525–543, 2001.

[8] Colagrossi, A. & Landrini, M., Numerical simulation of interfacial flows by smoothed particle hydrodynamics. *J. Comput. Phys.*, **191(2)**, pp. 448–475, 2003.

[9] Ciarlet, P.G., *The finite element method for elliptic problems*. North-Holland Publishing Company: Amsterdam, Holland, 1980.

[10] Crank, J. & Nicolson, P., A practical method for numerical evaluation of solutions of partial differential equations of the heat-conduction type. *Proc. Cambridge Philos. Soc.*, **43**, pp. 50–67, 1947.

[11] Brezzi, F. & Fortin, M., Mixed and hybrid finite element methods. *Springer Series in Computational Mathematics, 15*, Springer Verlag, 1991.

[12] HSL, *(Harwell Subroutine Library), a collection of Fortran codes for large scale scientific computation.* CCLRC, Oxfordshire, UK, 2004. URL http://www.cse.crlc.ac.uk/nag/hsl/.

[13] Lucy, L., A numerical approach to the testing of fission hypothesis. *Astronomical Journal*, **82(12)**, pp. 1013–1024, 1977.

[14] Morris, J., Fox, P. & Zhu, Y., Modeling low Reynolds number incompressible flows using SPH. *J. Computat. Phys.*, **1936**, pp. 214–226, 1997.

[15] Monaghan, J., Smoothed particle hydrodynamics. *Ann. Rev. Astron. Astroph.*, **30**, pp. 543–574, 1992.

[16] Monaghan, J., Simulating free-surface flows with SPH. *J. Computat. Phys.*, **110**, pp. 399–406, 1994.

[17] Monaghan, J., Particle methods for hydrodynamics. *Computer Physics Reports*, **3(2)**, pp. 71–124, 1985.

[18] Hernquist, L. & Katz, N., TREESPH: a unification of SPH with the hierarchical tree method. *Astrophys. J. Supp. Ser.*, **70**, pp. 419–446, 1989.

[19] Colicchio, G., Colagrossi, A., Greco, M. & M. L., Free-surface flow after a dam break: a comparative study. *Ship. Tech. Res.*, **49(3)**, 2002.

[20] Landrini, M., Colagrossi, A. & Faltinsen, O., Sloshing in 2-D flows by the SPH method. *The $8^{th}$ International Conference on Numerical Ship Hydrodynamics*, Busan, (Korea), 2003.

[21] Landrini, M., Colagrossi, A. & Tulin, M., Breaking bow and transom waves: Numerical simulations. *Proc. 16th International Workshop on Water Waves and Floating Body* (IWWWFB), Hiroshima, Japan, 2001.

[22] Bertolazzi, E. & Manzini, G., A mixed finite element solver for liquid-liquid impacts. *Communications in Numerical Methods in Engineering*, **20(8)**, pp. 595–606, 2003. 2003-36-PV.

[23] Shewchuk, J., Delaunay refinement algorithms for triangular mesh generation. *Computational Geometry: Theory and Applications*, **22**, pp. 21–74, 2002.

[24] Bertolazzi, E. & Manzini, G., Algorithm 817 P2MESH: generic object-oriented interface between 2-D unstructured meshes and FEM/FVM-based PDE solvers. *ACM TOMS*, **28(1)**, pp. 101–132, 2002.